I, Microbiologist

**A Discovery-Based Course in
Microbial Ecology and Molecular Evolution**

I, Microbiologist

A Discovery-Based Course in Microbial Ecology and Molecular Evolution

Erin R. Sanders and Jeffrey H. Miller
Department of Microbiology, Immunology, and Molecular Genetics
University of California, Los Angeles
Los Angeles, California

With contributions by
Craig Herbold, Krystle Ziebell, and Karen Flummerfelt

ASM
PRESS

Washington, DC

Address editorial correspondence to ASM Press, 1752 N St. NW, Washington, DC 20036-2904, USA

Send orders to ASM Press, P.O. Box 605, Herndon, VA 20172, USA
Phone: (800) 546-2416 or (703) 661-1593
Fax: (703) 661-1501
E-mail: books@asmusa.org
Online: estore.asm.org

Library of Congress Cataloging-in-Publication Data

Sanders, Erin R.
I, microbiologist : a discovery based course in microbial ecology and molecular evolution / Erin R.
Sanders and Jeffrey H. Miller ; with contributions by Craig Herbold, Krystle Ziebell, and Karen
Flummerfelt.
p. cm.
Includes bibliographical references.
ISBN 978-1-55581-470-0 (pbk.)
1. Microbial ecology—Textbooks. 2. Molecular evolution—Textbooks. I. Miller, Jeffrey H. II. Title.
QR100.S265 2009
579′.17—dc22

2009026843

Illustrations: Cori Sanders (iroc designs), Craig Herbold, and Erin R. Sanders
Cover and interior design: Susan Brown Schmidler

Cover images: Colony kaleidoscope. Colonies of *Escherichia coli* can be stained to reveal interesting color patterns. The plate second from the right in the top row shows colonies growing on LB medium. The plate in the third row from the top, far right, shows colonies on minimal medium with X-Gal (5-bromo-4-chloro-3-indolyl-β-D-galactopyranoside), a histochemical stain for β-galactosidase. The plate second from the left in the top row shows papillating Lac$^+$ colonies on X-Gal plates. The remaining panels show different enhanced color photographs of each plate. Photos courtesy of Elinne Becket and Jeffrey H. Miller, Department of Microbiology, Immunology, and Molecular Genetics, Molecular Biology Institute, and David Geffen School of Medicine, University of California, Los Angeles. (Rearranged from the cover of the *Journal of Bacteriology,* vol. 190, no. 15, 2008.)

To my patient and always supportive family: my husband, Todd Lorenz; my parents, Nancy and Alan Sanders; my sisters, Kari and Cori Sanders; and my grandmother, Joyce Sanders.

And in loving memory of my grandmother, Millie Smith.
ERIN R. SANDERS

To my wife, Kim Anh Miller, and to the memory of my parents, Irma and Jerome S. Miller.
JEFFREY H. MILLER

Contents

UNIT 6
Bioinformatics Analysis of 16S rRNA Genes 257

UNIT 7
Molecular Evolution: Phylogenetic Analysis of 16S rRNA Genes 325

Preface

Undergraduate science education should mirror the collaborative nature of research and reflect how scientific hypotheses are evaluated and results are communicated in the 21st century. Such an undertaking requires the development of teaching methods that actively engage students in the creative process of scientific inquiry, provide skills necessary for success in the modern research laboratory, and foster excitement about the discovery process central to scientific research. The "I, Microbiologist" research experience as presented in this textbook has been designed to enable educators to meet these goals in an instructional laboratory setting. The research project not only exposes students to cutting-edge topics in microbial ecology and molecular evolution but also provides opportunities for students to experience the interactive nature of the scientific method (Fig. 1).

Organization

The "I, Microbiologist" project is crafted around a single question in the field of microbial ecology: What is the extent of bacterial diversity in terrestrial ecosystems? The project described in this textbook focuses on the rhizosphere, the region of soil surrounding the roots of plants, but can be applied to the exploration of bacterial diversity in any soil sample. Each unit of the textbook contains experimental protocols preceded by background information to be used as a resource by students, who will formulate hypotheses, conduct experiments, gather data, and interpret results so that their hypotheses can be evaluated and conclusions reported in the form of oral and/or written assignments. The background reading is designed to show students how the tools and methodologies used in the "I, Microbiologist" research project can be applied to related fields. Students also will be challenged to develop the appropriate vocabulary and use it to accurately describe observations made and results analyzed during the course of the project. Instructors may use the background information to develop lectures or to seed discussion sessions. Note that information from several primary-literature articles is included in some of the units. Although key figures from these papers are provided in the textbook, the full-length articles must be downloaded by students through the journal publishers' websites. To

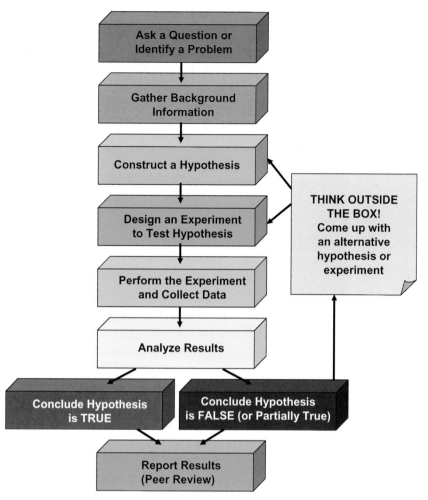

FIGURE 1 The scientific method, an iterative process used to acquire new knowledge. Illustration by Erin Sanders.

encourage students to read the material and actively participate in discussions about the material, homework assignments are incorporated into each unit.

As the instructor guides them through each phase of the project, students will have to draw upon their abilities to think critically, integrating knowledge gained from reading assignments and lectures into the development of hypotheses or the analysis of data. As "I, Microbiologist" participants, students will encounter project-specific nuances reflective of an experimental system (the soil) that is greatly understudied, and thus they must learn to navigate obstacles typical of any true research experience. At the same time, students should become proficient in basic techniques in microbiology and molecular biology and should develop competence with the computer-based analytical tools of bioinformaticists and phylogeneticists.

I, Microbiologist was devised for upper-division lecture and laboratory courses in microbiology for students who have a strong biology background and some previous laboratory experience. The entire project can be completed in as few as 10 weeks with as many as 12 to 18 students. The whole project can best be accomplished with groups of four students, with tasks divided up among members of the group. For smaller groups or less-experienced students, portions of the project can be omitted.

Special Features

Although students will be taking part in the "I, Microbiologist" research project at their home institution, there is the opportunity for students to become part of a collaborative network of colleges and universities. By contributing their results to the "I, Microbiologist" database (called the Consortium of Undergraduate Research Laboratories [CURL] Online Notebook [http://ugri.lsic.ucla.edu/cgi-bin/loginmimg.cgi]), students can share their discoveries with undergraduates at other participating institutions. Furthermore, the database is an online resource for students to utilize throughout the project.

A concern of instructors with regard to research projects and laboratory courses in general is the cost. The budget for consumable supplies all too often is a limiting factor in choosing which experiment to conduct or how many times to repeat it in the event of failure. For the "I, Microbiologist" project to be a true research experience, students must have the opportunity to troubleshoot experiments when unexpected or unusable results are obtained. The cost associated with the project could quickly get out of hand unless some restraint is imposed. To address this issue, student teams may be awarded a "grant" and be required to budget their expenditures throughout the project (see the budget worksheet in the introduction). In this way, the student experience mirrors that of a principal investigator in a research laboratory where decisions are dictated by the funding provided by external agencies such as the National Institutes of Health (http://grants.nih.gov/grants/oer.htm) or the National Science Foundation (http://www.nsf.gov/funding/). Students will decide whether they can repeat experiments on the basis of their allocated funding level, which cannot be exhausted before completion of the project. Instructors act as program directors, influencing the level of funding available to students.

Acknowledgments

The "I, Microbiologist" research project was first launched as a course by the Department of Microbiology, Immunology, and Molecular Genetics (MIMG) at the University of California, Los Angeles (UCLA), in January 2006. Its implementation served as a means for the department to integrate research experiences into the undergraduate instructional laboratory curriculum. We thank MIMG Professor and Chair Jeffery F. Miller for invaluable departmental support needed to implement and sustain the course. The UCLA Office of Instructional Development and Dean of Life Sciences Emil Reisler also have our gratitude for supporting the continued improvement of this course. We thank MIMG Professor and Life Sciences Core Curriculum Chair Robert Simons, as well Life Sciences Core instructors Cheryl Kerfeld, Debra Pires, and Gaston Pfluegl, who were instrumental in the creation of a collaborative, interdepartmental research environment for life science courses. Dr. Kerfeld was especially helpful during the initial offering of the course, giving valuable advice as well as sharing instructional resources and laboratory space.

Several people in Dr. Jeffrey H. Miller's laboratory assisted Krystle Ziebell with optimization of the original "I, Microbiologist" protocols: Grace Lee, Erika Wolf, Cindy Tamae, Jennifer Okada, Lavitania Bismart, Maxine Karimoto, and Tamar Sardarian. Working with Dr. Erin Sanders, UCLA undergraduate students To Hang (Shela) Lee, Brian Kirkpatrick, Areerat (Fah) Hansanugrum, Jeong-hee Ku, and Sin Il (Chris) Kang contributed to the development of new experiments and modified existing protocols for incorporation into the textbook. Professor Marcos García-Ojeda of the University of

California, Merced, helped with updates made to BLAST experiment instructions. The decision guides in Units 3 and 5 were modeled after similar diagrams in the *PGRI Resource Guide* crafted by Tuajuanda Jordan and Lucia Barker with the Howard Hughes Medical Institute's Science Education Alliance.

We are very grateful for the assistance of UCLA Professor Ann M. Hirsch, who has provided expertise and guidance in development of new experiments and who has participated as an enthusiastic guest lecturer for the course, sharing her knowledge and experience in rhizosphere biology and nitrogen fixation. We thank the many researchers, including former "I, Microbiologist" program participants at UCLA, who contributed their images, photographs, and sample results. We thank the Marc Levis-Fitzgerald, Moon Ko, Ed Ryan, and Lisa Millora of the UCLA Center for Educational Assessment as well as Julia Phelan of the National Center for Research on Evaluation, Standards, and Student Testing for their expert assistance in the conduct of assessment critical for acquisition of funds to support ongoing curriculum improvement projects. We are indebted to all the undergraduate researchers who offered constructive feedback and insightful suggestions to continually improve the "I, Microbiologist" program.

We thank Weihong Yan, program analyst at the Keck Bioinformatics User Center at UCLA for creating the CURL Online Notebook. The database was built with server space generously contributed by Cheryl Kerfeld, the former director of the Undergraduate Genomics Research Initiative.

We are grateful to numerous colleagues who offered encouragement and support while the research program and textbook were being developed. We are especially appreciative to the following reviewers for their thoughtful comments and insightful feedback:

Frederick M. Cohan, Wesleyan University, Middletown, CT

Ann M. Hirsch, University of California, Los Angeles

Peter H. Janssen, AgResearch Limited, Palmerston North, New Zealand

Reid C. Johnson, University of California, Los Angeles

Cheryl Kerfeld, DOE Joint Genome Institute, Walnut Creek, CA

Huiying Li, University of California, Los Angeles

Minghsun Liu, University of California, Los Angeles

Richard M. McCourt, National Science Foundation, Washington, DC

Debra Pires, University of California, Los Angeles

Finally, we offer special thanks to the ASM Press staff and freelancers who helped shape the textbook and bring the project to completion. We thank our editor, Gregory Payne; production manager, Ken April; assistant production editor, Cathy Balogh; cover and interior designer, Susan Schmidler; and copy editor, Yvonne Strong, for their invaluable assistance on this project.

Note to Readers

We thank you for selecting this textbook, which will enable you to implement the "I, Microbiologist" undergraduate research program at your institution. We welcome comments and would appreciate suggestions, especially if there are any errors in the text and figures or if there are alternative protocols or experimental strategies that we should consider including in the textbook. Please e-mail us at the following addresses:

Erin R. Sanders
erinsl@microbio.ucla.edu

Jeffrey H. Miller
jhmiller@microbio.ucla.edu

About the Authors and Contributors

The two primary authors, Erin R. Sanders and Jeffrey H. Miller, developed the textbook *I, Microbiologist: a Discovery-Based Course in Microbial Ecology and Molecular Evolution* with major contributions by Craig Herbold, Krystle Ziebell, and Karen Flummerfelt.

Erin R. Sanders received her B.S. in 1998 from DePaul University in Chicago, IL, and her Ph.D. in 2005 from the University of California, Los Angeles (UCLA), where she studied a bacterial site-specific DNA recombination system with Reid C. Johnson. She has taught undergraduate courses in microbiology, molecular genetics, and genomic biology for the Department of Microbiology, Immunology, and Molecular Genetics (MIMG) at UCLA. She converted the "I, Microbiologist" research project into a formal, research-based laboratory course which was successfully integrated into the MIMG undergraduate curriculum. Dr. Sanders has obtained funding from the National Science Foundation to conduct an ongoing assessment of the "I, Microbiologist" research program as well as to support development of a dedicated data management system (DUE 0737131). She also directs the design and implementation of new laboratory courses, collaborating with faculty and national education leaders to develop innovative instructional experiences for undergraduates. She is a member of the American Society for Microbiology, a faculty advisor for the DOE Joint Genome Institute Microbial Genome Annotation education program, an associate of the Howard Hughes Medical Institute's Science Education Alliance (2007–2008 cohort), and a National Academies Education Fellow in the Life Sciences.

Jeffrey H. Miller has made major contributions to molecular biology, particularly to the understanding of mechanisms of mutagenesis and DNA repair. After graduating with a B.A. in chemistry from the University of Rochester in 1966, he received his Ph.D. from Harvard University in 1970, working in Jon Beckwith's laboratory. Just 2 years later, Dr. Miller wrote the Cold Spring Harbor Laboratory manual *Experiments in Molecular Genetics* (1972). This was the first of a series representing a bible of techniques for several generations of graduate students, and it is still in use today in its revised and expanded version (1992). The manual included simple methods for selecting mutations, mutagenizing populations of bacteria with standard mutagens, and quantifying *lacZ* activity as β-galactosidase units. He has been a coauthor of the introductory genetics textbook *Introduction to Genetic Analysis* (W. H. Freeman) for editions 3 through 8 and is the author of the book *Discovering Molecular Genetics* (Cold Spring Harbor Laboratory Press, 1996). In 2001, he coedited, with Nobel laureate Sydney Brenner, the four-volume *Encyclopedia of Genetics* (Academic Press). He received the 1982 Friedrich Miescher Prize of the Swiss Biochemical Society and the 2007 Environmental Mutagen Society Award in recognition of outstanding research contributions in the area of environmental mutagenesis. He is currently a Distinguished Professor of Microbiology, Immunology, and Molecular Genetics at UCLA. His recent work addresses mutagenic cancer mechanisms, multidrug resistance systems, pathways of antibiotic action, and mutagenesis and repair in pathogens.

Craig Herbold received his Ph.D. in 2009 from UCLA, working in the laboratory of James Lake in the Molecular Biology Institute. His research is in molecular evolution and addresses the evolution of prokaryotes and the rooting of the universal tree of life. In collaboration with Erin Sanders, he contributed to the development of lecture materials and coauthored Units 6 and 7 of *I, Microbiologist*.

Krystle Ziebell is currently a medical student at the University of California, Irvine. She received her B.S. in microbiology, immunology, and molecular genetics from UCLA, where she was a research assistant in Jeffrey H. Miller's laboratory. Her senior honors thesis was on the development of streamlined protocols making up the original "I, Microbiologist" undergraduate research program. These procedures were modified and incorporated into a laboratory manual for the departmental course launched by Erin Sanders in 2006.

Karen Flummerfelt is a teaching associate with the Life Sciences Core Curriculum and the Department of MIMG at UCLA. She graduated with a B.S. from Mount St. Mary's College in Los Angeles, CA, and received an M.S. from UCLA, where she worked with Benhur Lee on the mechanism of human immunodeficiency virus attachment and entry into host cells. In 2001, she was awarded a Department of Education/GAANN Fellowship for Preparation of Future Faculty. Working with Erin Sanders, she contributed to the development of lecture materials and edited the "I, Microbiologist" experimental procedures to facilitate transformation of the undergraduate research program into a format appropriate for a laboratory course.

Introduction

Project Overview

Undergraduate research programs offer students a range of opportunities in many areas of the life sciences. However, a long apprenticeship on a highly specialized project is often an inevitability of such research programs, resulting in the delay of the discovery process for as long as 3 years and ultimately offering students only a narrow view of the entire life sciences field. The "I, Microbiologist" program was developed to enable undergraduates to participate in cutting-edge research within a laboratory course and to experience the thrill of discovery within only a few weeks of starting a project. Specifically, students learn to reconstruct the phylogeny, or evolutionary history, of a unique soil-based microbial community based on the analysis of 16S ribosomal RNA (rRNA) genes. The streamlined protocols are summarized in Fig. 1.

The experimental methodologies allow undergraduates to isolate the genomic DNA from microorganisms cultivated from the soil or to purify total DNA (the "metagenome") directly from the soil itself, to amplify 16S rRNA gene segments by polymerase chain reaction (PCR), and to build phylogenetic trees based on the DNA sequences. In addition, any number of biochemical or cytological assays may be adapted for the cultivation-dependent part of the project, allowing students to characterize the metabolic potential of soil bacteria. Since the enrichment strategies presented in this book have been devised to target the rhizosphere, students may screen their isolates for cellulase activity, the ability to break down cellulose, attributed to secretion of hydrolytic enzymes. Because soil microorganisms are prolific manufacturers of antibiotic substances, students may screen their isolates for antibiotic production as well as antibiotic resistance.

Students may collect soil samples from any terrestrial environment, depending on what appeals to them and is available locally. Each team (four students is the suggested team size) chooses a particular soil type and collects soil samples from around the base of a tree or plant (Fig. 2). Having some knowledge of the sampling site permits speculation about the microbial community in that habitat. For example, students might consider asking how different environments (e.g., the rhizosphere of different plants or soil differing in physical or chemical characteristics) may affect patterns of bacterial diversity.

Each project team should be subdivided into two groups. As schematized in the flowchart in Fig. 1, one group may use the soil sample to proceed with the cultivation-

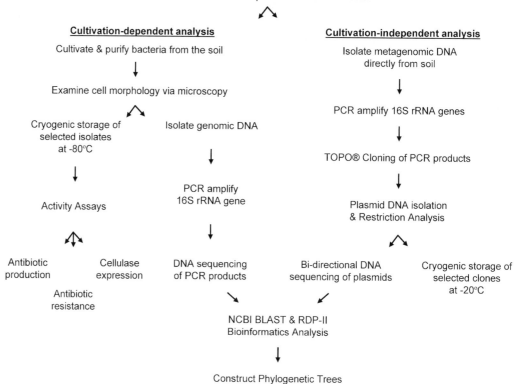

Collect soil samples from environment

Cultivation-dependent analysis

Cultivate & purify bacteria from the soil

Examine cell morphology via microscopy

Cryogenic storage of selected isolates at -80°C

Isolate genomic DNA

Activity Assays

PCR amplify 16S rRNA gene

Antibiotic production

Cellulase expression

DNA sequencing of PCR products

Antibiotic resistance

Cultivation-independent analysis

Isolate metagenomic DNA directly from soil

PCR amplify 16S rRNA genes

TOPO® Cloning of PCR products

Plasmid DNA isolation & Restriction Analysis

Bi-directional DNA sequencing of plasmids

Cryogenic storage of selected clones at -20°C

NCBI BLAST & RDP-II Bioinformatics Analysis

Construct Phylogenetic Trees

FIGURE 1 Overview of experiments involved in "I, Microbiologist" student research projects. Illustration by Erin Sanders.

FIGURE 2 Collection of soil samples from Mildred E. Mathias Botanical Garden at UCLA. An "I, Microbiologist" student project team in an area of chaparral gathers soil from the rhizosphere of *Gleditsia triacanthos* (honey locust, Fabaceae family), a native invasive plant in southern California. Photograph by Erin Sanders.

dependent part of the project and the other group may use the same soil sample and simultaneously proceed with the cultivation-independent part of the project. Each project team should generate two phylogenetic trees (Fig. 3 and 4). All four team members should discuss and debate the most appropriate interpretation of their DNA sequence data. Specifically, they should consider how well the isolates from the cultivation-dependent part of the project represent the diversity of microorganisms portrayed in the tree produced from the clones containing the 16S rRNA gene for the cultivation-independent part of the project. Which tree is predicted to contain a more phylogenetically diverse representation of the microbial community? Are the results consistent with expectations? We recommend that each team give an oral presentation to solicit feedback and constructive criticism from instructors and peers about their results and conclusions. However, we also suggest that students individually write their own research paper summarizing the work that they and their team members accomplished and providing a critical interpretation and analysis of both trees.

FIGURE 3 Phylogenetic tree representing cultivated members of the rhizosphere from the base of a conifer *(Podocarpus totara)* found in the UCLA Botanical Gardens. This phylogram was generated by using ClustalX and TreeView. Courtesy of Brian Kirkpatrick, a participant in the "I, Microbiologist" research program in the fall of 2006.

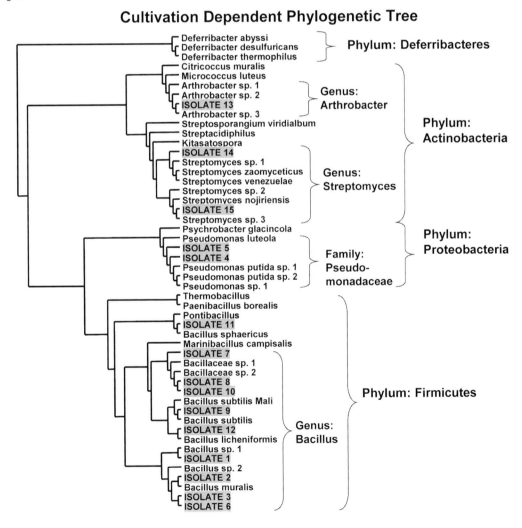

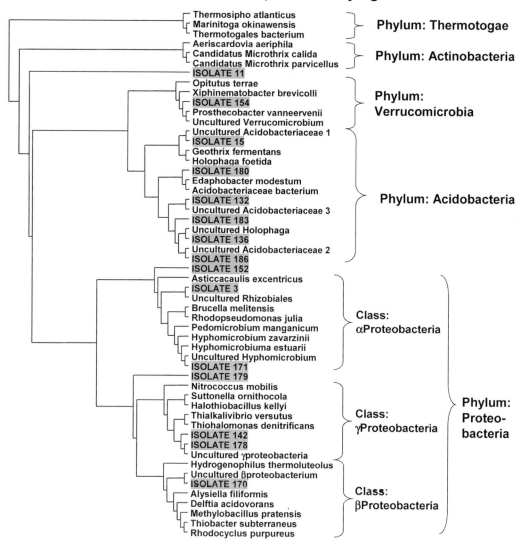

Cultivation Independent Phylogenetic Tree

FIGURE 4 Phylogenetic tree representing uncultivated members of the rhizosphere from the base of a conifer *(Podocarpus totara)* found in the UCLA Botanical Gardens. This phylogram was generated by using ClustalX and TreeView. Courtesy of Brian Kirkpatrick, a participant in the "I, Microbiologist" research program in the fall of 2006.

Timing Goals for Team Project Planned over a 10-Week Time Frame

Week	Group 1 goals	Group 2 goals
WEEK 1	Collect soil samples (Experiment 1.1); practice streak-plate procedure (Experiment 1.2); determine water content and pH of soil samples (Experiments 1.3 and 1.4)	
	Perform first enrichment (Experiment 2.1)	Begin metagenomic DNA isolation (Experiment 5.1)
WEEK 2	Take pictures of plates from first enrichment (Experiment 2.1); microscopic examination (Experiments 4.1–4.3); perform second enrichment (Experiment 2.2)	PCR amplification and gel electrophoresis of 16S rDNA amplicons (Experiments 5.2 and 5.3)
WEEK 3	Purify isolates (Experiment 2.3); microscopic examination and pictures (Experiments 4.1–4.3); cryogenic storage of isolates at −80°C (Experiment 2.4)	Purification and quantification of metagenomic 16S rDNA PCR products (Experiments 5.4 and 5.5)
WEEK 4	Genomic DNA preparation (Experiment 3.1); PCR amplification of isolates (Experiment 3.2)	Clone purified products into TOPO TA vector, transform ligation reaction, and streak purify candidate colonies (Experiment 5.6)
WEEK 5	Gel electrophoresis of 16S rDNA (Experiment 3.3); purify and quantify PCR products (Experiment 3.4 and 3.5)	Plasmid preparations (Experiment 5.7)
WEEK 6	Prepare and submit samples for sequencing (Experiments 3.6); start phenotypic screens (Experiments 4.4–4.6)	Begin secondary screening of TOPO clones by EcoRI restriction analysis (Experiment 5.8)
WEEK 7	Repeat phenotypic screens as necessary, taking pictures of plates and cells (Experiments 4.1–4.6)	Continue preparation and screening of TOPO clones (Experiments 5.6–5.9); start submitting samples for DNA sequencing (Experiment 5.9)
WEEK 8	Begin BLAST and RDP analyses of DNA sequences (Experiments 6.2 and 6.3)	Continue sequencing TOPO clones (Experiment 5.9); construct consensus sequences (Experiment 6.1); begin BLAST and RDP analysis of DNA sequences Experiments 6.2 and 6.3)
WEEK 9	Begin multiple sequence alignment and tree construction (Experiments 7.1 and 7.2)	
WEEK 10	Work with project team to complete data analysis; ensure that all sample and clone information is deposited into database (CURL Online Notebook)	

Note: Students must adjust the schedule depending on how long it takes isolates to grow (group 1) or how long it takes metagenomic DNA preparation and subsequent cloning steps to work (group 2).

Budget Worksheet

Congratulations! Your project team has been awarded a grant from the Consortium of Undergraduate Research Laboratories to support your exploration of microbial diversity in soil-based environmental samples. However, the program director (your instructor) requires you to submit a formal proposal describing your project plan, including a budget justifying anticipated expenditures associated with conduct of experiments.

Below is an overview of "costs" associated with each experiment listed in *I, Microbiologist* (Units 1 to 7). The currency used for this project will be the "phylobuck." The cost of each experiment is representative of the materials and/or services required for its successful completion under normal circumstances. Note that some experiments are "free" and others have a per-sample charge, while there are a few in which the cost is independent of the number of trials. Use these amounts as guidelines to generate your anticipated budget for successful completion of your project in the designated time frame; assume that none of the experiments are optional unless specified by your program director. It is helpful to keep in your notebook a running tally of the phylobucks used during each laboratory period. Be sure to consider potential problems that could be encountered during the project, necessitating certain experiments to be repeated; such deviations from specified amounts in each experimental category must be properly justified in your proposal. *Hint:* Your program director will be able to clue you into some unforeseen cost considerations if you ask. Each project team has been awarded a total of 1,500 phylobucks. Ideally, the project should require about 1,250 phylobucks. Distribute your expenditures wisely and as accurately as possible. Do not exhaust your funding before the project is finished!

All members of the project team should work together to devise a budget.

Experiment	Description	Cost (phylobucks)
1.1	Soil collection (one-time cost)	5
1.2	Streak plate practice (one-time cost)	0
1.3	Soil water content determination (one-time cost)	5
1.4	Soil pH determination (one-time cost)	5
2.1	Cultivation of soil bacteria (first enrichment)	1 (per plate)
2.2	Cultivation of soil bacteria (second enrichment)	1 (per isolate)
2.3	Purification of bacterial isolates	1 (per isolate)
	Propagation of selected isolates (every 2 weeks)	1 (per isolate)
2.4	Cryogenic storage of isolates (one-time cost)	20
3.1	Genomic DNA isolation	
	Methods A and F	1 (per isolate)
	Methods B and C	2 (per isolate)
	Methods D and E	3 (per isolate)
3.2	PCR (need negative control)	5 (per reaction)
3.3	Gel electrophoresis	10 (per gel)

(continued)

Experiment	Description	Cost (phylobucks)
3.4	Purification of PCR products	3 (per column)
3.5	Quantification of purified PCR products	
	Method A (qualitative gel comparisons)	10 (per gel)
	Method B (spectrophotometric)	2 (per scan)
3.6	DNA sequencing	5 (per reaction)
4.1	Microscopic examination (one-time cost)	10
4.2	Wet mount preparation (one-time cost)	3
4.3	Gram stains (one-time cost)	5
4.4	Testing for antibiotic production on four indicators	4 (per isolate)
4.5	Testing for antibiotic resistance	3 (per isolate)
4.6	Testing for cellulose activity	2 (per isolate)
5.1	Metagenomic DNA isolation	10 (per sample)
5.2	PCR (need negative control)	5 (per reaction)
5.3	Gel electrophoresis	10 (per gel)
5.4	PCR gel purification (electrophoresis step included)	5 (per column)
5.5	Quantification of purified PCR products	
	Method A (qualitative gel comparisons)	10 (per gel)
	Method B (spectrophotometric)	2 (per scan)
5.6	TOPO cloning	20 (per reaction)
	Streak purification of transformants	3 (per plate)
5.7	Plasmid DNA preparation (be sure to account for both A and B tubes per clone)	3 (per column)
5.8	Restriction analysis (screen only A tubes for each clone)	3 (per reaction)
5.9	DNA sequencing (be sure to account for two primers per clone)	5 (per reaction)
6.1	Assembly of consensus sequences	0
6.2	NCBI BLAST search	0
6.3	RDP-II search	0
7.1	RDP-II tree	0
7.2	MEGA4 tree	0

Laboratory Etiquette and Safety Regulations

Be aware:

1. Live microorganisms are used for the "I, Microbiologist" project. These microorganisms are generally considered safe to use for those with a normal, uncompromised immune system.

If you have any health concerns associated with exposure to the microorganisms used for this project, please alert your instructor or teaching assistant immediately.

2. Potential bioterrorism threats and the emergence of multidrug-resistant infectious diseases present a public health concern that can be circumvented with proper biosafety training, biocontainment resources, and security measures. The precautions outlined below meet the standards for working in a biosafety level 1 (BSL1) laboratory, although additional precautions are recommended to maximize the safety of all laboratory personnel and prevent escape of potential pathogens.

For a full description of safety procedures, equipment, and laboratory facilities pertaining to microbiological practices, please consult *Biosafety in Microbiological and Biomedical Laboratories (BMBL)*, 5th ed., published by the Centers for Disease Control and Prevention and the National Institutes of Health.

3. Soil may contain plant pests such as nematodes and insects as well as plant pathogens of fungal, bacterial, or viral origin. If inadvertently released into the environment, these pests and pathogens can have devastating effects on native plant species and the local agricultural industry. Soil permits may be required for transport and manipulation of certain soil samples.

Please consult the U.S. Department of Agriculture's Animal and Plant Health Inspection Service (http://www.aphis.usda.gov/) to ensure that you are in compliance with environmental regulations pertaining to plant health before transporting soil samples to the laboratory from anywhere other than a local collection site.

Follow these rules:

1. No food or drink in the laboratory! Always use proper aseptic techniques when handling cultures.

2. A laboratory coat is required, as it will help protect you and your clothing. It may not leave the laboratory under any circumstances. Your laboratory coat should be stored in the same room in which the "I, Microbiologist" project is being conducted.

3. Gloves are required for some of the experiments. Disposable gloves will be provided for you.

4. No open-toed shoes (e.g., sandals or flip-flops) are allowed while you are working in the laboratory.

5. Know the location of the fire extinguisher, eye wash, and shower.

6. Always clean up spills *immediately*. In the case of a bacterial spill, flood the spill with disinfectant, wait a few minutes, and wipe it up with a paper towel.

7. Use plastic test tube racks and plastic containers to organize your supplies.

8. You must be able to distinguish your media and reagents from those of your laboratory colleagues, so always *label* supplies with your name, date, and other identification as necessary.

9. Not all students will be on the same experimental schedule. Be sure to communicate with your instructor or teaching assistant about supplies you will need to perform upcoming experiments, as your needs may be different from those of the other students in the laboratory.

10. Take only the supplies required to perform your experiments on any given day. If you make a mistake and need to repeat a procedure, tell your instructor or teaching assistant before taking extra supplies.

11. When pipetting, always use a pipette bulb. *Do not pipette by mouth!*

12. It is your responsibility to make sure plates and flasks are incubated at the proper temperature and for the appropriate amount of time. Be sure to communicate your incubation schedule with your instructor or teaching assistant.

13. Do not store media or other personal items in the laboratory, especially if the space is shared with other laboratory personnel. Store materials only in designated locations.

14. *Students will be held responsible for any damage that occurs to P10, P20, P200, or P1000 pipettors due to carelessness.* Immediately report any equipment problems to your instructor or teaching assistant. Replacement cost for the pipettors exceed $200!

15. *Students will be held responsible for any damage that occurs to the microscopes and cameras due to carelessness.* Immediately report any equipment problems to your instructor or teaching assistant. The microscopes cost approximately $7,000 each, and costs for the cameras and accessories range from $800 to $5,000. Use care with this equipment, cleaning lenses with lens paper before and after each use. Avoid getting oil on any lens other than the 100× oil immersion lens. Keep this equipment away from Bunsen burners. Turn equipment *off* when finished to avoid burning out lightbulbs unnecessarily.

16. Wear gloves when handling dimethyl sulfoxide (DMSO). Use DMSO in a fume hood, and dispose of contaminated pipette tips in the proper waste container.

17. Wear gloves when handling phenol and chloroform. Use these reagents in a fume hood, and dispose of contaminated tips and organic waste in designated containers.

18. Wear gloves when handling ethidium bromide (EtBr). EtBr is a mutagen and suspected carcinogen. Dispose of contaminated pipette tips, Kimwipes, gloves, used agarose gels, and Tris-acetate-EDTA (TAE) buffer in the proper waste containers.

19. Wear gloves when handling Congo red, which is a toxic benzidine-based dye. It is suspected to metabolize to benzidine, a known human carcinogen. Dispose of contaminated liquid and solid waste in the proper containers.

20. Never pull the electrodes out of the power supply while it is running. Turn the power supply *off* before removing the electrodes or gel electrophoresis cover.

21. Computers may be available in the laboratory classrooms. These computers should be used by students only for project-related work, not for personal e-mail, Web surfing, or other activities not related to the project. Keep Bunsen burners away from computers and connecting wires or network cables.

22. Lab cleanup

    a. Discard all used plastic pipettes directly into a designated collection bin. Never leave dirty pipettes on the bench. If glass pipettes are used, place them directly into buckets containing disinfectant (10% bleach).

    b. Dispose of all consumable supplies in the proper waste disposal bin. Separate disposables, such as plastic petri dishes, contaminated paper towels, used swab sticks, and toothpicks, from broken glassware, including used microscope slides and Pasteur pipettes.

    c. All used regular glassware (flasks, beakers, test tubes, etc.) must be placed in a collection bin, the contents of which will be autoclaved, cleaned, and sterilized. *Remove all labeling tape and metal caps before discarding glassware.*

    d. All benches should be wiped down with disinfectant at the beginning and end of each laboratory session.

    e. Be sure to wash your hands with antiseptic soap before you leave the laboratory.

    f. Factor cleanup time into your laboratory work schedule.

COMPLIANCE CONTRACT

I have read and understand the Laboratory Etiquette and Safety Regulations for the "I, Microbiologist" research project, and I agree to abide by these rules and procedures. I understand that if I have any questions, I can ask my teaching assistant or the course instructor.

_____

Print Name

_____

Signature

_____

Date

UNIT 1

Soil Collection and Compositional Analysis

SECTION 1.1
INTRODUCTION TO MICROBIAL ECOLOGY IN THE RHIZOSPHERE

Ecologists span all sorts of disciplines—from microbiology to zoology to botany, as examples. The field of microbial ecology describes how microorganisms interact with each other and their environment. Ecology from this standpoint encompasses two major components—biodiversity ("what organisms reside there?") and organismal activity ("what are the organisms actually doing there?").

Soil: a terrestrial ecosystem at the frontier of microbial ecology

Soil is a very complex natural system, harboring an extensive community of microorganisms comprised of prokaryotes *(Bacteria* and *Archaea)* as well as mycorrhizal fungi and other eukaryotes, which coexist to create a unique terrestrial environment. In every gram of soil, there are at least 10^7 to 10^9 prokaryotic cells, which comprise between 10^3 and $>10^4$ species (Daniel, 2005, and references within; Torsvik et al., 2002). Microorganisms constitute the largest component of soil biomass (Hassink et al., 1993; Foster, 1988). The metabolic capabilities exhibited by soil communities supported by the terrestrial environment are incredibly diverse and may vary between different soil environments, giving each soil sample a unique population of microorganisms. The diversity of microorganisms present in terrestrial environments can be explored by reconstructing the phylogeny, or evolutionary history, of the community within soil samples. In fact, this method has led to the discovery of previously unknown microorganisms, adding to our scientific knowledge and appreciation of the microbial world (Amann et al., 1995; Joseph et al., 2003).

The degree of diversity observed in soil environments is necessary to achieve complex chemical transformations such as those that contribute to global biogeochemical cycles—

no single microorganism can accomplish all the reactions needed for the conversion of elements to forms required for maintenance of the biosphere (Fig. 1.1). For instance, both autotrophic and heterotrophic microbes play a central role in soil respiration, specifically contributing to the utilization of carbon (Bond-Lamberty et al., 2004). Understanding the bacterial processes that mediate carbon exchanges between the soil and the air, and how these exchanges react to climate change, is gaining considerable attention in the context of global warming awareness (Bardgett et al., 2008). As another example, the biogeochemical cycling of nitrogen, which makes up nearly 80% of Earth's atmosphere, is influenced by the overall composition of the microbial community and its associated metabolic activities since key chemical transformations such as biological nitrogen fixation (BNF) are mediated by specialized bacteria (Van der Heijden et al., 2008). There are microbial communities in which each microorganism performs perfunctory tasks, producing metabolic end products that become substrates or energy sources for other microorganisms within the community. The cooperative activities among members of a microbial community, whether in the soil or in the ocean or within the bodies of humans, are essential to maintain life on Earth.

The influence of soil structure on microbial communities

Soil comes in a variety of textures, with characteristics thought to influence the metabolic activities of microbes which reside in these habitats. Soil can be differentiated as being part of the rhizosphere, the contiguous region of soil surrounding plant roots, or simply as bulk soil, which is everything else. According to soil scientists at the U.S. Department of Agriculture, soil can be categorized based on composition and structural features (Soil Survey Division Staff, 1993) (Fig. 1.2a). For example, some soils tend to be well drained, aerated, and acidic with a high content of organic matter (e.g., sandy and loamy soils). Others exhibit high water-holding capacity but tend to be high in nutrients (e.g., silt and clay). Table 1.1 summarizes some of the features of soil that allow estimates of textural identification in the field. Structural characteristics that lead to variations in soil texture are attributed to the size, shape, and chemical composition of soil particles to which

FIGURE 1.1 Representation of Earth's biosphere, which is composed of several ecosystems, including the hydrosphere (ocean), lithosphere (soil and rock), atmosphere (air), and ecosphere (living things). Illustration by Cori Sanders, iroc designs.

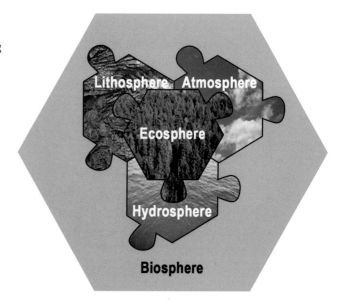

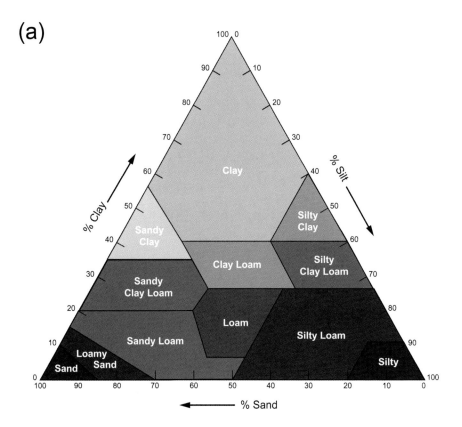

(a)

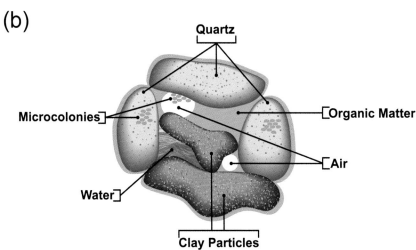

(b)

FIGURE 1.2 Composition and structural features of soil particles. (a) Triangular representation of soil texture describing the proportion of sand, silt, and clay within soil particles (based on p. 75 of *NRCS Soil Survey Manual* at http://soils.usda.gov/technical/manual). (b) Soil particle composed of mineral and organic components. Most soil microorganisms localize as microcolonies attached to the surface of soil particles (modified from Fig. 19.7 in Madigan and Martinko, 2006). Illustrations by Cori Sanders, iroc designs.

microorganisms adhere or adsorb (Daniel, 2005). As shown in Fig. 1.2b, microcolonies form on both the surfaces of soil aggregates as well as the pore spaces within and between soil particles (Foster, 1988).

Soil composition also is influenced by depth. The movement of organic (e.g., plant roots and decomposing organisms) and inorganic (e.g., minerals and water) materials results in the formation of layers, or soil horizons, which vary in the amount of microbial activity (Fig. 1.3). Microbial activity tends to be greatest in the organic-rich surface and subsurface layers of the soil (Madigan and Martinko, 2006). The amount of vegetation and leaf litter on the ground can provide clues about the soil profile, although mature soil typically takes hundreds of years to develop, depending on various abiotic factors.

Table 1.1 Summary of the determinants of soil texture

Texture class[a]	Description[b]
Sand	Loose, single-grained soil particles that feel gritty and do not stick together unless moistened; individual grains can be seen by the human eye; subclass (coarse, normal, fine, or very fine sand) determined by the relative proportion and size of sand grains
Loamy sand	Loose, single-grained soil particles containing slightly higher percentages of silt (up to 30%) and clay (up to 15%), which makes the particles somewhat cohesive when moist; most individual grains can be seen by the human eye; sand subclass (coarse, normal, fine or very fine sand) determined by the relative proportion and size of sand grains
Sandy loam	Contains sufficient amounts of silt (up to 50%) and clay (up to 20%) that soil readily sticks together; sand subclass (coarse, normal, fine or very fine sand) determined by the relative proportion and size of sand grains
Loam	Soft, fairly smooth and powder-like; contains up to 50% silt and up to 25% clay, which make it fairly sticky and capable of being molded (plastic) when moist
Sandy clay loam	Behavior dominated by sand and clay, giving it cohesive properties (stickiness and plasticity) when moist
Clay loam	Has a fairly even distribution of sand, clay, and silt but behaves and feels more like clay in that it is sticky and plastic
Silt	Smooth and rather silky soils that exhibit no stickiness when moistened
Silty loam	Forms large aggregates (clumps) that can be easily broken between fingers when dry, producing soft and powder-like particles; breaks into small bits when moistened
Silty clay loam	Smooth soil that is firm when moist but sticky and plastic when wet
Silty clay	Smooth soil that is very sticky and plastic when wet and forms very hard clumps when dry
Sandy clay	Similar to silty clay but with less silt and more sand, which may be difficult to detect
Clay	Very fine soil with smallest particle size, giving it immense surface area for physical and chemical activities; forms extremely hard aggregates and is extremely sticky when wet and capable of being molded into long, thin ribbons when moist
Peat or muck	Organic soils composed of accumulating plant and animal remains in various stages of decomposition; may be found in marshes, swamps, and lakes

[a] Class (except peat and muck) from USDA soil triangle depicted in Fig. 1.2.

[b] Descriptions derived from UF/IFAS Fact Sheet SL-29 (http://edis.ifas.ufl.edu/SS169; Brown, 2003).

How environmental factors affect microbial activities in the soil

Several major abiotic factors are thought to affect microbial activity in the soil. Water availability, which is a reflection of how well drained a particular terrestrial environment is, tends to be highly variable, often depending on the soil composition, amount of precipitation (rainfall) or drainage, and plant cover (Madigan and Martinko, 2006). The composition of soil microbial communities fluctuates as a consequence of cyclic changes

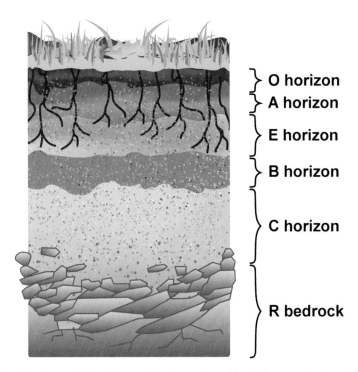

O horizon
A horizon
E horizon
B horizon
C horizon
R bedrock

FIGURE 1.3 Profile of mature soil depicting soil horizons defined by differences in chemical composition, color, and microbial activities. The O horizon is the topmost layer littered by undecomposed plant materials (leaf litter) such as leaves, needles, twigs, moss, and lichens. Some are saturated with water. Some have been saturated with water for long periods while others have either never been saturated or are artificially drained. The A horizon is the surface soil, which is high in humified organic matter, dark brown, and high in microbial activity. Since this is the layer where plants grow, it is tilled for agriculture. In addition to bacteria, which are often in close association with plant roots, other organisms such as fungi, nematodes, insects, and worms are in great abundance in this layer. The A horizon is depleted of soluble constituents like iron, clay aluminum, and organic compounds. The E horizon arises when depletion at the bottom of the A horizon is pronounced, resulting in a lighter-colored subsurface soil layer; this layer is present only in older, well-developed soils. The B horizon, extending below the A or E horizon, is the subsoil, which contains little organic matter but is loaded with minerals that have leached from the surface and subsurface layers and have accumulated here. There are typically detectable levels of microbial activity but at lower levels than in the surface and subsurface layers. The C horizon is the undeveloped soil base that forms directly above underlying bedrock. This layer generally has only low levels of microbial activity. The R horizon is composed of continuous masses of hard, partially weathered rock such as granite, quartz, limestone, and sandstone (p. 63–64 of *NRCS Soil Survey Manual* at http://soils.usda. gov/technical/manual). Illustration by Cori Sanders, iroc designs.

in water content, which contribute to the death of some proportion of community inhabitants (Kieft et al., 1987).

Oxygen availability is another major factor. Air tends to penetrate well-drained soils, leaving them with a high oxygen content that supports growth of aerobic bacteria (Madigan and Martinko, 2006). Waterlogged soils tend to become anoxic because the only oxygen present is that which has dissolved in the water. Dissolved oxygen is rapidly consumed by aerobic microorganisms, leaving an environment the supports the growth of anaerobes and facultative aerobes.

A third major factor is nutrient availability. Although carbon is not usually limiting in soil, it may not be readily available to microorganisms. Much of the organic matter derived from dead plants, animals, insects, nematodes, and microorganisms is converted to humus, which consists of stable materials that are refractory to further microbial decomposition processes. It turns out, however, that the availability of inorganic nutrients such as phosphorus and nitrogen tends to be most limiting in soil environments, and this can be detrimental to plant growth. Metabolic activities unique to the microbial world such as biological nitrogen fixation contribute to biogeochemical cycling of the key element nitrogen, facilitating its conversion to ammonia, a fixed form used by plants. The estimated contribution to this ecosystem by nitrogen-fixing bacteria is as much as 20% or even more (Van der Heijden et al., 2008).

A fourth environmental factor that contributes to microbial diversity is pH, the impact of which is discussed in Section 1.2 (Fierer and Jackson, 2006). Most soil microorganisms operate in an optimal pH range. Soil pH is influenced not only by the number of hydrogen ions (H^+) in water occupying pores within soil particles (Fig. 1.2) but also by the types and proportion of ions such as calcium (Ca^{2+}), magnesium (Mg^{2+}), and aluminum (Al^{3+}), which can be leached from the soil by rainwater (Cranfield University, 2007). If there are more calcium and magnesium ions than hydrogen and aluminum ions in the soil solution, the pH will be higher (alkaline, pH > 7). Conversely, if the proportion of hydrogen ions relative to calcium and magnesium ions increases, the soil pH becomes lower (acidic, pH < 7). The makeup of ions in the soil solution determines the pH, which is a critical indicator of soil health, affecting the ability of soil to support plant growth and microbial life.

The "I, Microbiologist" project is designed to be a discovery-based exploration of microbial communities derived from interesting soil environments. The media and enrichment strategies described in Unit 2 will enable students to specifically and easily address questions about the rhizosphere, an understudied experimental system with astounding microbial diversity. The soil area surrounding the roots of plants is a hot spot of microbial activity. Novel microorganisms await discovery in this particular environmental niche. It should be noted, however, that the techniques used in this project can be applied to any terrestrial environment where the goal is to characterize bacterial diversity.

KEY TERMS

Abiotic Nonliving; refers to physical factors in a particular environment.

Activity A measure of what microorganisms are doing in their habitats.

Aerobic Taking place in the presence of oxygen; the converse of *anaerobic* (taking place in the absence of oxygen).

Anaerobe An organism that does not employ oxygen as a terminal electron acceptor during respiration; the converse of an *aerobe*, which uses molecular oxygen as a terminal electron acceptor in an electron transport chain during respiration.

Anoxic Without oxygen; the converse of *oxic*, or containing oxygen.

Biodiversity The assortment of organisms living in a specific habitat or region.

Biogeochemical cycles Transformations catalyzed by biological and chemical agents during ecosystem movements of key elements of living systems such as carbon, nitrogen, sulfur, and iron.

Biosphere The global ecological system, which includes land (lithosphere), air (atmosphere), and water (hydrosphere), that supports all life on Earth.

Ecosystem The functional interaction among all biotic (plants, animals, and microorganisms) and abiotic (nonliving) factors within an environment.

Facultative aerobe An organism that can grow in either the presence or absence of oxygen.

Horizon Each layer or zone, approximately parallel to the soil surface, within the soil profile; horizons are distinguished by distinct properties produced by soil-forming processes.

Humus Partially decomposed heteropolymeric organic complexes such as lignin, humic and fulvic acids; *humified* (adjective) describes the organic materials transformed into humus.

Leaf litter Plant materials such as leaves, needles, twigs, moss, and lichens that are scattered throughout the O horizon at the soil surface.

Phylogeny Evolutionary relationships among a group of organisms.

Rhizosphere The contiguous region of soil surrounding the roots of plants.

REFERENCES

Amann, R. I., W. Ludwig, and K. H. Schleifer. 1995. Phylogenetics identification and in situ detection of individual microbial cells without cultivation. *Microbiol. Rev.* **59:**143–169.

Bardgett, R. D., C. Freeman, and N. J. Ostle. 2008. Microbial contributions to climate change through carbon cycle feedbacks. *ISME J.* **2:**805–814.

Bond-Lamberty, B., C. Wang, and S. T. Gower. 2004. A global relationship between the heterotrophic and autotrophic components of soil respiration? *Global Change Biol.* **10:**1756–1766.

Brown, R. B. 2003. University of Florida, Institute of Food and Agricultural Sciences (UF/IFAS) Fact Sheet SL-29. Soil and Water Science Department, Florida Cooperative Extension Service, Institute of Food and Agricultural Sciences, University of Florida, Gainesville, FL. Date first printed: April 1990. Reviewed: April 1998, September 2003. http://edis.ifas.ufl.edu/SS169.

Cranfield University. 2007. National Soil Resources Institute (NSRI) of Cranfield University, UK. Soil-Net project at www.soil-net.com.

Daniel, R. 2005. The metagenomics of soil. *Nat. Rev. Microbiol.* **3:**470–478.

Fierer, N., and R. B. Jackson. 2006. The diversity and biogeography of soil bacterial communities. *Proc. Natl. Acad. Sci. USA* **103:**626–631.

Foster, R. C. 1988. Microenvironments of soil microorganisms. *Biol. Fertil. Soils* **6:**189–203.

Hassink, J., L. A. Bouwman, K. B. Zwart, J. Bloem, and L. Brussaard. 1993. Relationships between soil texture, physical protection of organic matter, soil biota, and C and N mineralization in grassland soils. *Geoderma* **57:**105–128.

Joseph, S. J., P. Hugenholtz, P. Sangwan, C. A. Osborne, and P. H. Janssen. 2003. Laboratory cultivation of widespread and previously uncultured soil bacteria. *Appl. Environ. Microbiol.* **69:**7210–7215.

Kieft, T. L., E. Soroker, and M. R. Firestone. 1987. Microbial biomass response to a rapid change increase in water potential when dry soil is wetted. *Soil Biol. Biochem.* **19:**119–126.

Madigan, M. T., and J. M. Martinko. 2006. *Brock Biology of Microorganisms,* 11th ed. Pearson Prentice Hall, Pearson Education, Inc., Upper Saddle River, NJ.

Soil Survey Division Staff. 1993. *Soil Survey Manual.* USDA Handbook no. 18. Soil Conservation Service, U.S. Department of Agriculture, Washington, DC.

Torsvik, V., L. Øvreås, and T. F. Thingstad. 2002. Prokaryotic diversity: magnitude, dynamics, and controlling factors. *Science* **296:**1064–1066.

Van der Heijden, M. G. A., R. D. Bardgett, and N. M. van Straalen. 2008. The unseen majority: soil microbes as drivers of plant diversity and productivity in terrestrial ecosystems. *Ecol. Lett.* **11:**296–310.

Web Resources

Soil-Net Project http://www.soil-net.com/

University of Florida, Institute of Food and Agricultural Sciences (UF/IFAS) Soil Texture Descriptions http://edis.ifas.ufl.edu/SS169

USDA Soil Classification http://soils.usda.gov/technical/classification/

ANALYSIS OF THE PRIMARY LITERATURE: BIOGEOGRAPHY OF SOIL BACTERIAL COMMUNITIES

READING ASSIGNMENT
Fierer, N., and R. B. Jackson. 2006. The diversity and biogeography of soil bacterial communities. *Proc. Natl. Acad. Sci. USA* **103**:626–631.

Biogeography is the study of the distribution of macro- or microorganisms across various spatial scales, which can be local, continental, or global. The goal is to look for patterns that can be associated with certain abiotic factors and determine whether such factors are good predictors of biodiversity. Ecologists have been doing these sorts of analyses with plants and animals for ages, but its application to microbial communities is relatively new (Horner-Devine et al., 2004; Green and Bohannan, 2006). Moreover, ecological studies of microbes have become possible only due to the technological and methodological advances permitting cultivation-independent analyses of complex habitats such as soil.

The biogeography of microbes: is everything everywhere? Martinus Beijerinck was a 19th-century microbiologist and botanist who made numerous important contributions during this formative historical period for the field of microbiology. For instance, he showed that tobacco mosaic disease was caused by a filterable agent smaller than a bacterium, he discovered the bacterial metabolic processes nitrogen fixation and sulfate reduction, and he invented the enrichment culture (Chung and Ferris, 1996). However, this esteemed scientist also is credited for the ecological principle "everything is everywhere, but the environment selects" (O'Malley, 2007). In other words, although the diversity of microbes is immense (e.g., 10^3 to $>10^4$ species per g of soil), the composition and abundance of certain groups are influenced by environmental factors (Horner-Devine et al., 2004). Noah Fierer and Robert Jackson, the authors of the paper discussed in this section, challenge this hypothesis by investigating which, if any, abiotic factors contribute to bacterial biodiversity at the continental scale (Fierer and Jackson, 2006). The identification of spatial patterns will provide clues about how biogeography contributes to the distribution of microbial diversity.

Recall from Section 1.1 that water availability, oxygen levels, and nutrient status have been linked to microbial activities at the local scale. Of issue is whether these particular abiotic factors are reasonable predictors of biodiversity at the continental scale. Why do we care? Many stand to benefit from the knowledge gained about the distribution and abundance of microorganisms in the soil. For example, conservation biologists are interested in determining what role microbial communities play in facilitating the persistence of nonnative plant species, asking whether invasive plants alter the microbial community structure and thus allow these plant species to persist (Van der Putten et al., 2007). Ecologists are concerned about the long-term effects of land use on soil microbial communities with respect to ecosystem function (Fraterrigo et al., 2006). In particular, it has been shown that physical disruptions to soil such as agriculturally related tillage, forest clearing, and urban development actually change the abundance of particular microbial groups. This variation appeared to cause a decrease over time in the conversion of organic nitrogen (e.g., amino acids and urea) to mineral forms (e.g., nitrite, nitrate, and ammonia). Making efforts to maintain native biodiversity could potentially reduce the local effects of land use changes.

The biogeography of plants and animals provides a contextual model for its application to microbes. Before the concept of microbial biogeography is explored, it would be helpful to understand plant and animal biogeography since far more is known about the patterns of plant and animal diversity on continental scales (Gaston, 2000). Specifically, ecologists have established which abiotic factors are good predictors of plant and animal diversity; these traditionally include water availability in the form of annual precipitation (millimeters), temperature (degrees Celsius), latitude (degrees north or south of the equator), and elevation (meters). Notably, no single factor is universally predictive across all ecosystems, scales, or organisms.

By plotting the number of species against a particular abiotic factor, a spatial pattern may become evident, reflecting a relationship between the data plotted on each axis. A number of examples for species of plants and animals have been presented in a review article by Kevin Gaston; a subset of the plots are shown here (Gaston, 2000). For example, Fig. 1.4 is a plot of the spatial pattern detected for birds as a function of latitude. A bell-shaped curve is seen in which the greatest number of species resides at the equator (0° latitude), with the number gradually decreasing as the latitude approaches the north and south poles. A second example is shown in Fig. 1.5, revealing a spatial pattern for woody plants as a function of annual precipitation. An S-shaped curve is observed for plants. As the amount of precipitation increases, the number of species concomitantly increases, leveling off once the annual precipitation is measured at approximately 1,000 mm per year. These spatial patterns are easily recognizable and, more importantly, are reproducible. The particular patterns displayed for different organisms (e.g., birds versus plants) may not be the same, but a pattern exists nonetheless. The question that microbial ecologists are asking is whether bacterial communities exhibit biogeographical patterns on a continental scale in a manner similar to what is observed with plants and animals.

Overview of the study by Fierer and Jackson (2006). It was hypothesized that soil bacteria exhibit spatial patterns just like plants and animals do on a continental scale. Moreover, those variables previously cited as good predictors for plants and animals should also be good predictors for bacteria. Figure 1.6 outlines the experimental strategy used by Fierer and Jackson to conduct this study. Soil was collected from 98 sites across North

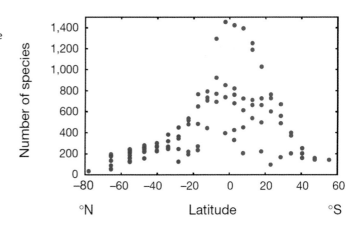

FIGURE 1.4 Spatial patterns for birds in approximately 611,111-km² grid cells across the New World, showing species-latitude relationship. Reprinted with permission from Gaston (2000).

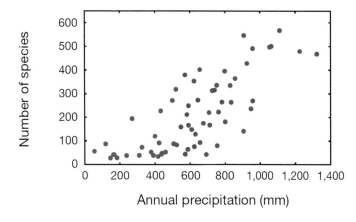

FIGURE 1.5 Spatial patterns for woody plants in 20,000-km² grid cells in southern Africa, showing species-precipitation relationship. Reprinted with permission from Gaston (2000); adapted with permission from O'Brien (1993).

and South America, representing a wide array of ecosystems (Table 1.2). Note that information such as the latitude, longitude, elevation, and dominant plant species was collected about each site. Furthermore, the physicochemical properties for each soil sample were determined, including the soil texture, pH and carbon content. The mean annual temperature and precipitation (MAT and MAP, respectively) also were described. This is an enormous number of samples with which to characterize the microbial community structure. The authors needed a practical yet informative way to analyze the samples and assess biodiversity as a function of the noted abiotic factors.

To simplify the analysis, the authors grouped the soil samples into six general categories: (i) tropical forest/grassland, (ii) boreal forest/tundra, (iii) humid temperate forest, (iv) humid temperate grassland, (v) dry forest, and (vi) dry grassland/shrubland. Next, the authors performed a cultivation-independent analysis of the soil bacterial communities in each group based on variations in restriction fragments generated by specific endonucleases. Referred to as T-RFLP, or terminal-restriction fragment length polymorphism, the method is used to evaluate bacterial diversity and community structure based on the length and abundance of unique restriction fragments, which represent phylotypes in each soil sample. The technique provides information about the total number and relative proportion of bacterial groups within a particular population but does not nec-

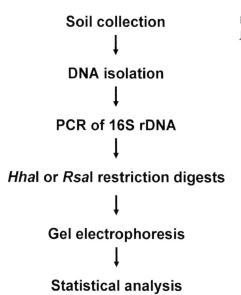

FIGURE 1.6 Overview of project conducted by Fierer and Jackson (2006). Illustration by Erin Sanders.

Table 1.2 Site information and physiochemical properties of a subset of the soils used in the Fierer and Jackson (2006) study[a]

Soil code	Location	Latitude[a]	Longitude[b]	Elevation (m)	Dominant plant species[c]	MAT (°C)	MAP mm	% Organic carbon[d]	Texture class[e]	pH[f]	Phylotype richness[g]	Phylotype diversity (H')[g]
AR1	Misiones, Argentina	27.73	55.68	150	Balfourodendron sp., Tabebuia sp.	23	1,400	2.2	Clay	6.0	31	3.3
BB1	Bear Brook, Maine	44.87	68.10	400	Picea rubens	6.1	1,200	12.8	Sandy loam	4.3	24	2.5
BZ1	Bonanza Creek LTER, Alaska	64.80	148.25	300	Picea glauca	−2.9	260	3	Silt loam	5.1	40	3.4
CA1	Cedar Mountain, Arizona	36.05	111.77	2,003	Pinus edulis	10.3	400	1.7	Silt loam	7.3	33	3.2
CM1	Clymer Meadow Preserve, Texas	33.30	96.23	200	Andropogon gerardii, Sorghastrum nutans, Schizachyrium scoparium	18.5	850	3	Silty clay	7.9	39	3.4
DF3	Duke Forest, North Carolina	35.97	79.08	150	Quercus alba	14.6	1,100	1.7	Loamy sand	5.1	24	2.9
GB6	Great Basin Experimental Range, Utah	39.33	111.45	3,750	Lupinus perennis, Bromus inermis, Achillea millifolium, Taraxicum officinales	2.0	400	2.2	Clay	7.2	44	3.6

IE3	Institute for Ecosystem Studies, New York	41.80	73.75	75	Galium aparine, Solidago sp., Phleum pratensis, Poa pratensis	8.6	1,200	6.4	Sandy loam	5.7	40	3.5
LQ2	Luquillo LTER, Puerto Rico	18.30	65.83	400	Dacryodes excelsa	21.5	3,500	4.1	Silty clay loam	5.0	29	3.0
MD3	Mojave Desert, California	35.20	115.87	776	Opuntia echinocarpa, Echinocactus polycephalus	21.0	150	0.12	Sandy loam	8.9	33	3.2
PE6	Manu National Park, Peru	12.63	71.27	860	Maieta sp., Ficus sp.	23	5,000	9.4	Clay loam	3.6	21	2.6
SK2	BOREAS site, Saskatchewan, Canada	53.60	106.20	601	Populus tremuloides	0.4	467	0.9	Loam	5.8	30	3.1

[a] Excerpted with permission from Table 3 (supporting data in online version) of Fierer and Jackson (2006). Original table copyright 2006 National Academy of Sciences, U.S.A. LTER, long-term ecological research; MAT, mean annual temperature; MAP, mean annual precipitation.

[b] All longitudes are west and all latitudes are north, with the exception of sites in Argentina and Peru.

[c] The dominant plant species at each site were determined in a qualitative manner at the time of sample collection. Dominant plants are described by genera or family when species identification was unclear.

[a] Soil organic carbon content was measured on a CE Elantech model NC2100 elemental analyzer (ThermoQuest Italia, Milan, Italy) with combustion at 900°C. Values are reported in grams per 100 grams of soil.

[c] Particle size analyses were conducted at the Division of Agriculture and Natural Resources Analytical Laboratory, University of California Cooperative Extension (Davis, CA) by using standard methods.

[f] Soil pH was measured after shaking a soil/water (1:1, weight/volume) suspension for 30 min.

[g] Phylotype richness and diversity (Shannon index, H') were estimated by using a terminal-restriction fragment length polymorphism analysis.

essarily reflect evolutionary history, which requires the DNA sequence data. The T-RFLP method also is commonly known as DNA fingerprinting and is frequently applied to crime scene investigations in forensic science (Varsha, 2006).

For the T-RFLP assay performed in the Fierer and Jackson study, genomic DNA was extracted simultaneously from the bacteria present in a particular soil sample. This composite genomic DNA sample is what we refer to as the metagenome of a particular soil bacterial community. Using primers designed to complementary base pair to specific regions at the 5′ and 3′ ends of the 16S ribosomal RNA (rRNA) gene thought to be conserved among members of the *Bacteria,* polymerase chain reaction (PCR) was performed using the soil metagenomic DNA as a template (Fig. 1.7). To perform T-RFLP analysis, one of the two primers must be tagged (e.g., covalently linked) to a fluorescent dye, in this case HEX, which emits green light when excited by a laser of appropriate wavelength. Thus, every amplicon obtained from a reaction using the HEX-8F–1492R primer combination will be fluorescent. Fluorescently tagged PCR products can be detected using specialized electrophoresis equipment (Fig. 1.8), allowing greater sensitivity than standard methods that depend upon secondary-staining procedures (e.g., with ethidium bromide). This system allows for detection of bacterial phylotypes that are in relatively low abundance, as well as those in very high abundance.

Following purification, the PCR products are treated with one of two restriction enzymes, HhaI or RsaI. These enzymes were selected for this study based on previous work, in which a number of enzymes were examined and their restriction patterns were analyzed—these two particular enzymes were shown to digest a broad representation of bacterial phyla in comparison to other enzymes examined in the study (Dunbar et al., 2001). These enzymes cut the 16S rRNA gene in locations that differ and with a frequency that varies from organism to organism, producing restriction fragments with a signature size characteristic of a particular phylotype. When electrophoresed on a special gel composed of an acrylamide-based matrix, the restriction fragments migrate through the matrix in response to an electric current and ultimately are resolved based on size. A simplified example of a T-RFLP result is shown in Fig. 1.9. Each of the lanes has a number of bands of different sizes resulting from a restriction digest of metagenomic DNA purified from six independent soil samples. The pattern for the population of DNA fragments varies for each sample, but the intensity of particular fragments in each lane also varies in comparison to that of other bands in the same lane. It turns out that intensity is a function of the number of 16S rRNA gene segments within that particular sample that were cut in an identical way, generating a bacterial phylotype that appears to be in higher abundance than others in the same sample. Taken together, the analysis provides two types of categorical measurements: (i) the number of restriction fragments, representing the relative amount of bacterial diversity, and (ii) the intensity of DNA fragments, reflecting the relative distribution of bacterial phylotypes within a particular population.

FIGURE 1.7 Diagram of the 16S rDNA region analyzed via T-RFLP in the Fierer and Jackson study. Illustration by Erin Sanders.

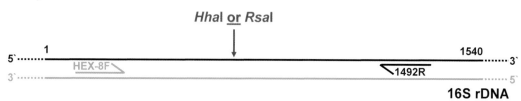

FIGURE 1.8 Student loading DNA samples into vertical gel on a LI-COR Biosciences 4300 DNA Analyzer. Photograph by Erin Sanders and Rachel Sauvageot (UCLA).

For a microbial ecologist, the patterns produced by these restriction fragments can be quantified and used to numerically report bacterial diversity as a function of the environmental factors under study.

To understand the data presented in the Fierer and Jackson study, some dissection of the statistical terms is required. The authors measure bacterial diversity using the Shannon index (H'), which is a common diversity index used by ecologists (Chao and Shen, 2003). The Shannon index enables the authors to translate the T-RFLP data (e.g., the number and intensity of bands on a gel) into numerical data points that can be analyzed, modeled, and evaluated using various statistical parameters. Specifically, the index treats phylotypes as symbols and their relative population size as a probability within an entropic distribution. It takes into account the number of phylotypes as well as the evenness, or uniformity in terms of band intensity on a gel, of phylotypes within a particular bacterial population. The value of the index increases as the number of phylotypes increases or by having greater phylotype evenness. Thus, there is a maximum possible H' for any given number of phylotypes, which occurs either when all unique phylotypes have been detected or when all phylotypes are present in a population in equal numbers. Biodiversity data also may be reported as simply "phylotype richness," which ignores the relative abundance of each phylotype and instead focuses only on the total number of

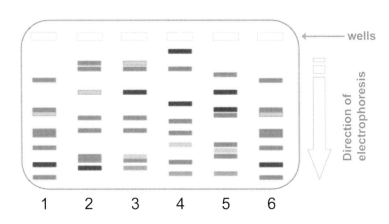

FIGURE 1.9 Results of T-RFLP performed on six hypothetical bacterial communities (lanes 1 to 6). Illustration by Craig Herbold.

wells

Direction of electrophoresis

1 2 3 4 5 6

unique phylotypes (e.g., bands on a gel) for a particular population. Both H' and phylotype richness can be plotted against values for a particular environmental parameter, and if a correlation exists between the two variables, a pattern will be detected relating bacterial biodiversity as a function of the abiotic factor under consideration.

Results and implications of the study by Fierer and Jackson (2006). As shown in Fig. 1.10, bacterial diversity for each of the 98 sites in the study was plotted against pH in a range between approximately 3.5 and almost 9. Based on these results, phylotype diversity is maximal in the neutral range and gradually decreases as the soil becomes either more acidic or more basic. Clearly, there is a pattern suggesting that soil pH may be a good predictor of bacterial diversity.

The authors also explore the relationship between soil bacterial diversity and a number of other environmental factors, including MAT and latitude, previously shown to be good predictors of plant and/or animal diversity on a continental scale. As shown in Fig. 1.11, there is no obvious relationship between bacterial phylotype diversity and either of these particular variables. Thus, those environmental factors that appear to be good predictors of plant and/or animal diversity do not seem to be good predictors of bacterial diversity, at least not at the continental scale. However, there is a caveat to the analysis.

It turns out that sampling area is a critical factor considered in plant and animal studies. Ecologists consider an area of 100 m² to be a local scale and greater than 500,000 km² to be a regional scale (Hawkins et al., 2003; Willis and Whittaker, 2002). If an analysis is performed at an inappropriate scale, biodiversity patterns cannot be detected;

FIGURE 1.10 Relationship between soil pH and bacterial phylotype diversity (Shannon index, H'). Reprinted from Fierer and Jackson (2006) with permission. Copyright 2006 National Academy of Sciences, U.S.A.

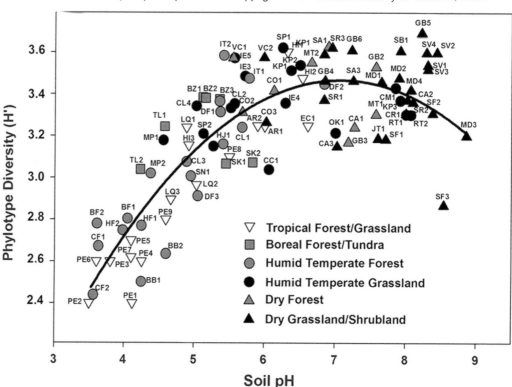

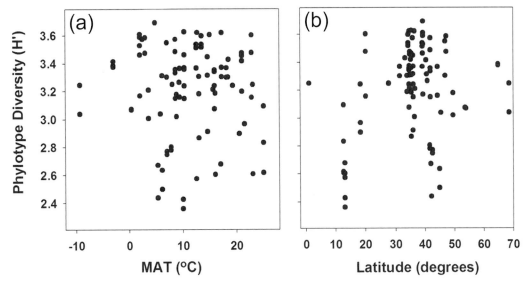

FIGURE 1.11 Relationships between phylotype diversity (Shannon index, H') and mean average temperature (MAT) (a) or latitude (b). Reprinted from Fierer and Jackson (2006) with permission. Copyright 2006 National Academy of Sciences, U.S.A.

the data would appear only as a random distribution. Taken from plots of approximately 100 m², the soil samples for the Fierer and Jackson study were collected from plots smaller than those used in regional, or large-scale, plant and/or animal biogeographical analyses. Thus, the soil collections made for the bacterial diversity studies were equivalent to a local scale in plant and animal biodiversity studies. However, Fierer and Jackson justify their choice of plot size by stating that the average size of bacterial cells within a population in any environment is orders of magnitude smaller than that of plants or animals, so the number of individuals per plot is comparable. The authors do acknowledge the potential consequences to their analysis if their assumption is incorrect. Specifically, they predict that the importance of local parameters such as pH would cause community composition to be overestimated whereas the importance of regional factors such as MAT and latitude would cause community composition to be underestimated.

Concluding thoughts. Not only did environmental factors cited to be good predictors of plant and animal diversity at continental scales have little effect on soil bacterial diversity, but the biogeographical patterns observed in soil bacterial communities appear to be fundamentally different from those observed in well-studied plant and animal communities. For example, phylotype diversity for bacteria is low in soil collected from the Peruvian Amazon (tropics). In contrast, this region exhibits the highest plant diversity (Myers et al., 2000). Instead, bacterial diversity appears highest in semiarid regions of the United States (Arizona, New Mexico, Nevada, Utah, California, Colorado, Idaho, Oregon, Montana, and Wyoming), which typically receive between 10 and 20 in. of rain annually and see temperature variations up to as much as 100°F (see NCDC Climate Monitoring Reports and Products in Web Resources below). It would seem that a larger number of bacterial phylotypes is seen in environments with greater fluctuations in temperature and precipitation than in environments in which the abiotic factors are more constant, such as the equatorial tropics or north and south poles. Seasonal or annual changes in the environment seem to promote diversification of bacterial communities, whereas plants and animal thrive within a much more restrictive range.

REFERENCES

Chao, A., and T.-J. Shen. 2003. Nonparametric estimation of Shannon's index of diversity when there are unseen species in sample. *Environ. Ecol. Stat.* **10**:429–443.

Chung, K., and D. H. Ferris. 1996. Martinus Willem Beijerinck (1851–1931): pioneer of general microbiology. *ASM News* **62**:539–543.

Dunbar, J., L. O. Ticknor, and C. R. Kuski. 2001. Phylogenetic specificity and reproducibility and new method for analysis of terminal restriction fragment profiles of 16S rRNA genes from bacterial communities. *Appl. Environ. Microbiol.* **67**:190–197.

Fraterrigo, J. M., T. C. Balser, and M. G. Turner. 2006. Microbial community variation and its relationship with nitrogen mineralization in historically altered forests. *Ecology* **87**:570–579.

Gaston, K. 2000. Global patterns in biodiversity. *Nature* **405**:220–227.

Green, J., and B. J. M. Bohannan. 2006. Spatial scaling of microbial biodiversity. *Trends Ecol. Evol.* **21**:501–507.

Hawkins, B. A., R. Field, H. V. Cornell, D. J. Currie, J.-F. Guégan, D. M. Kaufman, J. T. Kerr, G. G. Mittelbach, T. Oberdorff, E. M. O'Brien, E. E. Porter, and J. R. G. Turner. 2003. Energy, water, and broad-scale geographic patterns of species richness. *Ecology* **84**:3105–3117.

Horner-Devine, M. C., K. M. Carney, and B. J. M. Bohannan. 2004. An ecological perspective on bacterial biodiversity. *Proc. R. Soc. Lond. Ser. B* **271**:113–122.

Myers, N., R. A. Mittermeier, C. G. Mittermeier, G. A. B. da Fonseca, and J. Kent. 2000. Biodiversity hotspots for conservation priorities. *Nature* **403**:853–858.

O'Brien, E. M. 1993. Climatic gradients in woody plant species richness: towards an explanation based on an analysis of southern Africa's woody flora. *J. Biogeogr.* **20**:181–198.

O'Malley, M. A. 2007. The nineteenth century roots of 'everything is everywhere.' *Nat. Rev.* **5**:647–651.

Van der Putten, W. H., J. N. Klironomos, and D. A. Wardle. 2007. Microbial ecology of biological invasions. *ISME J.* **1**:28–37.

Varsha. 2006. DNA fingerprinting in the criminal justice system: an overview. *DNA Cell Biol.* **25**:181–188.

Willis, K. J., and R. J. Whittaker. 2002. Species diversity—scale matters. *Science* **295**:1245–1248.

Web Resources

National Climatic Data Center (NCDC) Climate Monitoring Reports and Products http://www.ncdc.noaa.gov/oa/climate/research/monitoring.html

READING ASSESSMENT

1. Using the textural triangle for soil characterization (Fig. 1.2a) and descriptions in Table 1.1, determine which class best describes the soil sample collected by you and your team for the "I, Microbiologist" project. Briefly explain your choice.

2. In addition to the structural features of your soil sample, what other abiotic factors could potentially influence the composition of the residing bacterial community? Predict the range of diversity you expect to find in your soil sample for this class. Which abiotic factors will likely be most important in making your predictions? Consider what you know about the rhizosphere sampled by your project team: what kind of plant was it, and where was it growing?

3. Which of the following statements defines a scientific hypothesis (as opposed to a theory or a fact)?

    a. a proposal that explains observations

    b. well-substantiated principles that can be used to make predictions

    c. an explanation that has been experimentally verified

    d. a tentative explanation that guides experimental investigations

    e. a concept that has been proven and is known to be true based on experience or observation

4. In selecting restriction enzymes for use in T-RFLP, why is it important to choose enzymes that do not over- or underdigest the region of interest in the genome?

5. One limitation of T-RFLP is that the method may underestimate the total bacterial diversity within an environmental sample. Why?

6. Does a phylotype represent all bacteria in a population comprising a single species?

The following questions pertain to the Fierer and Jackson (2006) study.

7. What is the purpose or goal of this study?

8. Find the hypothesis or hypotheses tested in this study, and rephrase it (them) as a conditional proposition (i.e., an "If...then..." statement).

9. Do the results support or refute the hypothesis or hypotheses?

10. Identify and briefly explain the key result that enabled you to draw the conclusion stated in your answer to question 9 (if there was more than one hypothesis, there will likely be more than one result to discuss; specify which result addressed which hypothesis).

11. Based on your own evaluation of the data, do you agree with the conclusions of the authors? Why or why not? Identify problems or ambiguities in their results that could lead you to question their analysis.

12. Thinking about future directions, suggest one experiment the authors should do next as a follow-up to this study.

13. Why was T-RFLP analysis appropriate for this particular study? Why would a sequence-based analysis of 98 soil samples be considered impractical?

UNIT 1

EXPERIMENTAL OVERVIEW

In Experiments 1.1 through 1.4, each student team will collect a soil sample from a specific location and perform some preliminary analyses of its physical composition, including water content and pH. Detailed observations about the collection site and soil texture also should be recorded. This sample will be used as the source of bacteria for the cultivation-dependent and cultivation-independent strategies used to assess community composition in subsequent units of this textbook.

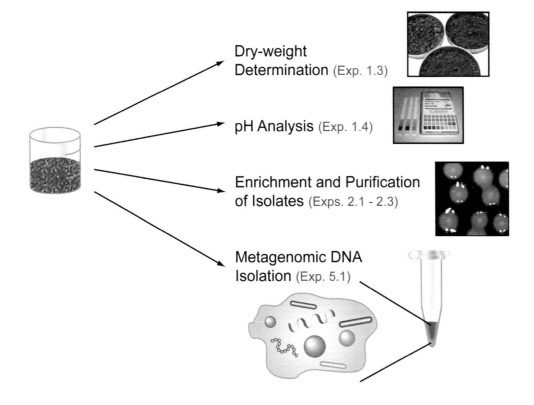

Dry-weight Determination (Exp. 1.3)

pH Analysis (Exp. 1.4)

Enrichment and Purification of Isolates (Exps. 2.1 - 2.3)

Metagenomic DNA Isolation (Exp. 5.1)

MATERIALS

Sterile 500-ml glass beakers (two)   Sterile large-bladed spatulas (two)
Sterile wire mesh colander or sieve   Digital camera

METHODS

Use aseptic technique when handling soil samples and throughout all subsequent cultivation procedures. Wear gloves except when using a Bunsen burner.

Hint: Evidence of how well the techniques were mastered will be provided by whether any of the isolates or clones turns out to be *Staphylococcus epidermidis,* a contaminant from your skin.

1. Use one 500-ml sterile beaker and one sterile large-bladed spatula.

2. Go outside with your teaching assistant (TA) or instructor. Your team should choose a site near the base of a tree or plant from which to collect soil. Record observations about the collection site in your lab notebook, including the date and time of the collection, the location of the site and taxonomic identification of the tree or plant, the weather conditions (temperature, humidity, precipitation, etc.), and any other characteristics that may be important determinants of bacterial diversity.

3. Have your TA or instructor take your team's picture next to the collection site.

4. Using gloved hands, clear the ground of leaf litter (O horizon) before you begin digging up soil, then quickly uncover your beaker and use the spatula to scoop soil into the beaker until it is approximately three-quarters full. Try to obtain soil particles from top 2- to 3-cm range (A horizon), though not too close to the base of the plant or tree because you want a soil sample near young roots, where the plant is actively growing and thus stands to benefit from potential soil microbe interactions. Be sure to note observations about soil composition (are there roots, rocks, or insects present?) and texture (silt, sand, clay, loam?). Make sure the dirt is not too wet, as excessive moisture will inhibit subsequent experimental steps in the project.

5. Once you have finished collecting the soil sample, immediately re-cover your beaker before transporting it back to the laboratory. Return the soil collection site to the condition in which you found it by sweeping the leaf litter back around the collection area. Return to the laboratory, bringing your used gloves and other collection materials with you for proper disposal.

6. Use a fresh pair of gloves and a second sterile spatula. With a clean, sterile wire mesh colander, sieve the soil into a second 500-ml sterile beaker to separate out the rocks and organic matter (roots, twigs). It may help to have one team member use the sterile handle of the spatula to crush larger soil aggregates while a second team member is sieving the sample. Once finished, re-cover the second beaker with aluminum foil.

7. Label the beaker with the date of collection, collection site, tree identification (species and native country of origin), and collector's initials (or team name).

8. Store soil samples at 4°C for the duration of the "I, Microbiologist" project.

MATERIALS

Serratia marcescens (UCLA lab strain 1010)   30°C incubator
LB plates   Sterile flat toothpicks

The streak plate procedure is a method designed to isolate pure cultures of bacteria from mixed populations by simple mechanical separation. Cells derived from a bacterial colony are spread over the surface of a solidified medium in a petri dish such that fewer and fewer bacterial cells are deposited as the streaking proceeds. Ultimately, single cells should be deposited at widely separated points on the surface of the medium, and after incubation, they should develop into isolated colonies. It is generally assumed that well-isolated colonies arise from single cells, an assumption that is frequently, but not always, true. Many streaking patterns will give the desired result if properly done. Practice these techniques as illustrated below.

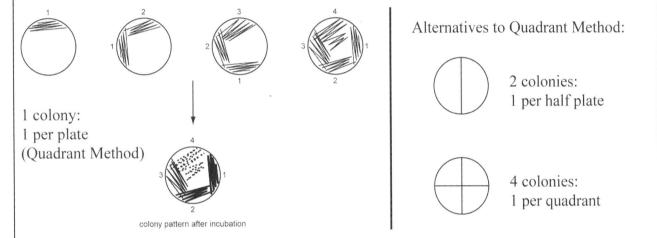

1 colony:
1 per plate
(Quadrant Method)

colony pattern after incubation

Alternatives to Quadrant Method:

2 colonies:
1 per half plate

4 colonies:
1 per quadrant

Quadrant method

For the first quadrant, the sample is picked up with the wide end of a sterile flat toothpick and gently spread over about one-quarter of the surface of the medium by using a rapid, smooth, back-and-forth motion. The toothpick should be held gently between your thumb and ring finger at a slight angle (10 to 20°) to the medium. Streak the sample back and forth across the plate many times, moving towards the center, as illustrated in the diagram. The toothpick should not dig into the agar. For the second quadrant, flip the toothpick over, then turn the petri dish 90° to streak the second quadrant, again using the back-and-forth pattern as shown. The streaks should cross over about the last half of the streaks in the first quadrant. The toothpick must not go back into the first streak of the first quadrant, along which most of the original colony was deposited. For the third and fourth quadrants, use the wide end of a new toothpick to streak the quadrants. Repeat the streaking steps as illustrated, crossing over the streaks in the preceding quadrant. Avoid going into the first quadrant when streaking the fourth quadrant.

PERIOD 1

Examine the bright pink *S. marcescens* colonies on the LB plate provided by your TA or instructor. Practice streak plating from a single colony using toothpicks onto a new LB plate (quadrant method). Next, divide a second LB plate in half by drawing a line with a marker

Experiment continues

down the center of the bottom of the plate, then streak purify two colonies onto the same plate (alternative method). With practice, you should be able to purify up to four colonies on a single plate. Incubate the plates at 30°C for 48 hours.

PERIOD 2

Examine the streak plates for single colonies in the appropriate quadrant, half, or quarter of the plates. Record the results in your lab notebook, noting which method works best for you. For subsequent experiments, you will want to use the method that results in single colonies. If you were not successful during this training exercise, you should repeat the procedure until you obtain single colonies.

MATERIALS

Fresh sieved soil

Drying oven preheated to 105°C

Desiccator

Aluminum weighing dishes

Analytical scale

Crucible tongs

METHODS

PERIOD 1

Weigh soil, then dry it at 105°C.

1. Obtain two aluminum weighing dishes. Ensure that each dish is labeled with a number.

2. Weigh the first dish. Be sure to tare the scale, then fill the dish with soil (20 to 30 g).

   Caution: Avoid spilling soil; do not overfill the weighing dishes. Record the dish number and corresponding soil sample in your lab notebook. Reweigh and calculate the mass of soil added by difference:

   (Mass of dish + soil) − (Mass of empty dish) = Mass of soil sample (grams)

3. Repeat step 2 with a second dish and another 20 to 30 g of your soil sample.

4. Place the dishes containing soil in a drying oven at 105°C for at least 24 hours.

PERIOD 2

Reweigh soil after drying.

1. Wearing gloves, use tongs to carefully remove the soil-filled dishes from the drying oven one at a time.

2. Place dishes in a desiccator, and allow them to cool for 1 hour or more if needed.

3. Once they have cooled to room temperature, remove them from the desiccator and reweigh:

   Mass of dish + dry soil = _____

4. Return the dishes containing soil to the 105°C drying oven for at least another 24 hours.

PERIOD 3

Reweigh the soil after drying it a second time, and calculate the water content based on dry weight.

1. Wearing gloves, use tongs to carefully remove the soil-filled dishes from the drying oven one at a time.

2. Place dishes in a desiccator, and allow them to cool for 1 hour or more if needed.

3. Once they have cooled to room temperature, remove them from the desiccator and reweigh:

   Mass of dish + dry soil = _____

Experiment continues

Dry-weight measurements for soil must be consistent (less than 5% difference) for two consecutive trials, indicating that all the water has evaporated. If the results are not consistent, the procedure for Period 3 must be repeated until consistent measurements are obtained.

4. Calculate the gravimetric moisture content of each soil sample by using the following equation:

$$\theta_g = (m - d)/d \times 100$$

where θ_g represents the gravimetric moisture content (%), m is the mass of soil prior to drying, and d is the mass of the same soil sample following drying.

5. Report the average percent moisture content based on the replicate values obtained from two soil samples.

6. Discard your dried soil samples when measurements have been completed. Weighing dishes can be cleaned, autoclaved, and reused. Do not leave samples in the drying oven or desiccator.

REFERENCE
Pepper, I. L., and G. P. Gerba. 2004. *Environmental Microbiology: a Laboratory Manual*, 2nd ed. Elsevier Academic Press, Burlington, MA.

EXPERIMENT 1.4 Determine the pH of Soil

The number and types of microorganisms in soil depend to some extent on the pH of the soil (Fierer and Jackson, 2006). Moist, acidic soil or dry, basic soil harbors fewer and different microbes than neutral soil, which tends to be rich with minerals that support various kinds of microbial life.

MATERIALS

Fresh sieved soil	Sterile large-bladed spatula
Sterile distilled water	Sterile 15-ml conical tube with cap
pH strips (or blue and red litmus paper)	Rack for 15-ml tubes
Vortex Genie II	Horizontal multitube vortex adapter
Clinical centrifuge	Digital camera

METHODS

1. Aseptically transfer a small portion of sieved soil into a 15-ml conical tube (the amount should not exceed the 3.0-ml graduation mark on the side of the tube). Repeat with a second tube.

2. For the first tube, add sterile distilled water to the soil, filling the tube to approximately the 9-ml graduation mark, effectively making a 1:3 dilution. Screw the cap onto the tube. Repeat with the second tube, but add tap water instead of distilled water.

3. Mix the soil and water by shaking or gently vortexing the tubes for 30 minutes. Then place the tubes upright in a rack on a flat surface, allowing the soil particles to settle to the bottom of each of the tubes. *Note:* **You may need to centrifuge at low speed briefly to ensure that the particles are completely settled.**

4. While the soil particles settle, prepare two control tubes, one containing 9 ml of sterile distilled water and the other containing 9 ml of tap water.

5. Using pH strips with colorimetric indicators, check the pH of the control tubes, one at a time, by dipping the end of a pH strip into the water. Observe the color changes on the pH strip right away, as the color intensity diminishes with time. Estimate the pH of the water controls by comparing the colors on each strip to the pH standards key on the box.

Repeat this procedure with each of the tubes containing soil plus water, being careful not to disturb the soil. Approximate the pH of the soil sample. Do the mixtures made with distilled and tap water agree?

6. Take pictures with the camera of all four pH strips for your lab notebook. Discard used pH strips. Return unused materials to TA or instructor when finished with the experiment.

Alternative procedures for pH determination

1. Remove 20 μl of water from settled mixture using a pipette, and dab water onto strips of blue and red litmus paper such that the strips absorb some of the water. With color changes occurring over a pH range of 4.5 to 8.3 (at 25°C), blue litmus

Experiment continues

paper turns red under acidic conditions and red litmus paper turns blue under basic conditions. Observe the color changes on the litmus paper and then predict the approximate pH of the sample (acidic or alkaline). Always include controls.

2. A more sophisticated way to measure soil pH is with the use of a handheld pH meter, which can be purchased at garden shops and nurseries or from a variety of standard scientific-equipment vendors. Such devices may also permit concurrent measurements of temperature or ion concentrations.

REFERENCES

Fierer, N., and R. B. Jackson. 2006. The diversity and biogeography of soil bacterial communities. *Proc. Natl. Acad. Sci. USA* **103**:626–631.

University of South Carolina Center for Science Education http://www.cas.sc.edu/cse/detersoil.html

UNIT 2

Cultivation-Dependent Community Analysis

SECTION 2.1
ENRICHMENT STRATEGIES THAT TARGET THE MICROBIAL INHABITANTS OF THE RHIZOSPHERE

Microbial ecologists use a variety of strategies to study and measure the activities of the microflora in the environment. Both cultivation-dependent and cultivation-independent techniques are used to help ecologists understand the roles of individual microorganisms in their natural settings and formulate hypotheses about metabolic interactions between and among microorganisms. If a microorganism is available for study in pure culture, its genetic information becomes accessible for further study. Unfortunately, fewer than 5% of microorganisms present in many natural environments are readily culturable, representing a bottleneck to microbial ecologists (Torsvik et al., 1990; Torsvik and Øvreås, 2002; Amann et al., 1995) (Fig. 2.1). The impact of this bottleneck is profound, meaning the vast majority of microorganisms are not accessible for basic research (e.g., ecological studies of microbial activities or physiological contributions to biogeochemical cycles) and are not available for biotechnological and pharmaceutical advancement. Nevertheless, bioprospecting, or discovery of novel products, has advanced with the use of cultivation-independent approaches (discussed in Unit 5) that bypass the need to first isolate bacteria in the laboratory.

A variety of methods have been developed to cultivate a broad and diverse representation of soil bacteria, a goal that requires knowledge of the nutritional requirements of microorganisms as well as the environmental conditions needed to promote and sustain their growth. This information can be applied to control growth under artificial laboratory conditions that simulate a natural environment, allowing differentiation between types of bacteria during the isolation process as well as their characterization based on biochemical and physiological properties (Joseph et al., 2003; Kaeberlein et al., 2002; Zengler et al., 2002). Before selecting growth media for inoculation, one should formulate

FIGURE 2.1 Representation of the "bottleneck" to microbial ecology. Illustration by Cori Sanders (iroc designs) based on an image from Campbell and Reece (2005, with permission).

hypotheses about the types of microorganisms that would be expected to grow based on their nutritional requirements and cultivation conditions.

Microbial metabolism "in a nutshell"

When cultivating microorganisms, you need both an energy source and a carbon source. An energy source supplies electrons needed to make adenosine triphosphate (ATP), which can be generated by one of two general processes: fermentation produces ATP via substrate-level phosphorylation, whereas cellular respiration couples the proton motive force generated by an electron transport chain with an ATP synthase to make ATP via oxidative phosphorylation or photophosphorylation. Some respiring bacteria derive energy in an aerobic manner, using oxygen (O_2) as a terminal electron acceptor during respiration, whereas others use inorganic compounds other than O_2 (e.g., nitrate or sulfate) as a terminal electron acceptor, producing ATP under anaerobic conditions.

Organisms can be classified according to the source from which they derive energy, which may involve NAD(P)H as a reductant for biosynthetic reactions in the cell. Phototrophs such as the cyanobacteria use light as an energy source and require reducing power for the conversion of CO_2 to biomass. Phototrophs use a source of electrons with high redox potential, such as those derived from the oxidation of water (H_2O), to produce NAD(P)H. The purple and green sulfur bacteria also use light as an energy source but generate reducing power for CO_2 assimilation by using hydrogen sulfide (H_2S) or hydrogen (H_2) in place of water.

Chemoorganotrophs and chemolithotrophs use organic or inorganic chemicals, respectively, that are capable of donating electrons as an energy source. Chemolithotrophic metabolism is typically aerobic, whereas chemoorganotrophic metabolism can be either aerobic or anaerobic, although less energy is released under anoxic conditions. Chemoorganotrophs are responsible for degradation of substrates such as those derived from plants (cellulose, hemicellulose, lignin, etc.). Some microorganisms even decompose toxic hydrocarbons such as petroleum; this metabolic function can be exploited in bioremediation projects involving oil spills (Head et al., 2006). Fermentative chemoorganotrophs couple the oxidization of NADH to the reduction of pyruvate into fermentation products (lactate, ethanol, butyrate, H_2, etc.). Respiring chemoorganotrophs, on the other hand, oxidize NADH in generating a proton motive force to drive ATP synthesis. In chemolithotrophs, reducing power is obtained either directly from inorganic compounds with low redox potential or by reverse electron transport reactions.

A carbon source provides the carbon atoms used in biosynthesis of organic compounds. Again, organisms can be classified according to the type of carbon source used in the anabolic pathways. Autotrophs use carbon dioxide (CO_2) and are referred to as primary producers in the overall food chain. Heterotrophs are consumers, assimilating organic compounds produced by other organisms; in the microbial world, these organisms are the scavengers and decomposers of organic matter (cellulose, hemicellulose, lignin, chitin, etc.). A more thorough treatment of microbial metabolism is beyond the scope of this textbook; however, interested students are encouraged to consult microbiology textbooks such as Staley et al. (2007) or Madigan and Martinko (2006) as well as online resources such as *Todar's Online Textbook of Bacteriology* (see Web Resources at the end of this section) to explore this topic in more detail.

Microbial growth requirements

Major elements other than carbon are required for growth. Water, as the "universal solvent" for nutrients in culture media, provides hydrogen and oxygen for chemical reactions. For example, the oxygen in CO_2 formed during catabolism of most substrates comes from H_2O. Note, however, that the O_2 used in respiration is taken up as O_2 and is reduced to H_2O during respiration. Microbes also require sources of nitrogen, sulfur, phosphorus, potassium, magnesium, calcium, and other trace elements. Whereas prototrophs can produce the essential constituents to meet metabolic and structural needs, fastidious microorganisms and auxotrophs are not capable of synthesizing all necessary organic compounds and thus require growth factors as well.

As discussed in Unit 1, microbial growth is affected by a range of abiotic factors such as pH, temperature, light, oxygen, and water availability. Each organism has an optimal range for growth, and the habitat from which the organism is obtained should be considered when choosing environmental conditions suitable for promoting growth in the laboratory. Again, when thinking about the variations in environmental conditions used in the "I, Microbiologist" project, one should consider asking what types of microorganisms one would expect to find. The most important message to take away from this type of thought experiment is that *no single medium can support the growth of all microorganisms*. This statement summarizes one of the biggest challenges to microbial ecologists, in that there are so many microorganisms in the environment and they cannot be studied from a genetic, biochemical, or physiological standpoint because they all cannot be readily cultivated. Success is completely dependent upon the ability of a scientist to devise a medium that will support the growth of various nutritional classes of microorganisms, often at the expense of one group with respect to another. When one starts to consider the various lifestyles of microorganisms, especially those living on the edge of energetic sustenance (e.g., methanogens), those that depend upon syntrophic relationships with other organisms to sustain energetic requirements, and those that maintain symbiotic relationships with host cells (e.g., rhizobia), then the task begins to feel like an almost insurmountable challenge. And although coculture techniques have been developed to try to get around these problems, the reality is that microbes live within consortia that cannot be replicated under laboratory conditions.

Cultivation media for the "I, Microbiologist" project

The ingredients in culture media vary extensively, ranging from complex cellular materials to pure chemical compounds. Media may be in liquid (broth) form or, if agar is added to the broth, in solid form. Five different types of solid media have been selected for the "I, Microbiologist" research project targeting bacteria residing in the rhizosphere: RDM, R2A, ISP4, N$_2$-BAP, and VXylA. Remember, however, that the challenge is not only to characterize the diversity of microorganisms in the soil but also to find novel bacteria: those that no one else has isolated before. Think about devising a cultivation strategy accordingly, exercising some patience and allowing plates to incubate longer so that the slow-growing microbes will have a chance to form colonies.

Rhizobium defined medium (RDM) is a minimal medium designed to promote the growth of rhizobia, which are gram-negative, non-spore-forming soil bacteria (Fig. 2.2) (Vincent, 1970). There are two groups of rhizobia, one designated the α subclass with members in the family *Rhizobiaceae* in the *Alphaproteobacteria* (e.g., species of *Mesorhizobium, Azorhizobium, Sinorhizobium, Bradyrhizobium, Rhizobium,* and *Methylobacterium,* as well as *Blastobacter denitrificans* and *Devosia neptuniae*) and the other designated the β subclass with members in the *Betaproteobacteria* (e.g., some *Burkholderia* species, *Cupriavidus taiwanensis,* and *Herbaspirillum lusitanum*) (Lee and Hirsch, 2006). These bacteria are best known for their ability to form nitrogen-fixing nodules on the roots of a specific group of angiosperms called legumes (Hirsch, 1992). Nitrogen fixation occurs in the context of a symbiosis between particular rhizobacteria and certain plants, an interaction of importance to the agricultural industry since many legumes are food stocks (Table 2.1). Rhizobia capable of establishing a symbiotic existence with legumes

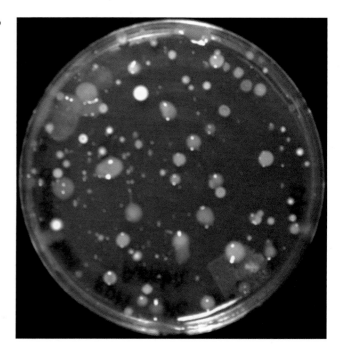

FIGURE 2.2 Sample enrichment on RDM agar incubated for 24 to 48 hours at 30°C. Photograph taken by Chao Xian (Jessica) Lin, Phong Pham, and Omar Sandoval, "I, Microbiologist" participants at the University of California, Los Angeles (UCLA), in spring 2008.

Table 2.1 Symbiotic nitrogen-fixing bacteria and their host plants[a]

Plant	Bacterial symbiont(s)
Leguminous plants	
Alfalfa *(Medicago sativa)*	*Sinorhizobium meliloti, S. medicae*
Lotus *(Lotus japonicus)*	*Mesorhizobium*
Pea *(Pisum sativum)*	*Rhizobium leguminosarum* bv. viciae
Siratro *(Macroptilium atropurpureum)*	*Azorhizobium*
Soybean *(Glycine max)*	*Bradyrhizobium japonicum*
Sweet clover *(Melilotus* spp.)	*Sinorhizobium meliloti*
Nonleguminous plants	
Alder tree (birch [Betulaceae] family)	
Alnus spp.	*Frankia* (actinomycete)
Bayberry tree (Myricaceae family)	
Sweet fern *(Comptonia peregrina)*	*Frankia* (actinomycete)
Sweet gale, bayberry *(Myrica* spp.)	*Frankia* (actinomycete)
Water fern *(Azolla)*	*Anabaena* (cyanobacterium)

[a]The list is not comprehensive.

do so by entering the cells of plant roots, a complex but well choreographed process that leads to the formation of root nodules containing modified bacterial cells called bacteroids that fix nitrogen (Lee and Hirsch, 2006). Infected plants assimilate the ammonia produced by the bacteroids into organic nitrogen-containing compounds such as the amino acids glutamine and asparagine. In exchange, the plant provides bacteroids with carbon compounds as an organic energy source that drives nitrogen fixation, as well as a microaerophilic environment in which N_2 is reduced to ammonia. As presented in Table 2.1, nitrogen-fixing symbioses also may occur in nonleguminous plants, but these associations involve bacteria other than rhizobia.

RDM also supports the growth of *Agrobacterium,* a genus of gram-negative plant pathogens that persist as biofilms, or surface-associated populations of cells, on living plant tissue (Ramey et al., 2004). Specifically, these bacteria use horizontal gene transfer to produce tumors in plants. For example, *Agrobacterium tumefaciens* is a well-studied species that causes crown gall disease upon transferring a DNA segment (T-DNA plasmid) from the bacterium to the plant (Francis and Spiker, 2005). The plasmid T-DNA integrates into the genome of the host plant cell, triggering the expression of virulence genes that alter the plant hormone balance, which results in unregulated cell division (tumors).

Like RDM, R2A agar is considered a minimal medium, but only in the sense that nutrient concentrations are relatively low compared to those in complex medium (Fig. 2.3) (Reasoner and Geldreich, 1979). R2A contains yeast extract, which provides microbes with a source of amino acids, vitamins, coenzymes, growth factors, and some trace minerals. It also contains peptone (digested beef muscle), providing a source of nitrogen, sulfur, carbon, and energy. Addition of Casamino Acids, glucose, and pyruvate to the medium provides preformed organic compounds necessary for heterotrophs to grow. When used in combination with low incubation temperature and longer incubation times, R2A supports the growth of fastidious microorganisms as well. Based on previous work with this medium, students are likely to isolate gram-positive representatives of the phyla *Firmicutes* (including *Bacillus*) and *Actinobacteria,* as well as several gram-negative

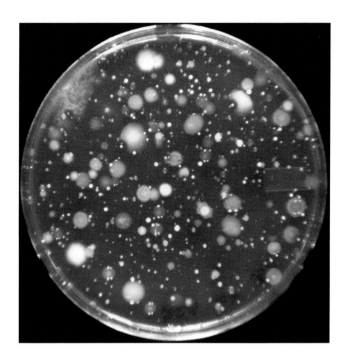

FIGURE 2.3 Sample enrichment on R2A agar incubated for 24 to 48 hours at 30°C. Photograph taken by Gary Chou, Kha Lai, Bac Nguyen, and Michael Nguyen, "I, Microbiologist" participants at UCLA in fall 2007.

members of the phyla *Bacteroidetes* and *Proteobacteria*, the latter of which includes nitrogen-fixing rhizobia. In addition, the microbes purified from R2A agar often display a variety of antibiotic production and/or resistance patterns.

ISP4 and N_2-BAP agars are considered selective media, selecting for *Streptomyces* and actinomycete nitrogen fixers, respectively (Fig. 2.4 and 2.5). Each medium contains nutrients that specifically promote the growth of a particular organism while inhibiting the growth of other types of organisms. Other common selective media contain antibiotics (e.g., ampicillin, kanamycin, and chloramphenicol), which inhibit the growth of microbes not expressing resistance genes necessary for survival on this type of medium. These

FIGURE 2.4 Sample enrichment on ISP4 agar incubated for 72 hours to 1 week at 30°C. Photograph taken by Gary Chou, Kha Lai, Bac Nguyen, and Michael Nguyen, "I, Microbiologist" participants at UCLA in fall 2007.

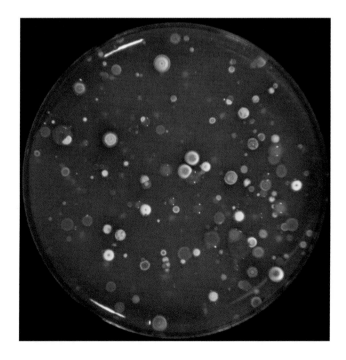

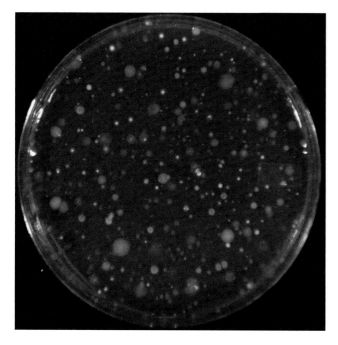

FIGURE 2.5 Sample enrichment on N$_2$-BAP agar incubated for 72 hours to 1 week at 30°C. Photograph taken by Gary Chou, Kha Lai, Bac Nguyen, and Michael Nguyen, "I, Microbiologist" participants at UCLA in fall 2007.

medium types are often used in bacterial transformation experiments whereby artificial plasmids that encode genes conferring antibiotic resistance are introduced (refer to Unit 5 for specific examples).

ISP4, which stands for International *Streptomyces* Project, is also referred to as inorganic salts starch agar (Fig. 2.4) (Shirling and Gottlieb, 1966). Although this medium was specifically developed for isolating *Streptomyces* species, it selects for other gram-positive bacteria, including coryneforms and actinomycetes. The coryneforms are non-sporulating, nonfilamentous rod-shaped bacteria. In contrast, the actinomycetes are filamentous, with many species that form aerial filaments (mycelia) containing spores called conidia (Madigan and Martinko, 2006). *Streptomyces* species are responsible for the "earthy odor" of soil due to the production of geosmins. Mature *Streptomyces* colonies are readily identifiable, being compact and having a powder-like appearance. The conidia are often pigmented, giving rise to colonies of diverse colors characteristic of a particular species. The other characteristic for which members of the genus *Streptomyces* are well known is their ability to produce antibiotics. More than 50% of commercially available antibiotics, including tetracycline, erythromycin, streptomycin, and chloramphenicol, are produced by *Streptomyces*. This property is best demonstrated when one examines plates with bacterial concentrations yielding confluent growth, in which a zone of growth inhibition around particular colonies can be seen, indicative of antibiotic production by those colonies (Fig. 2.6). Antibiotic production is linked to the developmental process of sporulation, which is triggered by nutrient depletion, providing a potential ecological rationale to explain why antibiotics are produced by *Streptomyces* in the first place (i.e., as a mechanism to inhibit the growth of other organisms competing for limited resources).

N$_2$-BAP agar is a defined medium specifically designed to select for certain bacteria that convert nitrogen (N$_2$), which is a major constituent of Earth's atmosphere, into a form that can be used for biosynthesis (Fig. 2.5) (Murry et al., 1984). Note that in the literature this medium is termed "BAP." We have added the "N$_2$" as a reminder to

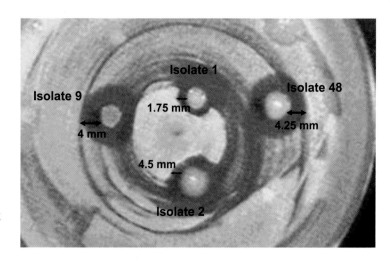

FIGURE 2.6 Antibiotic production against *M. luteus* bacterial indicator strain. Photograph taken by To Hang (Shela) Lee, "I, Microbiologist" participant at UCLA in winter 2006.

students of the selection that this medium facilitates. Nitrogen fixation comprises several steps catalyzed by an enzyme called nitrogenase. The overall reaction results in the reduction of atmospheric nitrogen (N_2) to ammonia (NH_3) as shown below:

$$N_2 + 8H^+ + 8e^- + 16ATP \rightarrow 2NH_3 + H_2 + 16ADP + 16P_i$$

This metabolic process commands a great deal of energy (requiring 16 ATPs) due to the stability of the relatively inert triple bond in dinitrogen ($N \equiv N$). Owing to the high energy demands, a microbe capable of nitrogen fixation does so only under conditions where nitrogenous compounds such as ammonia, nitrates, or amino acids are extremely scarce or completely lacking in the environment, making N_2 the sole source of nitrogen. Such environmental conditions are provided by the N_2-BAP agar, which supplies all essential elements except nitrogen.

In addition to *Rhizobium*, the soil contains free-living nitrogen fixers known as diazotrophs, some of which are listed in Table 2.2. Based on previous work with this medium, students are likely to isolate nitrogen-fixing pseudomonads, *Bacillus*, and *Streptomyces*. Depending on the soil collection site selected for this project, the plants may or may not have been treated with nitrogen-based fertilizers. In the absence of treatment, the only source of nitrogen available to plants is that produced by microbes. It is important to note whether the soil samples are collected from a nutrient-limited terrestrial environment. So again, when considering the types of microorganisms one could expect to find in the soil samples on the basis of their nutritional requirements, the source of the soil sample as well as the type of plant at the collection site should be considered when developing one's hypothesis. For example, rhizobia are expected to be isolated from the roots of legumes but can be isolated from the rhizosphere of other plants as well.

VXylA is a defined medium that allows the growth of a wide range of heterotrophic soil bacteria, including commonly isolated groups and those belonging to phylogenetic lineages represented by only a few cultivated members (Fig. 2.7) (Davis et al., 2005). This medium utilizes gellan as the solidifying agent, rather than agar, making the plate medium more translucent and delicate; it is easy to puncture with a colony spreader. As summarized in Table 2.3, in comparison to results using an all-purpose medium (tryptic soy agar [TSA]) in which the nutrient amounts have been reduced 10-fold (analogous to R2A in the "I, Microbiologist" project), 21% of the isolates obtained on VXylA agar

Table 2.2 Some examples of free-living aerobic nitrogen-fixing bacteria[a]

Phototrophs	Chemoorganotrophs	Chemolithotrophs
Cyanobacteria	*Azotobacter*	*Alcaligenes*
	Azomonas	*Thiobacillus*
	Azospirillum lipoferum	*Acidithiobacillus*
	Beijerinckia	*Cupriavidus*[b]
	Burkholderia unamae	
	Citrobacter freundii	
	Gluconoacetobacter diazotrophicus	
	Methylococcus	
	Methylomonas	
	Methylosinus	
	Mycobacterium flavum	
	Paenibacillus azotofixans	
	Pseudomonas stutzeri	

[a] Nitrogen fixation may occur in only one or a few species from the listed genera. This list is not comprehensive.

[b] This bacterium is a facultative aerobe.

were members of rarely isolated groups (e.g., *Acidobacteria* and *Rubrobacteridae*). The *Acidobacteria* are a particularly underrepresented group within the domain *Bacteria*, having only a few cultivated members despite their abundance in the soil (Sait et al., 2002). Members of this phylum comprise up to 14% of the total soil community as ascertained from cultivation-independent studies. The entire genome sequence for one acidobacterium isolated from an Australian pasture (Ellin345) has been determined recently by the DOE Joint Genome Institute and is available to the public in the Integrated Microbial Genomes database (Markowitz et al., 2007). Although phenotypic analysis of Ellin345

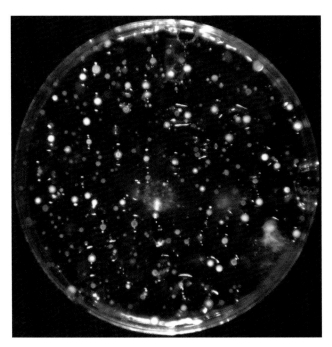

FIGURE 2.7 Sample enrichment on VXylA agar incubated for 2 to 3 weeks at 30°C. Photograph taken by Gary Chou, Kha Lai, Bac Nguyen, and Michael Nguyen, "I, Microbiologist" participants at UCLA in fall 2007.

Table 2.3 Bacterial isolates that formed visible colonies on plates with different media inoculated with soil[a]

Medium	Total no. of isolates identified	% of isolates affiliated with rarely isolated lineages	Lineage represented (no. of isolates)
VXylA	33	21 (7/33)	*Acidobacteria* (2), *Actinobacteridae* (14), *Proteobacteria* (12), *Rubrobacteridae* (5)
0.1X TSA	33	0	*Actinobacteridae* (18), *Bacteroidetes* (1), *Firmicutes* (5), *Proteobacteria* (9)

[a]Table from Davis et al. (2005), modified with permission of the authors.

has revealed a gram-negative, highly capsulated, aerobic heterotroph, annotation of the genome should produce information about the gene content of Ellin345, suggesting how the *Acidobacteria* contribute to the soil ecosystem.

The enrichment of *Acidobacteria* members on VXylA medium should not be surprising. VXylA has a pH of 5.5, providing an acidic growth medium typical of environments in which *Acidobacteria* members are most abundant (Sait et al., 2006). As shown in Fig. 2.8, when the DNA sequences from 16S rRNA genes from 25 different soil libraries were related to the pH of the soil from which the library was made, there was a strong correlation between the proportion of *Acidobacteria* 16S rRNA gene sequences in each library and the soil pH (Sait et al., 2006). The sequences were derived from members of a single subdivision (class) in the phylum *Acidobacteria* and appear to be numerically more abundant in soils with pH values below 6.

The same research group also demonstrated that colony counts increase with longer incubation times on VXylA medium, although 80% of the total number of colonies produced over a 12-week period are obtained by the fourth week (Fig. 2.9) (Janssen et

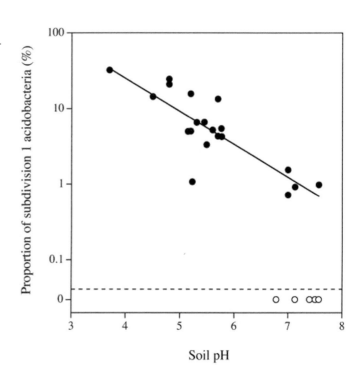

FIGURE 2.8 Proportion of *Acidobacteria* 16S rRNA genes detected in 25 libraries derived from soils with different pH values. Libraries in which *Acidobacteria* were detected (●) were used to calculate the best fit line. Libraries with no *Acidobacteria* (○) were omitted from the calculation. Reprinted from Sait et al., 2006, with permission.

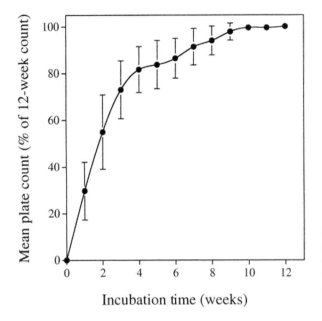

FIGURE 2.9 Effect of incubation time on the viable colony counts on enrichment agar, expressed as a percentage of the total colony count obtained after a 12-week incubation period. Each point (●) represents the mean of 12 triplicate plate count experiments, and the vertical bars indicate one standard deviation from the mean. Reprinted from Janssen et al., 2002, with permission.

al., 2002). The authors also found that using sonication, which produces sound waves that agitate and perturb soil clumps before plating, increased viable-cell counts.

The take-home message is that the bacteria belonging to phylogenetic lineages represented by only a few cultivated members tend to take a long time to grow—patience is a must! The goal is not only to enrich for rare or novel microorganisms but also to isolate them successfully in pure culture. Purification of soil isolates is not a trivial undertaking due to the challenges presented by diverse colony and cellular morphologies, which may complicate the manipulation of such microorganisms by standard laboratory techniques. A little creativity and attention to unusual morphological details are therefore required.

KEY TERMS

Abiotic factors Nonliving (physical) factors in a particular environment.

Aerobic Taking place in the presence of oxygen.

Agar A solidifying agent in microbiological growth media. Agar is an impure polysaccharide gum obtained from certain marine algae; it dissolves and melts around 100°C and solidifies around 43°C. It typically is not used as a nutrient by microorganisms.

Anaerobic Taking place in the absence of oxygen.

Angiosperms Flowering plants.

Anoxic Lacking oxygen; the converse of *oxic* (containing oxygen).

Autotroph A microorganism that obtains carbon from carbon dioxide (CO_2) for biosynthesis.

Auxotroph A bacterial strain unable to synthesize a particular compound essential for growth that is normally produced by prototrophic strains.

Bioprospecting Discovery of natural products; searching for novel enzymes or antibiotics produced by microorganisms.

Bioremediation Cleanup of toxic chemicals and environmental pollutants by microorganisms.

Chemolithotroph A microorganism that uses inorganic compounds to drive energy production; also called a lithotroph.

Chemoorganotroph A microorganism that uses organic compounds to drive energy production.

Coculture Cultivation of two or more microorganisms in single culture.

Complex medium A medium rich in a wide variety of nutrients (including growth factors), prepared from complex materials such as cellular extracts or protein digests (peptone) and tissues; exact chemical composition not known.

Confluent growth Growth where the entire surface of an agar plate is covered with bacterial cells.

Consortia Complex communities of microorganisms that interact and coordinate activities such that all members benefit from the association.

Cultivate To promote the growth of microbial cultures in the laboratory.

Defined medium A medium in which the exact amounts of nutritional ingredients are known.

Diazotroph A free-living bacterium capable of nitrogen fixation; synonym for nitrogen fixer.

Enrichment A method to increase the relative number of microorganisms demonstrating desired properties, growth characteristics, or behaviors.

Fastidious Having complex nutritional requirements which must be met via uptake of particular compounds (e.g., amino acids, nucleotides, vitamins) that organisms cannot synthesize themselves from the environment.

Gellan A solidifying agent used in microbiological growth media as an alternative to agar. Gellan tends to be broken down by a wide variety of microorganisms, so colonies growing on the surface of the media may appear depressed. It is softer than agar; thus, gellan-based medium is more easily punctured with toothpicks during streak plating.

Heterotroph A microorganism that uses preformed organic carbon compounds, which it is unable to synthesize itself, from the environment for biosynthesis.

Horizontal gene transfer Movement of a gene or group of genes from one organism to another by mechanisms other than vertical transmission. It may occur among bacteria via transduction, transformation, or conjugation but not by binary fission. It occurs between *Agrobacterium* and plants via the conjugative tumor-inducing (TI) plasmid. Also called lateral gene transfer.

Legumes Flowering plants in the family Fabaceae, which are known for their ability to fix atmospheric nitrogen due to their symbiotic association with rhizobia.

Microflora The microbial inhabitants of a specified area.

Minimal medium A medium that supplies only the basic nutritional requirements of prototrophic organisms.

Phototroph A microorganism that uses light-driven electron transport for energy production.

Prototroph A strain of bacteria capable of synthesizing all nutrients required for growth on a particular medium; considered the wild-type strain of a particular species.

Purification A method used to obtain a pure and genetically uniform culture of microorganisms.

Rhizobium A genus of bacteria that fixes nitrogen after establishing a symbiotic association inside root nodules of certain plants.

Rhizosphere The contiguous region of soil surrounding the roots of plants.

Selective medium A medium that supports growth of the desired organisms while inhibiting growth of many or most of the unwanted ones, either by containing one or more selective agents which "poison" certain types of organisms or by containing or not containing certain nutrients such that the desired organisms are able to grow.

Sonication The application of ultrasonic energy to a sample using a laboratory device called a sonicator, causing the agitation and dispersal of particles such as soil clumps.

Symbiosis Two or more dissimilar organisms intimately living together, facilitating mutually beneficial interactions between the organisms.

Syntrophic relationship A nutritional scenario in which two or more microorganisms combine their metabolic capabilities to catabolize a substance that cannot be catabolized by either microorganism on its own.

REFERENCES

Amann, R. I., W. Ludwig, and K. H. Schleifer. 1995. Phylogenetic identification and in situ detection of individual microbial cells without cultivation. *Microbiol. Rev.* **59**:143–169.

Campbell, N. A., and J. B. Reece. 2005. *Biology,* 7th ed. Pearson Education, Inc., San Francisco, CA.

Davis, K. E. R., S. J. Joseph, and P. H. Janssen. 2005. Effects of growth medium, inoculum size, and incubation time on culturability and isolation of soil bacteria. *Appl. Environ. Microbiol.* **7**: 826–834.

Francis, K. E., and S. Spiker. 2005. Identification of *Arabidopsis thaliana* transformants without selection reveals a high occurrence of silenced T-DNA integrations. *Plant J.* **41**:464–477.

Head, I. M., D. M. Jones, and W. F. Röling. 2006. Marine microorganisms make a meal of oil. *Nat. Rev. Microbiol.* **4**:173–182.

Hirsch, A. M. 1992. Tansley Review no. 40. Developmental biology of legume nodulation. *New Phytol.* **122**:211–237.

Janssen, P. H., P. S. Yates, B. E. Grinton, P. M. Taylor, and M. Sait. 2002. Improved culturability of soil bacteria and isolation in pure culture of novel members of the divisions *Acidobacteria, Actinobacteria, Proteobacteria,* and *Verrucomicrobia. Appl. Environ. Microbiol.* **68**:2391–2396.

Joseph, S. J., P. Hugenholtz, P. Sangwan, C. A. Osborne, and P. H. Janssen. 2003. Laboratory cultivation of widespread and previously uncultured soil bacteria. *Appl. Environ. Microbiol.* **69**:7210–7215.

Kaeberlein, T., K. Lewis, and S. Epstein. 2002. Isolating "uncultivatable" microorganisms in a simulated natural environment. *Science* **296**:1127–1129.

Lee, A., and A. M. Hirsch. 2006. Choreographing the complex interaction between legumes and α- and β-rhizobia. *Plant Signaling Behav.* **1**:161–168.

Madigan, M. T., and J. M. Martinko. 2006. *Brock Biology of Microorganisms,* 11th ed. Pearson Prentice Hall, Pearson Education, Inc., Upper Saddle River, NJ.

Markowitz, V. M., E. Szeto, K. Palaniappan, Y. Grechkin, K. Chu, I. A. Chen, I. Dubchak, I. Anderson, A. Lykidis, K. Mavromatis, N. N. Ivanova, and N. C. Kyrpides. 2007. The integrated microbial genomes (IMG) system in 2007: data content and analysis tool extensions. *Nucleic Acids Res.* **36**(Database issue)**:**D528–D533.

Murry, M. A., M. S. Fontaine, and J. G. Torrey. 1984. Growth kinetics and nitrogenase reduction in *Frankia* sp. HFP ArI3 grown in batch culture. *Plant Soil* **78:**61–78.

Ramey, B. E., M. Koutsoudis, S. B. von Bodman, and C. Fuqua. 2004. Biofilm formation in plant-microbe associations. *Curr. Opin. Microbiol.* **7:**602–609.

Reasoner, D. J., and E. E. Geldreich. 1979. A new medium for the enumeration and subculture of bacteria from potable water, abstr. N7. *Abstr. 79th Annu. Meet. Am. Soc. Microbiol.* American Society for Microbiology, Washington, DC.

Sait, M., P. Hugenholtz, and P. H. Janssen. 2002. Cultivation of globally distributed soil bacteria from phylogenetic lineages previously only detected in cultivation-independent surveys. *Environ. Microbiol.* **4:**654–666.

Sait, M., K. E. R. Davis, and P. H. Janssen. 2006. Effect of pH on isolation and distribution of members of subdivision 1 of the phylum *Acidobacteria* occurring in soil. *Appl. Environ. Microbiol.* **72:** 1852–1857.

Shirling, E. B., and D. Gottlieb. 1966. Methods for characterization of *Streptomyces* species. *Int. J. Syst. Bacteriol.* **16:**313–340.

Staley, J. T., R. P. Gunsalus, S. Lory, and J. J. Perry. 2007. *Microbial Life,* 2nd ed. Sinauer Associates, Inc., Sunderland, MA.

Torsvik, V., J. Goksøyr, and F. L. Daae. 1990. High diversity in DNA of soil bacteria. *Appl. Environ. Microbiol.* **56:**782–787.

Torsvik, V., and L. Øvreås. 2002. Microbial diversity and function in soil: from genes to ecosystems. *Curr. Opin. Microbiol.* **5:**240–245.

Vincent, J. M. 1970. *A Manual for the Practical Study of Root Nodule Bacteria.* IBP Handbook no. 15. Blackwell Scientific Publications, Oxford, United Kingdom.

Zengler, K., G. Toledo, M. Rappe, J. Elkins, E. J. Mathur, J. M. Short, and M. Keller. 2002. Cultivating the uncultured. *Proc. Natl. Acad. Sci. USA* **99:**15681–15686.

Web Resources

Department of Energy Joint Genome Institute Integrated Microbial Genomes database http://img.jgi.doe.gov/cgi-bin/pub/main.cgi

Todar's Online Textbook of Bacteriology (© 2008 by Kenneth Todar) http://www.textbookofbacteriology.net/ (refer to section on Nutrition and Growth of Bacteria)

SECTION 2.2
ANALYSIS OF THE PRIMARY LITERATURE: OVERCOMING THE MICROBIAL BOTTLENECK OF THE UNCULTIVATED MAJORITY

READING ASSIGNMENT
Sait, M., P. Hugenholz, and P. H. Janssen. 2002. Cultivation of globally distributed soil bacteria from phylogenetic lineages previously only detected in cultivation-independent surveys. *Environ. Microbiol.* **4:**654–666.

This paper exemplifies exactly what the cultivation-dependent part of the "I, Microbiologist" project sets out to accomplish, providing a classic example of how the work being done by undergraduates can be completed and published, thus offering the scientific community information that is novel.

The purpose of the study by Sait et al. (2002) was to isolate representatives of soil bacteria belonging to phylogenetic lineages represented only by 16S rRNA gene sequences and/or only by a few cultivated members (e.g., "the underrepresented majority"). Since the culturable portion of the soil microbiome is unrepresentative of the total phylogenetic diversity of the community, the strategy was to devise culturing methods that capture a more diverse representation of the microbial community and thus overcome the bottleneck to microbial ecology introduced in Section 2.1.

Looking at the flow chart in Fig. 2.10 describing the overall series of experiments in this paper, the similarity to what is being done for the cultivation-dependent part of the "I, Microbiologist" project becomes obvious—even the primers used for polymerase chain reactin (PCR) and initial sequencing are identical to those used in the study by Sait et al. (2002). Thus, close attention should be paid to not only the results but also the manner in which they are analyzed and presented, as this may serve as a model for the writing assignments associated with the "I, Microbiologist" project.

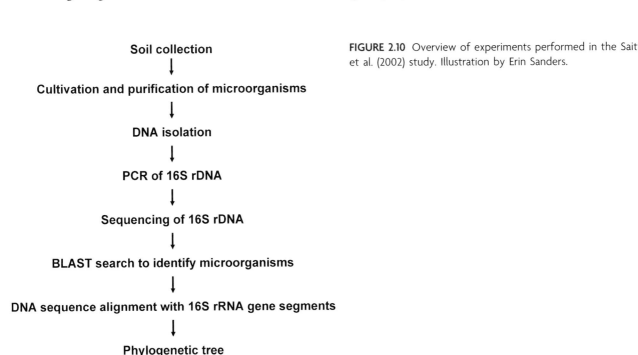

FIGURE 2.10 Overview of experiments performed in the Sait et al. (2002) study. Illustration by Erin Sanders.

What the scientific community in the field knew before the study by Sait et al. Using cultivation-independent approaches, which will be discussed in Unit 5, others had established that there were four bacterial lineages, or phylogenetic groups, common to almost all soil types. These groups are considered to be globally distributed since the soil studies were performed with soil collections from around the world. The four ubiquitous classes were identified as the *Alphaproteobacteria* (class), *Actinobacteria* (phylum), *Acidobacteria* (phylum), and *Verrucomicrobia* (phylum). The taxonomic assignments follow a nomenclature system outlined in several online resources listed at the end of Unit 7 (see Web Resources).

An attempt to cultivate members of the four globally distributed lineages. The soil for the cultivation study was collected from rotationally grazed pasture lands containing perennial rye grass *(Lolium perenne)* and white clover *(Trifolium repens)*. Rather than sampling strictly from the top surface layers, Sait and colleagues obtained a soil core 25 mm in diameter by 100 mm deep, which was transported from the field back to the laboratory in a sealed plastic bag. To reduce the effect of diluting bacterial cells found in deeper soil horizons by the larger number of cells expected in carbon-rich upper horizons, the soil core was sliced into 2-cm sections with a sterile scalpel and then sifted through a sterile sieve to remove stones and large organic debris. The sieved soil sections were subsequently used for dry-weight analysis, microscopic cell counts, and cultivation experiments as follows.

The dry weight of each soil section was determined so that the total number of cells or viable-cell counts (colony-forming units [CFU]) could be normalized and expressed as cells or CFU per gram of dry soil. The drying process significantly reduces the number of viable cells, so enumeration and cultivation experiments must be done using "wet" soil. Because the moisture content could vary depending on the depth of soil or between soil cores collected from different sites, the dry-weight analysis provides a conversion factor for wet to dry weight of soil.

The total number of cells in each section was counted by staining fixed cells with fluorescent dyes such as 4′,6-diamidino-2-phenylindole (DAPI) and acridine orange that bind DNA. Since all intact cells contain chromosomal DNA, the total number of cells in the soil samples may be viewed under a fluorescence microscope and counted on the basis of their fluorescent DNA content.

Viable-cell counts for each section were obtained after serial dilutions of soil suspensions were spread plated onto a growth medium similar to the VXylA used in the "I, Microbiologist" project. Like VXylA, the growth medium used in the Sait et al. (2002) study contains the heteropolysaccharide xylan as the carbon source. The authors hypothesize that use of complex sugar polymers, instead of simple sugars such as glucose, as the sole carbon source may reduce substrate-accelerated death observed when bacteria naturally growing under oligotrophic conditions (environments with nutrients in low concentration) are transferred to artificial environments with high substrate concentrations (e.g., rich media). Polymers such as xylan must first be hydrolyzed by enzymes secreted by cells into simple sugars that can be transported into the cytoplasm of cells; thus, the local concentration of simple sugars is regulated by the cells themselves. Moreover, it is unlikely that the natural substrate encountered by bacteria in the soil environment would be glucose or sucrose. It is more likely to be a complex polysaccharide such as xylan, cellulose, or hemicellulose, especially in the rhizosphere, if one considers what constituents make up plant roots (see Unit 4, section 4.2). These complex poly-saccharides first must be broken down into smaller components before being metabolized by bacterial cells. To ensure that the maximum number of colonies was allowed to develop, the plates were incubated for 12 weeks at ambient temperature in the dark (Janssen et al., 2002).

The left panel of Fig. 2.11 is a plot of the total number of either microscopically counted or viable cells as a function of soil depth (in centimeters). Overall, the total number of cells per gram of dry soil appears to decrease with increasing soil depth. The viable-cell count, which represents the total number of culturable cells in the soil sample, follows the same pattern, although not surprisingly the viable-cell count is lower than the total cell count in each section of the soil core. The viable-cell count for the top (0- to 2-cm) section was highest (1.8×10^9 CFU/g), whereas the count for the 8- to 10-cm section was reduced 100-fold (4.6×10^7 CFU/g). The right panel of this figure shows moisture content as a percentage of wet soil; this percentage appears to be pretty constant (~30%) throughout the entire soil core.

The authors purified 210 colonies from the spread plates. Although efforts were made to obtain representatives of phenotypically distinct morphological types, most colonies were small, white, and fairly unremarkable (P. H. Janssen, personal communication). The authors repeatedly streak purified colonies onto fresh VXylA plates until isolates were obtained (i.e., all colonies derived from a single cell were of identical morphology). Although the spread plates from the initial enrichment experiments were incubated for 12 weeks, purified isolates typically grew to form visible colonies within only 1 week. Of the 210 attempts, only 71 isolates were successfully purified, although all five 2-cm sections were represented in the final group.

Crude lysates containing genomic DNA, as well as other cellular debris, were prepared from each of the 71 isolates and used as the template for PCR; see Unit 3 for a detailed

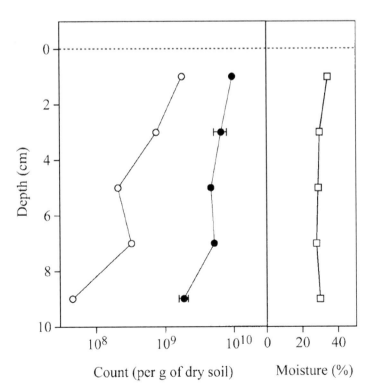

FIGURE 2.11 Distribution of viable colony-forming units (○) and total cells counted microscopically (●) with depth of soil core used for the cultivation experiments. The soil moisture content (□) is expressed as a percentage (by weight) of the freshly sieved soil. The standard deviations are indicated as error bars. The dashed line represents the soil surface. Reprinted from Sait et al., 2002, with permission.

discussion about the methodology. The authors used the primers 27F and 1492R, the same as those used in the "I, Microbiologist" project, which complementary base pair to internal regions within the 16S rRNA gene near the 5′ and 3′ ends, respectively (Fig. 2.12 and Table 2.4). For members of the *Bacteria,* the full length of the 16S rRNA gene is approximately 1,540 base pairs (bp). The primers amplify a region within the gene resulting in a product of about 1,500 bp, which can be verified by gel electrophoresis with a size standard—the amplicon should run at a mobility consistent with the size product you expect for the DNA target. The authors also used another internal primer called 519R for DNA sequencing of the amplified products; this primer anneals to the top strand (plus strand) and, as a reverse primer, amplifies approximately 500 bases at the 5′ end of the gene. Again, this is the same sequencing primer used for the cultivation-dependent part of the "I, Microbiologist" project.

The authors initially tried to identify the 71 isolates using NCBI-BLAST, which compares the DNA sequence for the 16S rRNA gene of an isolate to all other sequences in the GenBank database (Baxevanis, 2005; Benson et al., 2007). This algorithm, which is discussed in extensive detail in Unit 6, allowed the authors to generate hypotheses about the particular taxonomic group to which any given isolate could be affiliated. All 71 isolates appeared to be members of the domain *Bacteria.* Members of four phyla were identified: 45 *Proteobacteria,* the majority of which were in the *Alphaproteobacteria* class; 14 *Actinobacteria;* 10 *Acidobacteria;* and 2 *Bacteroidetes.* Recall that the authors were trying to develop enrichment conditions that would permit the growth of bacteria representing the four globally distributed lineages. According to the BLAST results, three of the four ubiquitous groups were identified.

It is important to realize that the results of a BLAST search are limited by the number of organisms within the database itself and are skewed by the search parameters actually used during the search. Thus, a BLAST result should be viewed only as a hypothesis

FIGURE 2.12 Amplification and sequencing of the 16S rRNA gene from genomes of 71 isolates in the Sait et al. (2002) study. (Top) Based on the primary structure of the *E. coli* 16S rRNA gene, the full-length gene for environmental isolates is expected to be approximately 1,540 bp. (Middle) After amplification with 27F and 1492R primers, the PCR product should be around 1,500 bp. (Bottom) The 519R primer reveals the nucleotide sequence for approximately the first 500 bp. Note that the DNA sequence will be in the reverse orientation by conventional standards; the reverse complement will have to be obtained prior to bioinformatics analysis in BLAST. Illustration by Erin Sanders.

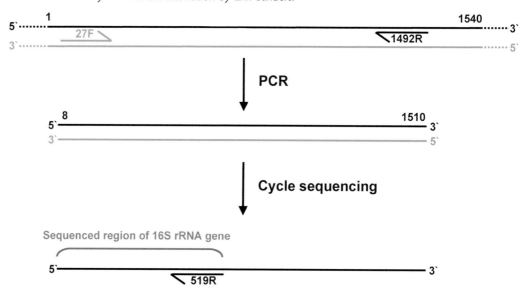

Table 2.4 Primers used to amplify the 16S rRNA gene for the Sait et al. (2002) study[a]

Target gene	Primer name	PCR primer (5′ to 3′)[b]
16S rRNA	16S_27F	AGAGTTTGATCMTGGCTCAG
16S rRNA	16S_1492R	GGTTACCTTGTTACGACTT

[a]Primer sequences originally from Lane (1991).

[b]M = A or C according to standard IUPAC ambiguity codes.

about the identity of a particular organism, not the absolute truth. Moreover, a BLAST search should not be mindless and should not rely upon default parameters. Instead, the search parameters should be selected within a logical experimental framework (e.g., if one is looking to resolve evolutionarily distant relationships, one should not constrain the search results such that only close relationships can be produced). Further annotation and analysis of the data set should be done to verify the taxonomic assignment proposed using only BLAST results.

Construction of a phylogenetic tree using DNA sequences for the 16S rRNA gene from the isolates, and appending the data set with additional sequences from organisms thought to be related to those isolates, is a common way to test a particular hypothesis put forth by BLAST analysis. The authors determined the phylogenetic placement of 31 of the 71 isolates by using a distance method coupled to bootstrap analysis, the details of which are discussed in Unit 7. Regions of the 16S rRNA gene other than the first 500 bp also are phylogenetically informative. The authors therefore sequenced additional regions of the gene to generate partial 16S rRNA sequences longer than 1,300 bp for use in the phylogenetic analysis (Fig. 2.13). Sequences for the 16S rRNA gene from related organisms of known identity were obtained from the public databases (e.g., GenBank) and appended to the alignment used to construct phylogenetic trees for the soil isolates. By inspecting the phylogenetic trees resulting from the comparative analysis of these sequences, the authors were able to deduce the taxonomic identity of their isolates to various degrees within the context of a nested hierarchy presented in a tree.

The isolates that clustered with bacterial lineages in the phylum *Actinobacteria* are depicted in the phylogenetic tree in Fig. 2.14. The authors not only confirmed that they had purified members of rarely isolated lineages from all soil depths, but also revealed

FIGURE 2.13 Extended 16S rRNA gene sequence information used to determine the phylogenetic placement of 31 isolates in the Sait et al. (2002) study. The map shows the approximate location of sequencing primer annealing sites within the 16S rRNA gene. The blue bar depicts the approximately 500-bp partial sequence of the 16S rRNA gene obtained from the original 519R sequencing primer (positions 1 to 519). The orange bar depicts the region amplified by additional sequencing primers. Together, the sequences produced for phylogenetic analysis were all longer than 1,300 bp. Illustration by Erin Sanders.

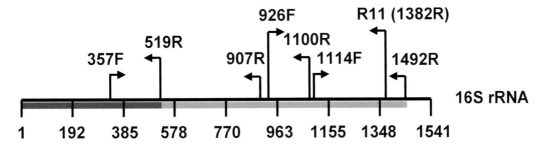

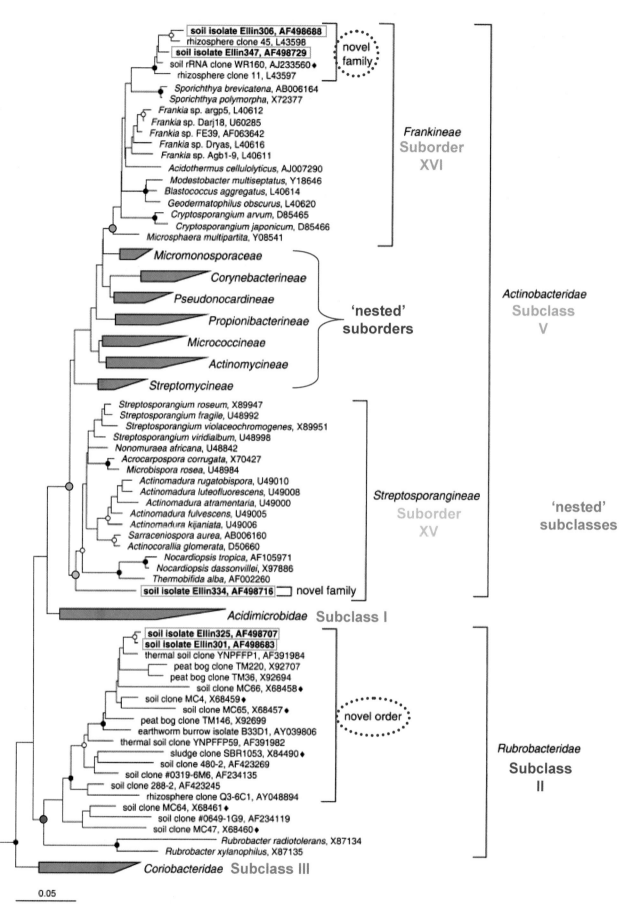

FIGURE 2.14 Evolutionary distance dendrogram of the bacterial phylum *Actinobacteria* based on comparative analysis of sequence data for the 16S rRNA gene in the study by Sait et al. (2002). Modified figure reproduced by permission of the authors.

that some represented novel clades within well-established higher-order lineages. For example, soil isolates Ellin306 (AF498688) and Ellin347 (AF498729) are members of a novel family within the suborder *Frankineae,* soil isolate Ellin334 (AF498716) is the sole member representing a new family within the suborder *Streptosporangineae,* and soil isolates Ellin325 (AF498707) and Ellin301 (AF498683) are members of a novel order within the subclass *Rubrobacteridae.* Overall, finer taxonomic relationships can be resolved with greater confidence using a phylogenetic analysis than can be done with BLAST-mediated database comparisons alone.

The cultivation method used for this study did not appear to select for novel groups, but it extended the range of cultivated bacteria to include additional members of "the underrepresented majority." For instance, the authors isolated two members of the phylum *Acidobacteria,* which had no known cultivated member prior to this study. Surprisingly, the authors did not see a trend in the incidence of particular groups within certain sections of the soil core, despite there being a difference in the total number of cells (Fig. 2.11). Furthermore, only three of the four globally distributed bacterial lineages were detected in each of the soil sections—no members of the phylum *Verrucomicrobia* were cultivated in this study when the VXylA medium or prolonged incubation time was used. It turns out that the *Verrucomicrobia* form colonies only rarely, since some members are easily inhibited by the presence of other colonies on the plate (Sangwan et al., 2005). The *Verrucomicrobia* form visible colonies only on plates with less than 10 colonies.

KEY TERMS

Isolate A purified bacterial strain in which all cells in the population are genetically identical and display the same colony morphology (and other growth characteristics).

Microbiome The genomes of all microbial community members living within a particular environment; a collective genome representing all community members; also called metagenomic DNA.

Oligotrophic conditions Environmental conditions in which the nutrient concentrations are very low.

REFERENCES

Baxevanis, A. D. 2005. Assessing sequence similarity: BLAST and FASTA, p. 295–324. *In* A. D. Baxevanis and B. F. Francis Ouellette (ed.), *Bioinformatics: a Practical Guide to the Analysis of Genes and Proteins,* 3rd ed. John Wiley & Sons, Inc., Hoboken, NJ.

Benson, D. A., I. Karsch-Mizrachi, D. J. Lipman, J. Ostell, and D. L. Wheeler. 2007. GenBank. *Nucleic Acids Res.* **35**(Database issue):D21–D25. doi:10.1093/nar/gkl986.

Janssen, P. H., P. S. Yates, B. E. Grinton, P. M. Taylor, and M. Sait. 2002. Improved culturability of soil bacteria and isolation in pure culture of novel members of the divisions *Acidobacteria, Ac-*

tinobacteria, Proteobacteria, and *Verrucomicrobia. Appl. Environ. Microbiol.* **68**:2391–2396.

Lane, D. J. 1991. 16S/23S rRNA sequencing, p. 115–175. *In* E. Stackebrandt and M. Goodfellow (ed.), *Nucleic Acid Techniques in Bacterial Systematics.* John Wiley & Sons, Inc., Chichester, United Kingdom.

Sangwan, P., S. Kovac, K. E. R. Davis, M. Sait, and P. H. Janssen. 2005. Detection and cultivation of soil verrucomicrobia. *Appl. Environ. Microbiol.* **71**:8402–8410.

READING ASSESSMENT

1. What is meant by the "bottleneck" to microbial ecology?

2. Why is it important to ensure that all colonies on each streak plate have the same morphology before proceeding with molecular analysis (Unit 3)?

3. What are the advantages of agar versus liquid cultivation methods? Why might you use liquid culture cultivation methods over agar cultivation methods?

4. Which aspects of cellular morphology might present challenges to obtaining a single colony representing a pure culture of a particular isolate when working with natural samples?

5. Think about what was discussed in the reading assignment in terms of media components and environmental factors, and develop hypotheses about what types of bacteria you expect to find after plating soil samples on each medium type (RDM, R2A, ISP4, N_2-BAP, and VXylA) and incubating the plates under the environmental conditions specified for the experiments in Unit 2. *Hint:* Consider aerobic versus anaerobic growth, long versus short incubation times, various dilutions plated, with and without ambient light, temperature, etc. For example, do you anticipate finding microorganisms that produce antibiotics, fix nitrogen, or break down cellulose?

   Aside *When analyzing your results at the end of the project, address whether your isolates fulfill your expectations or whether you obtained results that surprised you. Importantly, can you attribute your results to particular aspects of the media or growth conditions you employed specifically in your conduct of the experiment? Be specific.*

6. You are studying a microorganism that was collected from a hot spring and that uses organic carbon as an energy and carbon source. The microorganism is a facultative anaerobe and can use nitrate as a terminal electron acceptor. Your lab partner inadvertently mixes your isolate with a number of other microorganisms in the lab. To continue your research, you need to reisolate the hot-spring microbe.
   a. Develop an enrichment medium and conditions to support the growth of your microbe.
   b. Discuss the steps you would take to recover your original microorganism.

7. How might the "I, Microbiologist" enrichment procedures be expanded to include the *Archaea* or eukaryotes?

The following questions pertain to the study by Sait et al. (2002).

8. What is the purpose or goal of this study?

9. Which specific aspect of VXylA renders it a good medium to use in microbial diversity studies specifically targeting the rhizosphere?
   a. It contains glucose, which is readily metabolized by soil bacteria.
   b. It contains xylan, an abundant carbon source in this habitat.
   c. It deters the growth of bacteria that typically grow in oligotrophic conditions typical of soil environments.
   d. It reduces substrate-accelerated death caused by the decomposition of complex polysaccharides.
   e. It is a complex medium that supports the growth of a wide range of heterotrophic microorganisms.

10. Why was it necessary to grow purified isolates only for 1 week whereas initial cultivation experiments were allowed to progress for a full 12 weeks?

11. Find the hypothesis (or hypotheses) tested in this study and rephrase it (or them) as a conditional proposition (i.e., an "If...then..." statement).

12. Do the results support or refute the hypothesis or hypotheses?

13. Identify and briefly explain the key result that enabled you to draw your conclusion to question 12. (If there was more than one hypothesis, then there will likely be more than one result to discuss; specify which result addressed which hypothesis.)

14. On the basis of your own evaluation of the data, do you agree with the conclusions of the authors? Why or why not? Identify problems or ambiguities in their results that could lead you to question their analysis.

15. Thinking about future directions, suggest one experiment the authors should do next as a follow-up to this study.

16. Do you expect to find any or all of the members of the four groups representing globally distributed lineages in the soil sample you obtained for the "I, Microbiologist" project? Do you expect to find them using the cultivation-dependent approach used in this class? Why or why not?

UNIT 2

EXPERIMENTAL OVERVIEW

In Experiments 2.1 through 2.3, students will enrich for and eventually isolate microorganisms according to the scheme below using five different types of media: RDM, R2A, ISP medium 4, N_2-BAP, and VXylA. Once purified, the genomic DNA will be isolated from each of these organisms by one of the methods described in Unit 3.

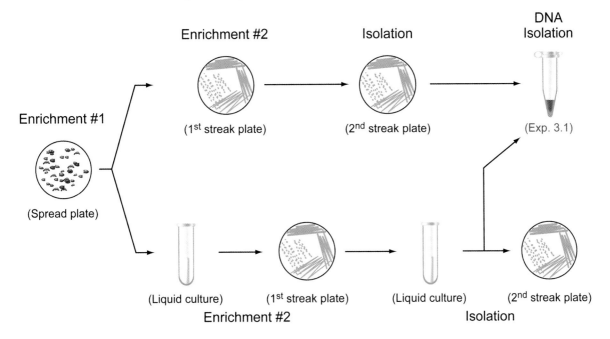

In this experiment, students will grow soil bacteria on a variety of media and under different environmental conditions. Specifically, plates will be inoculated with multiple dilutions of soil and will be incubated at different temperatures, for different lengths of time, under both light and dark conditions, or in the presence and absence of oxygen. The goal is to obtain a diverse representation of culturable bacteria present in the soil.

MATERIALS

RDM plates	Weighing paper
VXylA[hex] plates	Analytical scale
R2A plates	Sterile 5- and 10-ml plastic pipettes
N_2-BAP plates	P1000 and P200 pipettors
ISP medium 4[hex, ben] plates	Vortexer
Large sterile test tubes	Sterile glass beads (~12 per 13-mm tube)
Sterile distilled water	Ice bucket
Sterile 50-ml glass beaker	Aluminum foil
Anaerobic incubation chamber	Sachets for anaerobic chamber

Label the plates used for the first enrichment with the following information: (i) name, (ii) date, (iii) medium type, (iv) dilution, (v) temperature, and (vi) time (duration) of the incubation. For soil samples, plate a dilution that will result in no more than 10 to 30 colonies per plate (Davis et al., 2005). Too many colonies on a plate will inhibit the growth of slower-growing colonies. The dilutions to be plated for this experiment are 10^{-2} through 10^{-8}. Each medium type is to be incubated at 30°C under conditions as follows:

Cultivation Conditions

Medium type	No. of plates	Temp (°C)	Incubation time	Additional cultivation notes[a]
RDM	7	30	24–48 h	Longer incubation time is better[b]
R2A (14 total)	7	30	24–48 h	Incubation with oxygen $(+O_2)$
	7	30	5–7 days	Incubation without oxygen $(-O_2)$
ISP medium 4	7	30	72 h–1 wk	Longer incubation time is better[b]
N_2-BAP	7	30	72 h–1 wk	Longer incubation time is better[b]
VXylA (14 total)	7	30	2–3 wk (maybe longer)	Incubation in the dark[c]
	7	30	2–3 wk (maybe longer)	Incubation in the light[d]

[a] Most conditions described in the table are suited for cultivation of aerobic bacteria. To prevent plates from drying out during a prolonged incubation period, stack the plates but include an empty petri dish at the bottom of the stack. Place the stack in a zip-locked storage bag during the incubation period. The empty dish protects the inoculated medium from contact with the condensed water that accumulates in the bag over time.

[b] To cultivate anaerobes and facultative aerobes, plates should be placed in an anaerobic incubation chamber. Obligate anaerobes cannot be propagated beyond the first enrichment unless plates are manipulated in a specialized anaerobic workstation that prevents exposure of the cultures to oxygen. Facultative aerobes tolerate exposure to oxygen.

[c] Cover plates with aluminum foil to prevent exposure to any light during incubation period.

[d] No special care is taken to prevent exposure of plates to ambient light during daylight hours.

Technical aside: We are working in an environment (the soil) packed with spore-forming bacteria. There is a very high probability that spores reside in the soil you collected. To avoid cross-contaminating the laboratory surfaces or subsequent plates with spores, the method

Experiment continues

involves a special plating technique to help control the potential and indiscriminate spread of spores during the cultivation steps. Because high heat and pressure are required to kill spores, sterile glass beads should be used instead of ethanol-flamed hockey sticks to spread plate soil suspensions. The glass beads are collected after plating and reautoclaved so they are sterilized properly for subsequent use. Disposable toothpicks should be used instead of wire loops for streak plating to avoid unnecessary aerosolization of the cultured bacterial strains.

REFERENCE

Davis, K. E. R., S. J. Joseph, and P. H. Janssen. 2005. Effects of growth medium, inoculum size, and incubation time on the culturability and isolation of soil bacteria. *Appl. Environ. Microbiol.* **71**:826–834.

METHODS

Use aseptic technique throughout the cultivation procedures. Wear gloves except when using the Bunsen burners.

To minimize cross-contamination, avoid sharing media, water, and other reagents. Each team should have its own supplies, which should be stored in designated areas in the class-room between uses. *Do not store materials in drawers or cupboards at benches,* as these areas may be shared work areas.

To avoid contaminating reagent stocks when pipetting small volumes (≤ 1 ml), transfer 10 to 15 ml to a sterile tube, which can be used as a working stock during subsequent manipulations. This temporary reagent source can be discarded at the end of each lab period, whereas the permanent stocks can be saved for future use.

Part I Cultivating Microorganisms on ISP medium 4 To Isolate Strains from the Order *Actinomycetales*, including *Streptomyces* and *Arthrobacter* Strains

Note: There will be seven plates of ISP medium 4. These plates consist of 10^{-2}, 10^{-3}, 10^{-4}, 10^{-5}, 10^{-6}, 10^{-7}, and 10^{-8} dilutions to be incubated at 30°C.

1. Working individually, place 1.0 g of sieved soil into a 50-ml sterile beaker. Using a sterile 10-ml plastic pipette, add 9 ml of sterile distilled water to the beaker. This suspension will be your 10^0 solution for ISP medium 4 dilutions only. Place the beaker in an ice bucket, and shake the bucket at low to medium speed for 1 hour.

   Note: Do not flame presterilized plastic pipettes. They should be used once and then discarded. The flaming procedure should be done only if glass pipettes are used instead of plastic pipettes. Be sure to check the ice level every 15 to 20 minutes while the soil suspension is shaking. If too much ice melts, the beaker will become contaminated with ice water. If this should occur, you must start the experiment over.

2. Label eight sterile test tubes as follows: 10^{-1}, 10^{-2}, 10^{-3}, 10^{-4}, 10^{-5}, 10^{-6}, 10^{-7}, and 10^{-8}. Using a 5-ml pipette, fill each tube with 4.5 ml of sterile water.

3. Using proper aseptic technique, make 10^{-2} to 10^{-8} serial dilutions of the 10^0 soil suspension (e.g., add 0.5 ml of 10^0 suspension to 4.5 ml of sterile water to make 10^{-1}, then add 0.5 ml of 10^{-1} to 4.5 ml of sterile water to make 10^{-2}, etc.).

4. Carefully pour 10 to 12 sterile glass beads onto an agar plate. Aliquot 50 µl of each dilution onto the center of ISP medium 4 plates containing 20 µg of cycloheximide per ml and 50 µg of benomyl per ml. Supplement each aliquot of diluted soil sus-

pension with another 100 μl of sterile water to facilitate even spreading of cells. Close the lid of the plate, and spread cells using the glass bead shaking technique:

- Gently shake beads across the surface of the agar six or seven times. To ensure that cells spread evenly, use a horizontal shaking motion. *Do not swirl* the beads, or else all the cells will end up at the edge of the plate. (Hint: The procedure sounds like "shaking maracas" if done properly). If unsure, have your teaching assistant (TA) or instructor demonstrate the technique.

- Rotate the plate 60°, and then horizontally shake again six to seven times. Rotate the plate 60° a third time, and horizontally shake again. By now, you should achieve even spreading of cells across the agar surface.

- When you have finished spread plating, pour off contaminated beads into a marked collection container. *Do not discard beads in the trash.* The used beads will be autoclaved, washed, and resterilized for repeat usage.

Note: If the agar surface is still wet after being shaken three times, allow the plate to sit for several minutes while the liquid is absorbed by the agar, then repeat the shaking steps until the plate surface appears dry.

5. Incubate the plates at 30°C for at least 72 hours and up to 1 week.

Part II Cultivating Microorganisms on RDM, R2A, N₂-BAP, and VXylA

Note 1 There will be 7 plates each of RDM and N$_2$-BAP media for a total of 14 plates. Plate the 10^{-2}, 10^{-3}, 10^{-4}, 10^{-5}, 10^{-6}, 10^{-7}, and 10^{-8} dilutions, and incubate all plates at 30°C.

Note 2 There will be 14 plates of VXylA medium containing 20 μg of cycloheximide per ml, with two subsets as follows:

Subset A consists of 10^{-2}, 10^{-3}, 10^{-4}, 10^{-5}, 10^{-6}, 10^{-7}, and 10^{-8} dilutions to be incubated at 30°C in the light.

Subset B consists of 10^{-2}, 10^{-3}, 10^{-4}, 10^{-5}, 10^{-6}, 10^{-7}, and 10^{-8} dilutions to be incubated at 30°C in the dark (wrapped in aluminum foil).

Note 3 There will be 14 plates of R2A medium, with two subsets as follows:

Subset A consists of 10^{-2}, 10^{-3}, 10^{-4}, 10^{-5}, 10^{-6}, 10^{-7}, and 10^{-8} dilutions to be incubated aerobically at 30°C.

Subset B consists of 10^{-2}, 10^{-3}, 10^{-4}, 10^{-5}, 10^{-6}, 10^{-7}, and 10^{-8} dilutions to be incubated anaerobically at 30°C.

1. Working individually, label nine sterile test tubes as follows: 10^0, 10^{-1}, 10^{-2}, 10^{-3}, 10^{-4}, 10^{-5}, 10^{-6}, 10^{-7}, and 10^{-8}. Using a sterile plastic pipette, fill the 10^0 tube with 5 ml of sterile water and the 10^{-1} through 10^{-8} tubes each with 4.5 ml of sterile water. *Reminder: Do not flame presterilized plastic pipettes!*

2. Weigh out approximately 5 g of your soil sample, and add it to the test tube labeled 10^0. Vortex vigorously for 1 minute.

3. Using proper aseptic technique, make 10^{-1} through 10^{-8} serial dilutions of the 10^0 soil suspension (e.g., add 0.5 ml of 10^0 suspension to 4.5 ml of sterile water to make 10^{-1}, then add 0.5 ml of 10^{-1} suspension to 4.5 ml of sterile water to make 10^{-2}, etc.).

Experiment continues

Note: If the 10^0 solution is too muddy to pipette, add an additional 2 ml of sterile distilled water, vortex thoroughly, and then let settle for approximately 5 minutes before continuing with preparation of the dilution series. Be sure to record this deviation from the protocol in your lab notebook.

4. Carefully pour 10 to 12 sterile glass beads onto agar plates. Aliquot 50 µl of each dilution to the center of the appropriate medium plate. Supplement each aliquot of diluted soil suspension with another 100 µl of sterile water to facilitate even spreading of cells. Close the lid of the plate, and spread the cells using the glass bead shaking technique described in Part I.

5. Incubate the plates as described in the table at the beginning of this experiment. For incubation under anoxic conditions, place the plates in an anaerobic growth chamber, add an anaerobic pack (sachet), seal the chamber, and incubate at 30°C. Note that a new pack must be used every time the chamber seal is broken. You will need to use some judgment about the appropriate length of time of incubation for your isolates. Bear in mind that if you isolate only those that grow quickly and easily, your results will lack diversity, and that will be attributed to your cultivation bias. You must also consider the limitations imposed by the total number of weeks over which the "I, Microbiologist" project will take place when planning your cultivation procedures. Remember, it is important to determine which concentration of the soil suspension results in 10 to 30 colonies on a single plate.

ACKNOWLEDGMENT

The ISP medium 4 enrichment protocol was modified from one obtained from Julian Davies, professor emeritus at the University of British Columbia.

REFERENCE

Axelrood, P. E., M. L. Chow, C. S. Arnold, K. Lu, J. M. McDermott, and J. Davies. 2002. Cultivation-dependent characterization of bacterial diversity from British Columbia forest soils subjected to disturbance. *Can. J. Microbiol.* **48**:643–654.

Recommended supplier for anaerobic growth chamber
Mitsubishi Gas Chemical Co., Inc., AnaeroPack System
    7.0-liter chamber (catalog no. 50-70; requires three sachets per chamber)
    2.5-liter chamber (catalog no. 50-25; requires one sachet per chamber)
    AnaeroPack sachets (catalog no. 10-01; 20 sachets per box)

EXPERIMENT 2.2 Cultivation of Soil Microorganisms: Second Enrichment

MATERIALS

RDM plates	RDM broth
VXylA plates	VXylA broth
R2A plates	R2A broth
N$_2$-BAP plates	N$_2$-BAP broth
ISP medium 4$^{hex, ben}$ plates	Glucose+MSB
Sterile 250-ml baffled flasks	Sterile sticks
Sterile 250-ml Erlenmeyer flasks with stir bar	Sterile swabs
18-mm test tubes	Sterile flat toothpicks
Light microscope with camera	Camera for plate pictures
Anaerobic incubation chamber	Sachets for anaerobic chamber
	Aluminum foil

METHODS

Use aseptic technique throughout cultivation procedures. Wear gloves except when using the Bunsen burners.

1. Working with your partner, inspect the plates and find the dilutions that produced between 10 and 30 colonies on a single plate. Which dilution(s) worked best?

    Check "crowded" plates for antibiotic zones of growth inhibition, phage plaques, or evidence of fungal growth. Which dilution(s) produced confluent growth of bacterial colonies?

    Choose a total of 48 well-isolated, phenotypically distinct colonies from plates representing each of the seven incubation conditions (e.g., six or seven colonies per condition). *It is strongly recommended that students select colonies from plates with only 10 to 30 colonies unless they are choosing colonies from crowded plates with interesting phenotypes (e.g., antibiotic producers).* You will have two sets of plates to choose from, since you and your partner each plated separately during the previous laboratory period.

    Before proceeding to the next steps, you and your partner *must* take pictures of your selected plates with a digital camera. You will take a total of at least seven plate pictures, one of each medium type/incubation condition. These pictures should be printed and included in your notebooks, presentations, and writing assignments.

2. You and your partner will need plates or broth for a total of 48 potential isolates as designated above. Subsequent enrichment (Experiment 2.2) and purification (Experiment 2.3) steps either on solid or in liquid media will be determined empirically, depending on the morphology of the selected colony or cells.

    Some bacterial colony types (e.g., filamentous, gummy, encapsulated, or calcified) may prove difficult to purify by the standard streak plate technique. These colony types should be visually inspected by *phase-contrast microscopy* (see Experiments 2.2 and 2.3 for protocols to prepare wet mounts and operate the microscope) to confirm suspected morphological challenges (e.g., chains or clusters of cells) before beginning the second enrichment step (step 4 or 5). To obtain isolates of these particular bacterial types, you may need to first grow the bacteria under conditions that will not promote the development of the problematic colony morphologies, which are refractory to isolation on solid media. Instead, use liquid media to cultivate your isolates.

Experiment continues

For instance, as shown in the figure, on solid media, colonies representing species of *Actinomycetales* (*Streptomyces* spp.) appear calcified and slightly fuzzy and can range in color from white to green to gray or pink. If you choose to isolate such colonies, it is plausible that liquid culture will have to be used for the second enrichment.

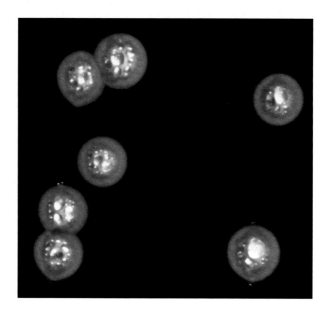

Image of colonies representing isolate ER__37R-1, which was assigned to the genus *Streptomyces* after phylogenetic analysis of the 16S rRNA gene, on R2A medium following incubation at 37°C. Note that phylogenetic results are consistent with the colony morphology expected for microorganisms from this genus. Photograph taken by Gary Chou, Kha Lai, Bac Nguyen, and Michael Nguyen, "I, Microbiologist" participants at UCLA in fall 2007.

3. Solid-medium cultivation. Using sterile flat toothpicks, pick colonies from the original dilution plates to streak onto plates with the corresponding medium type (i.e., use the same medium for streak plates as was used for dilution plates in the first enrichment). Try to select colonies from the RDM, R2A, N$_2$-BAP, and VXylA plates that are phenotypically distinct. Incubate your plates at 30°C as needed.

4. Liquid cultivation. Do not shy away from using the liquid culture enrichment procedure. Remember that you are trying to obtain the most diverse representation of your soil community as possible by avoiding cultivation bias (e.g., picking only the easy ones).

   Broth cultures of bacteria demonstrate a variety of growth patterns. Motile strains such as *Escherichia coli* typically display uniform fine turbidity (UFT). Bacteria such as *Streptococcus* produce flocculent growth in which cells appear to clump together. Some nonmotile microorganisms such as *Mycobacterium smegmatis* generate a waxy cell wall called a pellicle that enables them to float on top of the medium, while others such as *Staphylococcus aureus* descend to the bottom of the growth vessel, forming sediment. Microorganisms with growth characteristics that deviate from UFT may require incubation on a shaker in a baffled flask or on a stir plate in an Erlenmeyer flask with a sterile stir bar.

   Other bacteria form biofilms in flowing liquid environments, growing encased in a porous slime, attached to a solid surface. Such bacteria, when grown in liquid culture, may need a solid substrate upon which to grow. Including a sterile toothpick or long stick in the medium may serve to catalyze biofilm formation and thus promote growth of the organism.

Be prepared to troubleshoot your liquid cultures as necessary.

a. Period 1. Using long sterile sticks (*not* toothpicks or pipette tips), inoculate each tube of 3 to 5 ml of broth with a single colony. Use the same medium for liquid cultures as used for dilution plates in the first enrichment, except as follows:

- Avoid using Luria broth (LB) for liquid cultivation since it promotes the growth of common laboratory contaminants such as *Bacillus* species.

- ISP4 broth forms a precipitate that prevents liquid cultivation in this medium. Instead, use a minimal medium such as glucose+MSB (glucose minimal salts broth).

To confirm that the cultures of environmental isolates were not contaminated during this step, for each medium used you should always prepare one additional tube containing uninoculated medium (negative control).

For aerobic growth, place the tubes in a shaker or rotator at 30°C and incubate for 24 hours (or longer if needed; you should check for growth by assessing relative turbidity, flocculence, sedimentation, pellicle, or biofilm formation). No agitation is needed to grow anaerobes; however, it is not recommended that liquid cultivation techniques be used for anaerobes since gaseous metabolic by-products are produced, creating a hazard due to the buildup of pressure inside the tubes and growth chamber.

b. Period 2. After the culture demonstrates sufficient growth, use it to aseptically swab streak a plate of the corresponding medium as described below. In doing so, you will accomplish two things: (i) you will confirm that you still have a pure isolate after liquid cultivation (if not, you will have to repeat steps 4a and 4b until you obtain the necessary purification), and (ii) you will regenerate single colonies that will be needed as the inoculum for a second liquid culture (period 3). This step, in which you verify the colony morphology of your isolates, is critical if you wish to avoid propagation of contaminants.

Follow these steps comprising the "swab streak technique":

i. Take a sterile cotton swab and dip it into the culture tube, thoroughly moistening the cotton with the cell suspension.

ii. Roll the swab against the sides of the tube to squeeze out excess liquid.

iii. Use the swab to streak quadrant 1 of an agar plate, using the quadrant method as shown below.

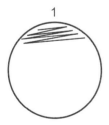 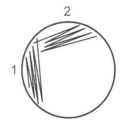 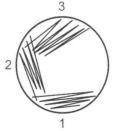 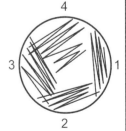

iv. Use a long sterile stick or toothpick to streak quadrants 2, 3, and 4.

v. Flip the plate upside down, and incubate at 30°C for 24 hours (or longer if needed).

MATERIALS

RDM plates	RDM broth
VXylA plates	VXylA broth
R2A plates	R2A broth
N₂-BAP plates	N₂-BAP broth
ISP medium 4[hex, ben] plates	Glucose+MSB
Sterile 250-ml baffled flasks	Sterile sticks
Sterile 250-ml Erlenmeyer flasks with stir bar	Sterile swabs
18-mm test tubes	Sterile flat toothpicks
Light microscope with camera	Camera for plate pictures
Anaerobic incubation chamber	Aluminum foil
	Sachets for anaerobic chamber

METHODS

Use aseptic technique throughout cultivation procedures. Wear gloves except when using the Bunsen burners.

1. You and your partner will need plates or broth for a total of 48 potential isolates. Examine your streak plates from the second enrichment. Has the colony morphology changed from your initial observations? Record observations in your laboratory notebook.

2. Solid-medium cultivation. Using sterile flat toothpicks, pick colonies from the second enrichment plates to streak again onto a plate containing the corresponding medium. Incubate your plates at 30°C as needed.

3. Liquid cultivation. Period 1: Using long sterile sticks (not toothpicks or pipette tips), inoculate each tube of 3 to 5 ml of broth with a single colony from the second enrichment plates derived from the initial liquid cultures. Use the same medium as was used in the first and second enrichments. Place the tubes in a shaker or rotator at 30°C, and incubate for 24 hours (or longer if necessary).

   Period 2: To verify that you still have a pure isolate after liquid cultivation, use this culture to aseptically swab streak a plate of the corresponding medium. *Do not discard liquid cultures.* After streak plating, use this culture and go on to Experiments 2.4 and 3.1. At this point, you may store your culture at 4°C for no more than 1 week.

4. Once the plates have been incubated for sufficient time, inspect them to confirm that the colonies have a uniform morphology consistent with that expected for a genetically pure culture. You should take pictures of the streak plates with a digital camera. These pictures should be printed and included in your notebooks, presentations, and writing assignments.

5. From a single colony, prepare a wet mount and examine cells under phase optics (Experiments 4.1 and 4.2) to verify uniform cellular morphology consistent with that expected for a pure culture. You also may Gram stain your isolates (Experiment 4.3). All microscope observations should be recorded in a laboratory notebook. You should take pictures with a digital microscope camera, documenting these observations for inclusion in laboratory notebooks, presentations, and writing assignments.

Experiment continues

6. If students will be contributing their project results to the "I, Microbiologist" database, then the plate and microscope pictures taken in steps 4 and 5 should be saved in a JPEG file format. The database is called the Consortium of Undergraduate Research Laboratories (CURL) Online Lab Notebook and can be accessed at http://ugri.lsic.ucla.edu/cgi-bin/loginmimg.cgi.

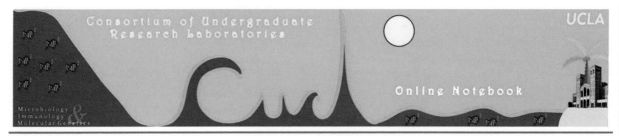

Please enter your user information
Username:
Password:

Login   Reset

Support provided by:

© 2007 Bioinformatics User Facility

DUE 0737131

Note: Before data can be uploaded to the database, a user login and password are required for access to the site. To set up an account, please contact the database administrator. Instructions are provided on the website.

MATERIALS

Sterile 80% glycerol (store at 4°C)   Cryogenic storage vials and caps
Liquid culture of each isolate   Sterile long sticks
5.0-ml pipettes   13-mm sterile test tubes

METHODS

Use aseptic technique throughout procedures. Wear gloves except when using the Bunsen burner. Pipettors are not sterile, so you must use sterile 5.0-ml pipettes for medium and culture transfers.

PERIOD 1
Prepare Samples for Cryogenic Freezing (Part I)

For each of your isolates cultivated on solid media, prepare overnight cultures by using sterile long sticks to aseptically inoculate tubes containing exactly 3.0 ml of broth of the appropriate medium with a single colony.

Note 1 Cultures to be used as long-term frozen stocks should not be allowed to grow and then sit for long periods in stationary phase. Instead, inoculate cultures and then place the tubes at 4°C until ready to begin incubation at 30°C for the appropriate duration.

Note 2 Some environmental isolates cannot be grown in broth cultures. Instead, prepare a fresh streak plate of each isolate, with only one isolate per plate. Incubate the plates at 30°C as needed.

PERIOD 2
Prepare Samples for Cryogenic Freezing (Part II)

1.   For each isolate, label the sides of two cryogenic tubes with quarter, course number, laboratory section, your initials, and isolate ID information (refer to Experiment 3.1 for sample ID format). In addition, label one tube W, for working stock, and the other tube P, for permanent stock. Instructors and TAs should be able to cross-reference your stored sample with the information in the course database, your laboratory notebooks, presentation slides, and writing assignments; make sure the identifier information for each isolate is identical.

2.   For isolates for which overnight liquid cultures were obtained, aseptically add 1.0 ml of sterile 80% glycerol to the test tube to give a final concentration of 20% glycerol in a 4.0-ml total volume. Vortex gently to mix.

       For isolates for which fresh streak plates were prepared, remove the lid and then aseptically add 1.5 ml of liquid medium across the top of the agar. Gently aspirate and dispense the broth across the surface of the plate, dislodging the colonies from the agar. Transfer the cell suspension to a sterile 13-mm test tube. Add 0.5 ml of sterile 80% glycerol to the test tube to give a final concentration of 20% glycerol in a 2.0-ml total volume. Vortex gently to mix.

Experiment continues

3. For each isolate, aseptically transfer 1.5 ml of culture or 1.0 ml of cell suspension into the cryogenic tube marked P. Close the cap, and place the tube in the box designated for permanent storage at −80°C.

   Note The P tubes will not be touched again during the project unless the viability tests fail in period 3. Otherwise, the P tubes containing the 16 strains selected for DNA sequencing will be maintained as part of a permanent collection of environmental isolates.

   Next, aseptically transfer 1.5 ml of culture or 1.0 ml of cell suspension into the cryogenic tube marked W. Close the cap, and place the tube in the box designated for storage of working stocks at −80°C.

   Note Do not throw away your plates or cultures used for the most recent propagation step yet. You must first confirm that your cell stocks are viable following freezing (see Period 3 below).

PERIOD 3
Test Viability of Cell Stocks from Part II

1. After a minimum of 24 hours at −80°C, recover the cells from the W glycerol stocks only by streak plating each cell sample onto the appropriate medium type, using a sterile stick or toothpick to scrape frozen cells from cryogenic tubes. Do not leave the vials out at ambient temperature for more than a few minutes as the frozen cells will begin to melt and their viability will decrease due to formation of ice crystals upon refreezing of the samples.

   Make sure you have advised your instructor or TA of the type and number of plates you will require for viability testing in advance. You should be able to streak out two samples per plate.

2. Incubate the streak plates at 30°C for 24 to 72 hours.

3. Check the plates for colony growth.

   a. If there is growth, and the colony morphology is identical to what you observed on plates throughout the enrichment, purification, and propagation steps, you may now discard your propagation plates.

   b. If there is no growth, repeat the streak plating procedure (steps 1 and 2 of Period 3).

   • If there is still no visible colony growth after the second attempt, you may assume that there are no viable cells in the glycerol stock culture. Discard both the P and W stocks. Repeat steps in Periods 1 and 2 for these samples, and retest the viability of the W stock culture as described in Period 3.

   • If there is still no visible colony growth from the second glycerol stock culture, see the instructor or TA for alternative strategies for preparing frozen cell stock cultures (e.g., try 10% DMSO instead of 20% glycerol).

4. After purification and cryogenic freezing, you and your partner are responsible for maintaining and propagating viable cultures of your isolates for the remainder of the project. You will need to obtain fresh streak plates directly from the W glycerol stocks about every 2 weeks to keep the cultures fresh and maintain purity. For this procedure, you must use the same medium from which the original isolate was obtained and incubate at the appropriate temperature and for the appropriate time, utilizing liquid-culture cultivation steps as required.

UNIT 3

Molecular Analysis of Cultivated Bacterial Communities

SECTION 3.1
PCR AND SEQUENCING OF THE 16S rRNA GENE

Principles of PCR

Developed a little more than 20 years ago, the polymerase chain reaction (PCR) is a powerful molecular technique used for the rapid amplification of a DNA target, whether that target is a single gene, part of a gene, or a noncoding sequence. This procedure can generate millions of copies or more of a specific segment of DNA starting with only one or a few copies. In 1993 Kary Mullis won the Nobel Prize in Chemistry for inventing PCR, now a fundamental protocol for most biological research laboratories (Mullis, 1990, 1993). This method is essential to the "I, Microbiologist" project, which comprises a biodiversity study on microbial communities in natural soil samples as well as a DNA-based phylogenetic analysis of community members, using the gene encoding the small subunit of the bacterial ribosome (16S ribosomal RNA [rRNA]).

PCR amplification of the 16S rRNA gene (16S rDNA), the target gene for this study, requires use of a heat-stable DNA polymerase called *Taq*. Discovered in 1976, this enzyme was purified from the extreme thermophile *Thermus aquaticus*, which resides in hot springs at temperatures ranging from 50 to 80°C (Chien et al., 1976). Conventional PCR takes place in a buffered salt solution with several reagents including a DNA template, two primers, and deoxynucleoside triphosphates (nucleotides, or dNTPs). PCR buffer is supplied by the manufacturer of *Taq* and is designed to provide a suitable and stable chemical environment for polymerase activity. In addition to containing a salt such as potassium chloride (KCl), PCR buffer contains magnesium chloride (MgCl$_2$). The divalent cations (Mg^{2+}) form complexes with the dNTPs, primers, and DNA template, held together by electrostatic interactions with the negatively charged phosphates (Nakano et al., 1999). In addition, the metal ions are part of the *Taq* polymerase active site (Li et al., 1998; Urs et al., 1999). The concentration of Mg^{2+} must be modulated: too low a concentration results in a low yield of PCR products, while too high a concentration promotes mutagenesis and synthesis of nonspecific products (Innis and Gelfand, 1994). Typically, between 0.1 and 1 μg of genomic (or metagenomic) DNA is used as the template (Landweber and Kreitman, 1993). Larger amounts tend to increase the yield of nonspecific DNA products. Two primers are necessary if amplification of a double-stranded DNA copy of the target is desired. Recall that a single strand of DNA is a polarized molecule, having 5′ and 3′ ends that correspond to the 5′ and 3′ carbons on the ribose sugar moiety of each terminal nucleotide in a chain, or linear polymer. A double-stranded DNA molecule, or helix, is comprised of two antiparallel chains that run in opposite directions. By convention, one chain, referred to as the plus strand (also known as the sense or nontemplate strand if the target DNA is part of a coding sequence), is aligned in the 5′-to-3′ direction while the second chain, referred to as the minus strand (also recognized as the antisense or template strand), is aligned in the 3′-to-5′ direction as shown below:

Plus strand (+): 5′ → 3′
Minus strand (−): 3′ → 5′

For PCR one primer, often called the forward primer, is required to amplify the minus strand while another primer, referred to as the reverse primer, is needed to amplify the plus strand. The dNTPs must be present in approximately equimolar amounts to prevent misincorporation of noncomplementary nucleotides, leading to mutagenesis of the PCR products. A single reaction mixture is typically only 25 to 100 μl in total volume.

PCR is an automated process, carried out in a machine that permits thermal cycling, which is necessary for the stepwise progression of cycles comprising this procedure. A single reaction proceeds through a series of 25 to 35 cycles, with three temperature-mediated steps within each cycle (Fig. 3.1). To minimize *Taq* polymerase errors, the smallest number of PCR amplification cycles should be employed (Acinas et al., 2005).

In the first step of a single PCR cycle, the reaction mixture is heated to 94 to 96°C for 20 to 60 seconds, causing denaturation of the double-stranded DNA template. This step results in two single-stranded DNA templates of opposite orientation, representing the plus and minus DNA strands.

Next, the reaction temperature is lowered to around 45 to 55°C for 20 to 30 seconds, which permits annealing of the primers to the single-stranded DNA templates. There are typically excess primers relative to the amount of target DNA, promoting hybridization between the primers and template DNA rather than reannealing of the plus and minus DNA template strands to each other. The exact annealing temperature is selected based on the calculated melting temperature (T_m) of the primers. One simple formula used to estimate the T_m of oligonucleotides is shown:

$$T_m = 64.9 + \{41 \times [(yG + zC - 16.4)/(wA + xT + yG + zC)]\}$$

where w, x, y, and z are the number of the bases A, T, G, and C in the sequence, respectively. However, this equation does not provide any adjustments for the concentration of monovalent cations (e.g., K^+ and Na^+) or salt, divalent cation concentration (Mg^{2+}), or other thermodynamic variables (e.g., pH, temperature, primer and target DNA ratio) that ultimately influence the T_m (Wallace et al., 1979; Sambrook and Russell, 2001). Because these calculations can become cumbersome and complicated, several T_m

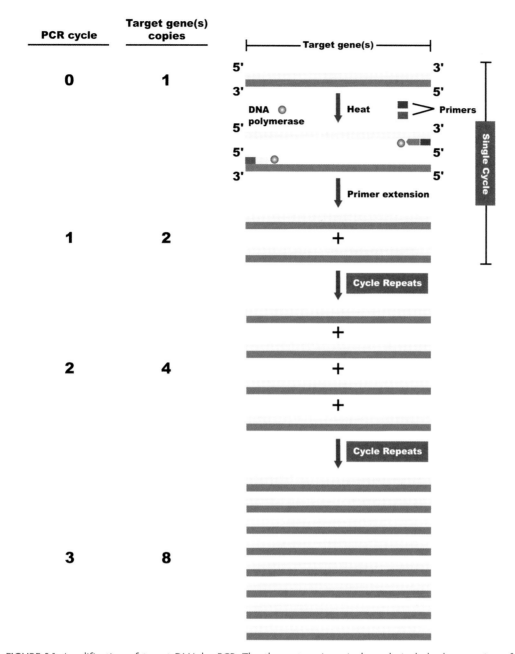

PCR cycle	Target gene(s) copies
0	1
1	2
2	4
3	8

FIGURE 3.1 Amplification of target DNA by PCR. The three steps in a single cycle include denaturation of DNA template, annealing of primers, and extension of the target DNA copy by DNA polymerase. There may be up to 35 cycles for a single reaction. By convention, the plus strand is written in the 5′-to-3′ direction whereas the minus strand, also called the template strand for mRNA synthesis, is written in the 3′-to-5′ direction. Illustration by Cori Sanders (iroc designs).

calculators are available online, some of which are listed in the Web Resources at the end of this section.

For PCR, the annealing temperature ideally should be approximately 5°C below the T_m of the primers so that hydrogen bonds form only when the sequences of the primer and DNA target closely match. This restriction is one of many ways in which to increase the stringency of the reaction. Under high-stringency conditions only two perfectly complementary DNA strands hybridize, whereas under lower-stringency conditions mismatches between the two strands are tolerated. One concern about performing PCR under lower-stringency conditions in microbial diversity studies is the potential to form chimeras and heteroduplex molecules, two PCR artifacts that complicate analysis of the sequence data generated later on (Acinas et al., 2005).

Primers are typically 15 to 30 nucleotides in length for optimal specificity and are designed to have a G+C content between 45 and 60% (Dieffenbach et al., 1995; Abd-Elsalam, 2003). Care also is taken to generate primers that do not form primer dimers and that have a base composition which exhibits minimal secondary structure. Once the primers and DNA templates hybridize, *Taq* polymerase binds and initiates DNA synthesis.

Finally, the temperature is increased to 72 to 75°C, which is the optimal temperature for *Taq* activity (Chien et al., 1976; Lawyer et al., 1993). During this part of the cycle, *Taq* polymerase synthesizes new DNA strands, which are complementary to the plus and minus template DNA strands, by adding nucleotides to the 3′ end of the nascent DNA strands. The extension time depends on the length of the DNA target to be amplified. For *Taq* at the optimal temperature, at least 1 minute is required to extend the first 2 kb, with 1 minute needed for each additional kilobase to be amplified. Since the length of the 16S rRNA gene for bacteria is expected to be well conserved at approximately 1,500 bp, an extension time of 1 minute should be sufficient to yield the desired PCR products. Longer extension times should be avoided to minimize sequencing artifacts that could be attributed to the incorporation of incorrect dNTPs by *Taq* polymerase (Acinas et al., 2005). Alternative polymerase enzymes are commercially available and recommended for amplification of longer target DNA molecules.

The three-step thermal cycling procedure is sometimes preceded by an initial denaturation step, called a "hot start," in which all of the reaction components except *Taq* polymerase are mixed and incubated at 90 to 94°C for as long as 9 minutes (D'Aquila et al., 1991; Erlich et al., 1991; Mullis, 1991). At the end of this initialization step, *Taq* and dNTPs are added so that the three-step thermal cycling procedure may begin. Hot-start PCR not only facilitates complete denaturation of the double-stranded template DNA, which consists of a large chromosome (>>1 Mb) in which the G+C content may not be known, but also eliminates the number of nonspecific priming events that result in the formation of secondary products during the first PCR cycle. For instance, the template may hybridize to itself, primer dimers may form, and individual primers may form hairpin structures or may partially anneal to nonspecific sites in the DNA template (SuperArray Bioscience Corporation, 2006). These nonspecific priming events tend to occur when lower-stringency reaction mixes are prepared at room temperature rather than strictly on ice and are most problematic when there is a small amount of target DNA, as is the case with individual genomes within the metagenomic DNA samples (see Unit 5), where there may only be one or a few copies of any single DNA target. Amplification of secondary products during subsequent cycles unnecessarily consumes PCR reagents, causing a decrease in the efficiency of amplification of the desired products. Thus, hot-start PCR is used for the cultivation-independent part of the "I, Microbiologist" project.

Another variation, called "touchdown" PCR, can be employed to reduce nonspecific primer annealing (Don et al., 1991). With this strategy, the earliest PCR cycles have high annealing temperatures, permitting only exact base pairing between the primer and the template to occur (high-stringency conditions), and the annealing temperature is incrementally decreased for subsequent sets of cycles. The primers hybridize at the highest permissible temperature, which at the same time is the least tolerant of nonspecific primer annealing. Therefore, the first sequence amplified is the one that contains regions of maximum primer specificity. Most likely, the product, also called an amplicon, is the anticipated target DNA sequence, which will continue to be amplified during the following PCR cycles that take place at lower annealing temperatures (low-stringency conditions). However, because the target sequence was amplified first, in the final population of amplicons it will be more abundant than any undesirable products generated by nonspecific primer annealing at the lower temperatures.

The last cycle of PCR is typically followed by a final elongation step in which a temperature of 72 to 75°C is held for 5 to 15 minutes. This part of the procedure ensures that the remaining nascent DNA strands fully extend and that the terminal transferase activity of *Taq* adds the extra dATP to the 3′ ends. The latter activity, which is discussed in detail in Unit 5, is essential for subsequent cloning of metagenomic PCR products.

A terminal hold for an indefinite time at 4 to 10°C is occasionally used as a practical means of briefly storing the reaction mixtures until they can be transferred to a permanent storage location (e.g., laboratory freezer) or analyzed by gel electrophoresis.

During each round of PCR-based amplification of a target DNA, the amount of PCR product doubles (Fig. 3.1). Because the products of one primer extension serve as a template in the next cycle, this repeated process leads to an exponential increase in the yield of PCR product. Thus, only a few molecules of target DNA (<10) need be present to start the reaction, resulting in an increase in the amount of target sequence by more than six orders of magnitude (Fig. 3.2).

Optimizing "I, Microbiologist" PCR conditions

In practice, PCR conditions must be optimized depending on the source of the DNA template (e.g., genomic versus metagenomic) or the application for which the PCR product is being generated (e.g., cloning and DNA sequencing). For the "I, Microbiologist" project, with the exception of the initial hot-start denaturation step for the cultivation-independent reactions, the cycling conditions for the cultivation-dependent and the cultivation-independent approaches are identical, with a denaturation step at 94°C for 3 minutes followed by a three-step thermal cycling procedure repeated 35 times (denaturation at 94°C for 1 minute, annealing at 48°C for 30 seconds, and extension at 72°C for 1 minute) and then by a final extension step at 72°C for 7 minutes. If troubleshooting

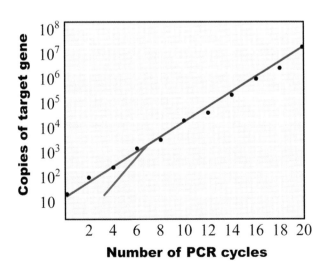

FIGURE 3.2 Graphical representation of the results after running 20 cycles of PCR with a genomic DNA preparation in which there were 10 original copies of the target gene. The tally of DNA copies of the target sequence made with each PCR cycle is shown (red line). These products contain sequences that flank the target gene due to extension along the original template DNA beyond the primer-annealing sites. It is not until the third PCR cycle that DNA fragments of the desired length first appear as products of the reaction (blue line). These products result from amplification of target gene copies, rather than the original template, and their length is defined by the location at which the forward and reverse primers anneal to the template. Beyond seven or eight cycles, the number of original copies and target copies is nearly equal (purple line). To illustrate the exponential function of the PCR process, the data are plotted on a semilogarithmic scale. Illustration by Cori Sanders (iroc designs).

becomes necessary, touchdown PCR cycling conditions may be programmed into the thermocycler and used to optimize primer annealing conditions.

As shown in Table 3.1, the components in the reaction mix for the cultivation-dependent (Experiment 3.2) and cultivation-independent (Experiment 5.2) approaches are not the same; in fact, each reaction mix contains additional reagents from those described for conventional PCR.

Of note for the cultivation-dependent reaction mix is the presence of $MgCl_2$ and dimethyl sulfoxide (DMSO). The 10× PCR buffer already contains 15 mM $MgCl_2$ such that a 50-μl reaction mixture contains a final concentration of 1.5 mM $MgCl_2$. In an attempt to increase the final yield of PCR product, the reaction is supplemented with an additional 1 mM $MgCl_2$, bringing up the final concentration to 2.5 mM. The additive DMSO facilitates DNA denaturation, which is important for genomes with high G+C content. Up to 10% (vol/vol) DMSO in the reaction is tolerated (note that the "I, Microbiologist" protocol only calls for 5%). Other additives that serve as functionally equivalent alternatives to DMSO include 10 to 15% (vol/vol) glycerol and 5% (vol/vol) formamide. The disadvantage of including DMSO in the reaction mix is that it inhibits *Taq* activity by approximately 50%, so an increased amount of enzyme must be used. Specifically, a typical reaction mixture contains between 1 and 1.5 units of *Taq*, whereas between 2 and 3 units of *Taq* are required for equivalent levels of activity in the presence of the inhibitor DMSO. Another inhibitor of *Taq* activity is phenol, which may be present in trace amounts if glass bead lysis (Experiment 3.1, method E) is used to prepare the genomic DNA lysate.

Of note for the cultivation-independent reaction mix besides the DMSO, which also functions to promote denaturation of the metagenomic DNA samples during PCR, is the presence of an additional salt, potassium chloride (KCl). The 10× PCR buffer already contains 500 mM KCl, such that a 50-μl reaction mixture contains a final concentration of 50 mM KCl. Supplementing the reaction mix with an additional 50 mM KCl, bringing up the final concentration to 100 mM, increases the stringency of the reaction by raising the effective melting temperature (T_m) of the primers. This modification to the reaction mix was done to somewhat counter the effects of adding DMSO, which decreases the T_m of the primer-template DNA duplex. Like in the cultivation-dependent PCR mix, the number of units of *Taq* must be increased relative to that in a typical reaction to counter the inhibitory effects DMSO on *Taq* activity. It also should be noted that the metagenomic DNA preparations are expected to be relatively pure compared to the genomic

Table 3.1 Comparison of the key components within each reaction mix used for PCR by the cultivation-dependent part of the "I, Microbiologist" project (Experiment 3.2) and the cultivation-independent part of the project (Experiment 5.2)

Cultivation-dependent PCR mix (Experiment 3.2)	Cultivation-independent PCR mix (Experiment 5.2)
PCR buffer (with KCl and $MgCl_2$)	PCR buffer (with KCl and $MgCl_2$)
More $MgCl_2$	**More KCl**
DMSO	DMSO
27F_16S rDNA primer	27F_16S rDNA primer
1492R_16S rDNA primer	1492R_16S rDNA primer
Crude genomic DNA template	**Purified** genomic DNA template

DNA lysates due to the use of a kit which employs column chromatography and several wash steps coupled to phenol-chloroform extractions and potentially followed by an isopropanol precipitation (Experiment 5.1). However, if care was not taken during the extraction steps, trace amounts of phenol may be present in the final DNA sample and will inhibit *Taq* activity during PCR.

For both the cultivation-dependent and cultivation-independent parts of the project, one should always assemble the reaction mixtures on ice not only to prevent formation of secondary products during the first PCR cycle that are amplified in subsequent cycles but also to inhibit residual nuclease activity that may be present in the genomic DNA lysates. The latter concern is especially critical during DNA manipulations with the crude lysates prepared during the cultivation-dependent part of the project. Only genomic DNA isolation methods that incorporate a phenol-chloroform extraction will fully inactivate and remove protein contaminants such as nucleases that degrade the bacterial chromosomes if left at ambient temperatures.

Gel electrophoresis

A simple procedure employed to confirm that PCR produced the anticipated products (amplicons) is agarose gel electrophoresis. Agarose is a highly purified polysaccharide derived from seaweed; when hydrated with an appropriate buffer, it serves as a porous matrix through which DNA molecules migrate upon application of an electrical current (Fig. 3.3). This method separates DNA fragments based on their size, shape, and charge (Sambrook and Russell, 2001). In an electric field, negatively charged DNA molecules migrate through the gel, with smaller fragments migrating faster than larger fragments in the case of linear DNA molecules. For plasmids, the topological form of the DNA (i.e., the shape) also influences the rate of migration: a highly compact, supercoiled plasmid will migrate more rapidly than a linear DNA fragment of equivalent size. However, any two plasmids may be similar in size yet have totally different base compositions, as should be the case with the 16S rRNA genes derived from the metagenomic DNA

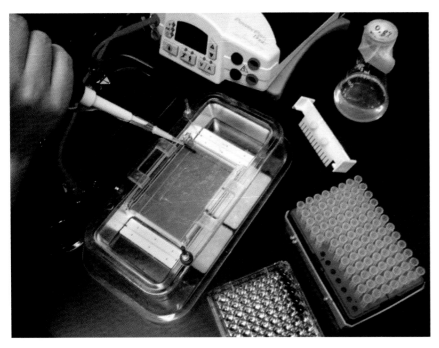

FIGURE 3.3 Setup for agarose gel electrophoresis. A pipetter is used to load DNA samples into individual wells of an agarose gel submerged in buffer within an electrophoresis chamber. The positive and negative electrodes are attached to cables running to the power supply, which generates a current, causing the DNA fragments within each well to migrate toward the positive electrode. Photograph by Cori Sanders (iroc designs) and Erin Sanders.

preparations after being cloned into vectors (Experiment 5.6). Therefore, the recombinant plasmids must be treated with restriction enzymes before gel electrophoresis. The location of the restriction sites within the gene may differ depending on the sequence of the gene, thus generating a pattern of DNA fragments characteristic of the gene for the organism from which it was obtained (Fig. 3.4). As discussed in Unit 1, this "DNA fingerprinting" technique can be applied to ecological as well as forensic science studies (Varsha, 2006). Following electrophoresis for an extended period, the gel can be stained with a compound that binds DNA. Ethidium bromide is a common stain used in research laboratories, as it is relatively inexpensive and sufficiently sensitive to detect products produced by conventional PCR. It intercalates between the nucleotide bases within the DNA strands and then fluoresces upon excitement with ultraviolet (UV) light, facilitating visualization of the DNA fragments within the gel. The sizes of the PCR products or restriction fragments may be determined by comparing their migration to that of the bands in a DNA marker, which contains multiple DNA fragments of known size. The marker should be loaded into an adjacent well and simultaneously electrophoresed alongside the PCR products.

DNA sequencing

The PCR products generated during the cultivation-dependent part of the project as well as the plasmids containing the 16S rDNA fragments produced during the cultivation-independent part of the project are subjected to another method of DNA analysis called DNA sequencing. This procedure is used to determine the order of nucleotides in a particular DNA fragment. The sequencing technique used for the "I, Microbiologist" project relies on the Sanger dideoxy method, which was originally developed in 1977 by Frederick Sanger and colleagues and permits analysis of the target DNA up to 1 kb from the site of primer hybridization (Sanger et al., 1977; Sambrook and Russell, 2001). Large-scale projects, such as sequencing entire microbial genomes, have created opportunities for the development of automated systems. As an automated process, now called cycle sequencing, the Sanger dideoxy method no longer relies upon the use of radioactivity. Instead, this procedure generates single-stranded DNA fragments terminating at each of

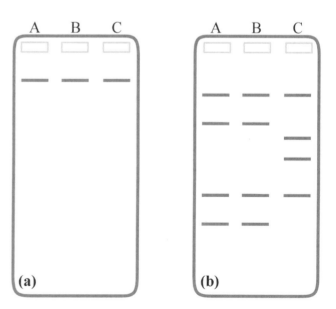

FIGURE 3.4 DNA analysis by gel electrophoresis. (a) Bands produced by three undigested, supercoiled DNA plasmids (A to C); (b) bands produced by the same three plasmids (A to C) after treatment with a restriction enzyme. Illustration by Cori Sanders (iroc designs).

the four nucleotides (G, A, T, and C), which are differentially labeled with fluorophores that emit light at different wavelengths upon excitation with a laser (Du et al., 1993; for a discussion of variations in labeling method, see Kaiser et al. [1993]). Being a variation of PCR, synthesis of the complementary DNA strand begins with annealing of a primer followed by the addition of dNTPs, a reaction catalyzed by DNA polymerase (Fig. 3.5).

The hallmark of cycle (Sanger) sequencing is that a nucleotide analog called a dideoxynucleoside triphosphate (ddNTP) is employed in each sequencing reaction together with the dNTPs (Fig. 3.6). The ddNTPs are missing the hydroxyl group on the 3′ carbon of the sugar moiety comprising the nucleotide analog. Thus, when this molecule is in-

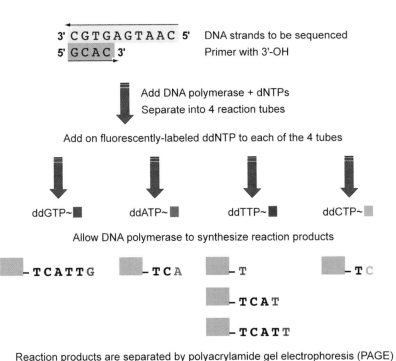

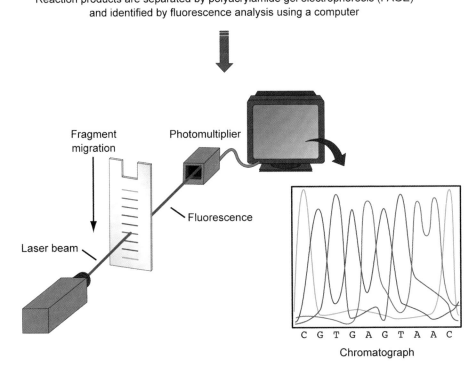

FIGURE 3.5 Overview of cycle sequencing methodology using fluorescently labeled dideoxynucleotides. See the text for details. Illustration by Cori Sanders (iroc designs).

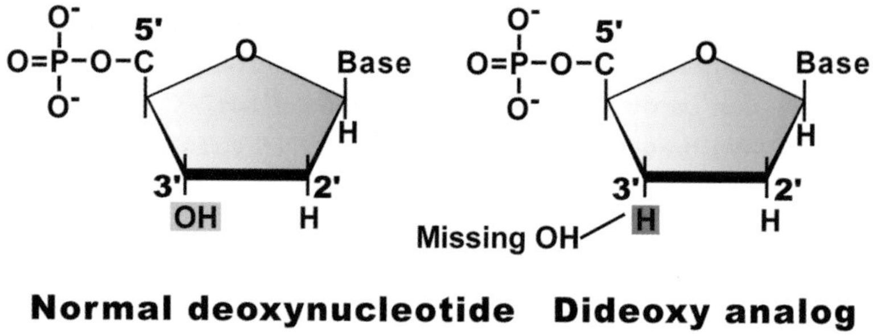

FIGURE 3.6 Structures of deoxynucleotides used for Sanger (cycle) sequencing. (Left) Deoxynucleotide (dNTP) with hydroxyl group on 3'-carbon of pentose sugar; (right) dideoxynucleotide (ddNTP) lacking the 3'-OH group. Illustration by Cori Sanders (iroc designs).

Normal deoxynucleotide   Dideoxy analog

serted during the extension of the complementary DNA strand, chain termination occurs at a defined location. As shown in Fig. 3.5, this reaction is carried out in four separate tubes, one for each base (ddGTP, ddATP, ddTTP, and ddCTP). Oligonucleotides of variable length are obtained in each tube. The products are denatured upon addition of a formamide-based load dye to the tubes followed by incubation at 95°C for 2 minutes. Then the single-stranded DNA fragments are separated by polyacrylamide gel electrophoresis (PAGE). This matrix is made from a small organic compound called acrylamide, which becomes cross-linked upon addition of an appropriate catalyst, forming a molecular sieve. In similar fashion to what is seen with agarose gel electrophoresis, PAGE separates negatively charged DNA fragments based on size when an electrical current is applied, with the smaller fragments migrating faster than larger ones. The resolution of PAGE far exceeds that of agarose gel electrophoresis, resolving DNA fragments that differ in length by only one nucleotide. On the basis of the size of the DNA fragment and the type of ddNTP responsible for terminating synthesis of that fragment, the computer software will generate a chromatograph reflecting the order of bases within the DNA target sequence in the 5'-to-3' direction from the site at which the primer hybridized to the DNA template. Only reaction products containing a ddNTP at their terminus are detected by the laser scanner. Moreover, because each of the four dyes emits fluorescence at a different wavelength, the four reaction products may be combined and analyzed in a single lane of a gel or, as is done more commonly, in a single threadlike capillary tube (Luckey et al., 1993; Madabhushi, 1998; Elkin et al., 2001).

Although capillary-based cycle sequencing is in widespread use today, several new technologies are being developed and optimized to facilitate rapid sequencing of entire genomes or environmental metagenomes (Hall, 2007; Schuster, 2008; Mardis, 2008; Medini et al., 2008). As solutions to the goal of broadening the applications of genomic information in biomedical research and health care, these state-of-the-art technologies hold promise with respect to reducing the cost of sequencing the human genome from between $10 million and $25 million per person to $1,000 or less (National Institutes of Health, 2004). Detailed descriptions of these processes can be found at the National Human Genome Research Institute (NHGRI) website (URL provided under Web Resources at the end of this section). Students also are encouraged to consult the literature (Hall, 2007; Schuster, 2008; Mardis, 2008; Medini et al., 2008).

Analysis of DNA sequence data

The directionality and antiparallel nature of the target DNA strands that are subjected to DNA sequencing must be considered before phylogenetic analysis of the sequence data is performed. All of the sequences used for the analysis must be in the same orientation. To keep the analysis simple and consistent with what is done in other research groups, only the plus strand should be used for all sequences derived from bacterial isolates and clones. Because the "I, Microbiologist" project employs primers that generate the sequence for the minus strand rather than the plus strand, the resulting sequence must be manipulated to deduce the sequence of the plus strand. Specifically, the process requires determination of the reverse complement of minus-strand sequence as demonstrated below:

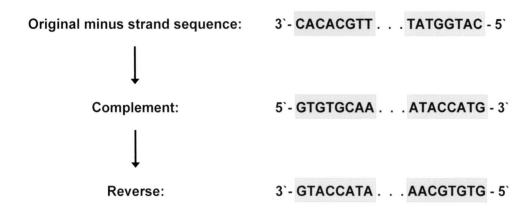

This procedure is essential for all 16S rDNA sequences generated for bacterial isolates by using the 519R primer. As shown in Fig. 3.7a, the approximately 500-base sequence

FIGURE 3.7 Deduction of the reverse complement from the sequence generated by the 519R primer in the cultivation-dependent part of the "I, Microbiologist" project. Note that the depicted DNA strands and primers are not to scale. (a) Region of the 16S rRNA gene amplified using the 519R primer. (b) The sequencing product reflects the minus-strand sequence, encompassing the first 500 bases of the 16S rRNA gene. The orientation of the sequencing read is 5' to 3'. (c) By convention, the plus strand in the 5'-to-3' orientation should be used for bioinformatics and phylogenetic analysis. This can be deduced from the sequencing product by generating the reverse complement. Illustration by Erin Sanders.

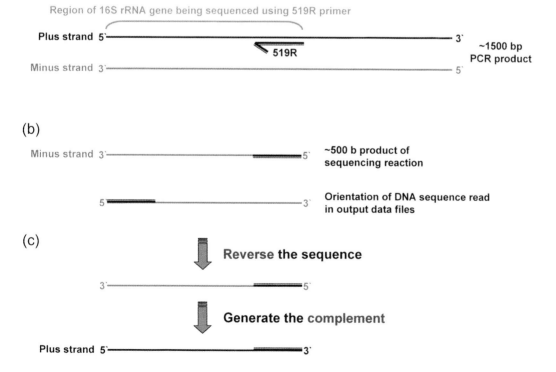

This way. . .

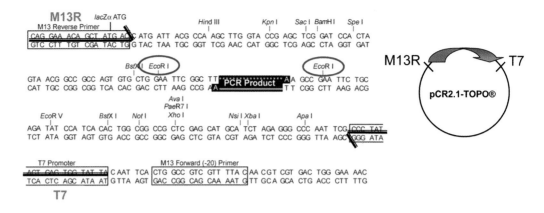

Or that way. . .

FIGURE 3.8 Deduction of the reverse complement from the sequences generated by the T7 and M13R primers in the cultivation-independent part of the "I, Microbiologist" project. The 16S rDNA PCR products may be cloned into the pCR2.1-TOPO vector in either orientation (compare the top and bottom panels). In one orientation (top panel), the sequence generated by the M13R primer produces the plus strand (orange dashed line), whereas in the opposite orientation (lower panel), the M13R primer produces the minus strand (orange solid line). In the latter scenario, the reverse complement would need to be generated to obtain the plus strand. On the right side of the figure are diagrammatic representations of the plasmids. The two possible orientations of the DNA insert are depicted as grey arrows within each plasmid (not to scale). Illustration of pCR2.1-TOPO vector sequence modified based on an image from *TOPO TA Cloning User Manual,* Invitrogen Corp., 2006.

product generated using the 519R primer reflects that of the minus strand. However, the sequence read itself is reported in the 5′-to-3′ orientation (Fig. 3.7b), with bases closest to the primer-annealing site determined first. Recall that base order is a function of the length of the DNA fragments produced from incorporation of the dideoxynucleotides, so the bases are reported for DNA fragments going from shortest to longest. As demonstrated in Fig. 3.7c, the sequence data file first must be reversed (3′-to-5′ orientation) and then the complementary sequence must be determined to generate the plus strand of the region of the 16S rRNA gene used for subsequent bioinformatic and phylogenetic analysis (Units 6 and 7). The plus-strand sequence is the reverse complement of the 519R sequencing read.

For the cultivation-independent part of the "I, Microbiologist" project, it is also necessary to obtain the reverse complement for one of the two sequences generated for 16S rDNA-containing plasmids by using the T7 and M13R primers. Because the DNA insert may be ligated into the vectors in either orientation, as displayed in Fig. 3.8, one cannot know which of the two sequences should be manipulated until the entire sequence is aligned with reference genes for which the orientation is known. The details for assembling the two sequence reads for a cloned 16S rRNA gene into a single, contiguous sequence are discussed in Unit 6.

Amplicon A DNA target of desired length in the resulting PCR product.

Annealing Association of two complementary DNA (cDNA) strands; see also **Hybridization.**

Chimera A PCR artifact that generates a sequence product comprised of two or more phylogenetically distinct parent sequences; thought to occur when a prematurely terminated amplicon reanneals to a DNA strand belonging to a different organism during a subsequent PCR cycle and is copied to completion in the second cycle, albeit using a foreign DNA template for the remaining sequence; also called a chimeric sequence.

Denaturation Physical separation, or melting, of the DNA strands comprising a single double helix; disrupting the hydrogen bonds between complementary bases in the DNA strands.

Deoxynucleoside triphosphates (dNTPs) The basic building blocks of DNA strands, with a single dNTP consisting of a five-carbon (pentose) deoxyribose sugar linked at the 5′ carbon to the three phosphate groups and at the 1′ carbon to one of four nitrogenous bases (the purines, adenine [A] or guanine [G], and the pyrimidines, thymine [T] or cytosine [C]). The four dNTPs are distinguished based on the attached base: deoxyadenosine triphosphate (dATP), deoxyguanosine triphosphate (dGTP), deoxythymidine triphosphate (dTTP), and deoxycytidine triphosphate (dCTP).

Extension The elongation step within the PCR cycle; occurs when the 3′-hydroxyl group on the pentose sugar of the nucleotide at the end of the nascent DNA strand initiates a nucleophilic attack of the 5′-phosphate group on the incoming dNTP that is complementary to the template at that position; the process repeats, extending the nascent DNA strand in the 5′-to-3′ direction one nucleotide at a time.

Heteroduplex molecules A PCR artifact in which two homologous DNA segments that differ by only a point mutation or a small insertion or deletion are amplified and then reanneal to each other rather than their native partner strand.

Hybridization Physical association of two DNA strands into a stable double-stranded DNA molecule resulting from the formation of hydrogen bonds between complementary base pairs.

Melting temperature (T_m) The temperature at which absorbance has increased to half the maximum absorbance. Because nitrogenous bases in DNA absorb UV light at 260 nm, the progress of DNA melting, or thermal denaturation, can be monitored by measuring changes in absorbance. As the temperature increases, a rise in absorbance is observed as double-stranded DNA gradually becomes denatured across its entire length. The higher the G+C content, the higher the T_m.

Nascent Newly synthesized or extending.

Nucleotide A nucleoside with one or more covalently attached phosphate groups.

Primer dimers A PCR artifact whereby primers act as their own template to make a small PCR product.

Primers DNA oligonucleotides required for initiation of DNA synthesis; complementary to regions at the 5′ and 3′ ends within specific strands of the DNA target.

Recombinant plasmids Plasmids that contain a DNA fragment insert interrupting the vector backbone; generated using molecular cloning procedures.

Restriction enzymes Endonucleases that recognize specific sequences, typically 4 to 6 bp long, and induce a double-stranded DNA break at that site.

Reverse complement A sequence constructed by writing the original DNA sequence backward and replacing the nucleotides by their complement when base paired.

Stringency The ability of two polynucleotide DNA strands to hybridize.

Template A source of DNA that contains target region to be amplified by PCR.

Thermal cycling Oscillation between specific temperatures; exposing a sample to regulated heating and cooling cycles.

REFERENCES

Abd-Elsalam, K. A. 2003. Bioinformatic tools and guidelines for PCR primer design. *African J. Biotechnol.* **2:**91–95.

Acinas, S. G., R. Sarma-Rupavtarm, V. Klepac-Caraj, and M. F. Polz. 2005. PCR-induced sequence artifacts and bias: insights from comparison of two 16S rRNA clone libraries constructed from the same sample. *Appl. Environ. Microbiol.* **71:**8966–8969.

Chien, A., D. B. Edgar, and J. M. Trela. 1976. Deoxyribonucleic acid polymerase from the extreme thermophile *Thermus aquaticus*. *J. Bacteriol.* **174:**1550–1557.

D'Aquila, R. T., L. J. Bechtel, J. A. Videler, J. J. Eron, P. Gorczyca, and J. C. Kaplan. 1991. Maximizing sensitivity and specificity of PCR by pre-amplification heating. *Nucleic Acids Res.* **19:**3749.

Dieffenbach, C. W., T. M. J. Lowe, and G. S. Dveksler. 1995. General concepts for PCR primer design, p. 133–155. *In* C. W. Dieffenbach and G. S. Dveksler (ed.), *PCR Primer: a Laboratory Manual.* Cold Spring Harbor Laboratory Press, Cold Spring Harbor, NY.

Don, R. H., P. T. Cox, B. J. Wainwright, K. Baker, and J. S. Mattick. 1991. 'Touchdown' PCR to circumvent spurious priming during gene amplification. *Nucleic Acids Res.* **19:**4008.

Du, Z., L. Hood, and R. K. Wilson. 1993. Automated fluorescent DNA sequencing of polymerase chain reaction products. *Methods Enzymol.* **218:**104–121.

Elkin, C. J., P. M. Richardson, H. M. Fourcade, N. M. Hammon, M. J. Pollard, P. F. Predki, T. Glavina, and T. L. Hawkins. 2001. High-throughput plasmid purification for capillary sequencing. *Genome Res.* **11:**1269–1274.

Erlich, H. A., D. Gelfand, and J. J. Sninsky. 1991. Recent advances in the polymerase chain reaction. *Science* **252:**1643–1651.

Hall, N. 2007. Advanced sequencing technologies and their wider impact in microbiology. *J. Exp. Biol.* **290:**1518–1525.

Innis, M. A., and D. H. Gelfand. 1994. Optimization of PCRs, p. 5–11. *In* M. A. Innis, D. H. Gelfand, J. J. Sninsky, and T. J. White (ed.), *PCR Protocols, a Guide to Methods and Applications.* CRC Press, London, United Kingdom.

Invitrogen. 2006. *TOPO TA Cloning User Manual,* version U, 25-0184. Invitrogen Corp., Carlsbad, CA.

Kaiser, R., R. Hunkapiller, C. Heiner, and L. Hood. 1993. Specific primer-directed DNA sequence analysis using automated fluorescence detection and labeled primers. *Methods Enzymol.* **218:**122–153.

Landweber, L. F., and M. Kreitman. 1993. Producing single-stranded DNA in polymerase chain reaction for direct genomic sequencing. *Methods Immunol.* **218:**17–26.

Lawyer, F. C., S. Stoffel, R. K. Saiki, S. Y. Chang, P. A. Landre, R. D. Abramson, and D. H. Gelfand. 1993. High-level expression, purification, and enzymatic characterization of full-length *Thermus aquaticus* DNA polymerase and a truncated form deficient in 5′ to 3′ exonuclease activity. *PCR Methods Appl.* **2:**275–287.

Li, Y., S. Korolev, and G. Waksman. 1998. Crystal structures of open and closed forms of binary and ternary complexes of the large fragment of *Thermus aquaticus* DNA polymerase I: structural basis for nucleotide incorporation. *EMBO J.* **17:**7514–7525.

Luckey, J. A., H. Drossman, T. Kostichka, and L. M. Smith. 1993. High-speed DNA sequencing by capillary gel electrophoresis. *Methods Enzymol.* **218:**154–172.

Madabhushi, R. S. 1998. Separation of 4-color DNA sequencing extension products in noncovalently coated capillaries using low viscosity polymer solutions. *Electrophoresis* **19:**224–230.

Mardis, E. R. 2008. The impact of next-generation sequencing technology on genetics. *Trends Genet.* **24:**133–141.

Medini, D., D. Serruto, J. Parkhill, D. A. Relman, C. Donati, R. Moxon, S. Falkow, and R. Rappuoli. 2008. Microbiology in the post-genomic era. *Nat. Rev. Microbiol.* **6:**419–430.

Mullis, K. 1990. The unusual origin of the polymerase chain reaction. *Sci. Am.* **262:**56–61, 64–65.

Mullis, K. 1991. The polymerase chain reaction in an anemic mode: how to avoid cold oligodeoxynuclear fusion. *PCR Methods Appl.* **1:**1–4.

Nakano, S., M. Fujimoto, H. Hara, and N. Sugimoto. 1999. Nucleic acid duplex stability: influence of base composition on cation effects. *Nucleic Acids Res.* **27:**2957–2965.

Sambrook, J., and D. W. Russell. 2001. *Molecular Cloning: a Laboratory Manual,* 3rd ed. Cold Spring Harbor Laboratory Press, Cold Spring Harbor, NY.

Sanger, F., S. Nicklen, and A. R. Coulson. 1977. DNA sequencing with chain-terminating inhibitors. *Proc. Natl. Acad. Sci. USA* **74:**5463–5467.

Schuster, S. C. 2008. Next-generation sequencing transforms today's biology. *Nat. Methods* **5:**16–18.

SuperArray Bioscience Corporation. 2006. *The Advantages of Hot-Start PCR Technology.* SuperArray Bioscience Corp., Frederick, MD. http://www.superarray.com/newsletter/hotstart.html.

Urs, U. K., R. Murali, and H. M. Krishna Murthy. 1999. Structure of Taq DNA polymerase shows a new orientation for the structure-specific nuclease domain. *Acta Crystallogr.* **D55:**1971–1977.

Varsha. 2006. DNA fingerprinting in the criminal justice system: an overview. *DNA Cell Biol.* **25:**181–188.

Wallace, R. B., J. Shaffer, R. F. Murphy, J. Bonner, T. Hirose, and K. Itakura. 1979. Hybridization of synthetic oligodeoxyribonucleotides to ΦX174 DNA: the effect of single base pair mismatch. *Nucleic Acids Res.* **6:**3543–3557.

Web Resources

Biology Animation Library

From Cold Spring Harbor's Dolan DNA Learning Center (Adobe Flash required): http://www.dnalc.org/ddnalc/resources/animations.html

We recommend viewing the following modules:
1. Gel Electrophoresis
2. Polymerase Chain Reaction
3. Cycle Sequencing

For a review of Sanger sequencing methodology, view the Sanger Sequencing module before the Cycle Sequencing module.

General Notes on Primer Design in PCR, compiled by Vincent R. Prezioso (BioSystems Laboratory, Eppendorf North America, Inc., Westbury, NY) http://www.eppendorfna.com/int/index.php?l=131&action=products&contentid=109

Melting-temperature (T_m) **calculators** https://www2.applied biosystems.com/support/techtools/tm_calculator.cfm

This freeware allows you to calculate the melting temperature of your primers, taking into account the base composition, primer concentration, and salt concentration (K^+ or Na^+) of your PCR solution.

Kibbe, W. A. 2007. Oligo Calc: an online oligonucleotide properties calculator. *Nucleic Acids Res.* **35:**w43–w46. http://www.basic.northwestern.edu/biotools/oligocalc.html (Oligo Calc).

Le Novere, N. 2001. MELTING: computing the melting temperature of nucleic acid duplex. *Bioinformatics* **17:**1226–1227. http://bioweb.pasteur.fr/seqanal/interfaces/melting.html.

Mullis, K. 1993. Nobel Lecture. http://nobelprize.org/nobel_prizes/chemistry/laureates/1993/mullis-lecture.html.

National Human Genome Research Institute (NHGRI) website http://www.genome.gov/12513162

National Institutes of Health. 2004. National Human Genomic Research Initiative (NHGRI) seeks next generation of sequencing technologies. http://www.genome.gov/12513210.

READING ASSESSMENT

1. Calculate the T_m for the following sequence, then determine the reverse complement of this sequence:

    5'- C T A A C G T T G C A A C G C T C A G T G -3'

2. Determine the melting temperature of the 16S rDNA primers used in the "I, Microbiologist" project if PCR is performed in a reaction mix containing KCl at 65 mM. (*Hint:* Use one of the suggested websites.)

3. Suggest three ways to increase the stringency of the reaction described in question 2.

4. What sort of artifacts result from performing PCR under low-stringency conditions at room temperature? Why are these artifacts especially problematic for metagenomic DNA samples?

5. If the DNA target you would like to amplify via PCR is expected to have a total length of 4.3 kb, how long must your extension time be during thermal cycling?

6. Spectrophotometric analysis confirmed that you successfully purified ample amounts of genomic DNA from a bacterial isolate. However, you have encountered problems at the PCR amplification step of the project. Despite repeated attempts using standard procedures, you have not been able to amplify the 16S rDNA product. You hypothesize that the G+C content in the genome of your isolate is unusually high, and thus refractory to PCR by conventional methods. Devise a PCR experiment in which you must change either the components of the reaction mix or the thermal cycling conditions to promote amplification of the 16S rRNA gene from this troublesome isolate. Briefly explain how the changes you make will affect the reaction.

7. What is the basis for the difference in resolving power between agarose and polyacrylamide gel electrophoresis?

UNIT 3

EXPERIMENTAL OVERVIEW

In Experiments 3.1 through 3.6, students will work with their purified isolates to obtain genomic DNA by using one of several methods (A to F), whichever produces the best results. The genomic DNA will serve as the template for PCR with primers specific for the 16S rRNA gene. If electrophoresis confirms that a product was produced that is consistent with the size expected for 16S rDNA, then the amplicons will be purified and submitted for DNA sequencing. The sequences will be used for bioinformatic and phylogenetic analyses as described in Units 6 and 7.

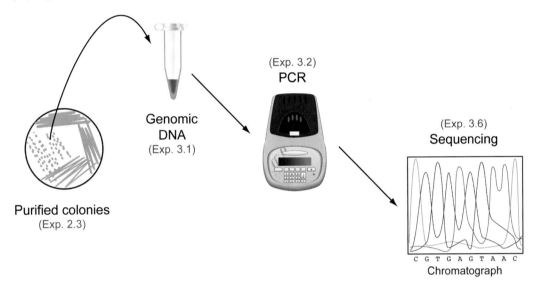

Purified colonies
(Exp. 2.3)

Genomic
DNA
(Exp. 3.1)

(Exp. 3.2)
PCR

(Exp. 3.6)
Sequencing

C G T G A G T A A C
Chromatograph

EXPERIMENT 3.1 Genomic DNA Isolation

The following series of protocols provide students with different strategies to obtain genomic DNA from their soil bacterial isolates. Some of these methods were developed for plasmid purification; however, enough of the bacterial chromosome typically can be recovered for PCR-based applications. Several genomic DNA purification kits are also commercially available, including the Wizard Genomic DNA Purification Kit (Promega), ChargeSwitch gDNA Mini Bacteria Kit (Invitrogen), Genomic DNA Purification Kit (Fermentas Life Sciences), and Generation Capture Column Kit (Qiagen). The manufacturer of each kit supplies instructions.

Method A Simple Boiling Lysis

MATERIALS

1.8-ml "boil-proof" microcentrifuge tubes	1.8-ml microcentrifuge tubes
Sterile distilled water	Sterile toothpicks
P200 pipetter	Microcentrifuge tube racks
P1000 pipetter	Microcentrifuge
100°C water bath	Ice bucket

METHODS

Wear gloves for all DNA manipulations. Always confirm that the water bath is at the proper temperature.

Working with your partner, choose your first 16 isolates. Between the two of you, there may be up to 48 isolates, either purified directly from plates or cultivated in liquid culture. For the "I, Microbiologist" project, a minimum of two colonies from each incubation condition is recommended. Furthermore, all isolates selected should be phenotypically distinct in some way (consider colony and cellular morphology).

Label the boil-proof microcentrifuge tubes as follows:

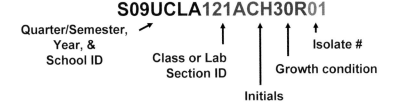

To designate incubation conditions in the sample ID, use a simple nomenclature as exemplified below:

30M isolate # (for colonies from the R̲D̲M̲ plates incubated at 30°C)

30RO isolate # (for colonies from the R̲2A plates incubated under o̲xic conditions at 30°C)

30RA isolate # (for colonies from the R̲2A plates incubated under a̲noxic conditions at 37°C)

30N isolate # (for colonies from the N̲₂-BAP plates incubated at 30°C)

30I isolate # (for colonies from the I̲SP medium 4 plates incubated at 30°C)

30VL isolate # (for colonies from the V̲XylA plates incubated at 30°C in the l̲ight)

30VD isolate # (for colonies from the V̲XylA plates incubated at 30°C in the d̲ark)

Experiment continues

For each isolate purified directly onto solid medium:

1. Add 200 μl of sterile water to a boil-proof microcentrifuge tube.

2. Using a sterile toothpick, scrape cell material from several isolated colonies, and suspend cells in the sterile water until the solution looks turbid or "milky."

3. Close the lid to the microcentrifuge tube tightly and place the sample into the 100°C water bath. Boil the cell suspension for 10 minutes.

4. Immediately place the sample on ice for 10 minutes.

5. Centrifuge the sample at 10,000 rpm for 10 minutes.

   Note: This centrifugation step will pellet the cell debris as well as the chromosome except what is sheared, which likely will be less than 10% of the total DNA in the cells. The fragmented genomic DNA is light enough (like a plasmid) to remain in the supernatant. The amount recovered is sufficient for PCR-based applications, which require only one or a few copies of the target DNA sequence.

6. Transfer 150 μl of the supernatant containing genomic DNA from your isolate into a clean microcentrifuge tube (boil-proof not necessary). Use new tips when collecting the supernatant for separate samples.

7. Keep genomic DNA samples on ice; nucleases present in the crude DNA preparation will degrade the bacterial chromosomes quickly if left at ambient temperature. Your subsequent PCR experiment will not work if you have no template DNA due to carelessness.

8. Be sure the tubes are properly labeled. The DNA samples must be stored in a freezer at −20°C.

For each isolate cultivated in liquid culture:

1. Add 200 μl of liquid culture to a boil-proof microcentrifuge tube. You may need to briefly vortex the culture to resuspend cells if they appear to clump or to have settled to the bottom of the tube.

2. Pellet the cells in microcentrifuge tubes by centrifugation at 10,000 rpm for 2 minutes.

3. Pour off the broth supernatant into waste containers (Hint: the supernatant should be clear if a cell pellet forms at the bottom of the microcentrifuge tube).

4. Resuspend each cell pellet in 200 μl of sterile water.

5. Close the lid to the microcentrifuge tube tightly, and place the sample in the 100°C water bath. Boil the cell suspension for 10 minutes.

6. Immediately place the sample on ice for 10 minutes.

7. Centrifuge the sample at 10,000 rpm for 10 minutes.

   Note: This centrifugation step will pellet the cell debris as well as the chromosome except what is sheared, which likely will be less than 10% of the total DNA in the cells. The fragmented genomic DNA is light enough (like a plasmid) to remain in the supernatant. The amount recovered is sufficient for PCR-based applications, which require only one or a few copies of the target DNA sequence.

8. Transfer 150 µl of the supernatant containing genomic DNA from your isolate into a clean microcentrifuge tube (boil-proof not necessary). Use new tips when collecting the supernatant for separate samples.

9. Keep genomic DNA samples on ice; nucleases present in the crude DNA preparation will degrade the bacterial chromosomes quickly if left at ambient temperature. Your subsequent PCR experiment will not work if you have no template DNA due to carelessness.

10. Be sure the tubes are properly labeled. The DNA samples must be stored in a freezer at $-20°C$.

Troubleshooting the boiling lysis genomic DNA isolation procedure

If you consistently have trouble isolating DNA by using the simple boiling lysis procedure described above, you should try the following two variations of this protocol instead.

Method A (Variation 1) Modified Boiling Lysis for Liquid Cultures

1. Grow a 3- to 5-ml liquid culture of the isolate, and then pellet the entire volume (use 1.8-ml boil-proof microcentrifuge tubes; you will need to sequentially pellet ~1.5 ml of cells two to four times by centrifugation at 10,000 rpm for 2 minutes each time).

2. Resuspend the pellet in 200 µl of sterile water.

3. Lyse the cells, and prepare genomic DNA by following protocols involving boiling lysis (method A), lysozyme coupled to boiling lysis (method B), guanidinium thiocyanate extraction (method C), microLYSIS-PLUS (method D), glass bead lysis (method E), or commercially available purification kits.

Method A (Variation 2) Modified Boiling Lysis for Isolates Propagated on Solid Media

ADDITIONAL MATERIALS

Glucose + MSB

METHODS

For all agar-based plates except VXylA (gelatin-based medium dissolves):

1. Streak out a purified culture of your isolate onto a fresh plate, and grow colonies under optimized conditions for that isolate.

2. Flood the plate with approximately 1.0 ml of sterile glucose + MSB. Let the plate sit for several minutes and then use a sterile flat toothpick to gently scrape all the colonies from the surface of the agar plate, collecting the suspension in a boil-proof microcentrifuge tube.

3. Centrifuge the cell suspension at 10,000 rpm for 2 minutes, and resuspend the cell pellet in 200 µl of sterile water.

4. Lyse the cells, and prepare genomic DNA by following protocols involving boiling lysis (method A), lysozyme coupled to boiling lysis (method B), guanidinium thiocyanate extraction (method C), microLYSIS-PLUS (method D), glass bead lysis (method E), or commercially available purification kits.

Experiment continues

Alternative genomic DNA isolation procedures

If you consistently have trouble isolating DNA by using the boiling lysis procedures described for method A (including variations 1 and 2), try any or all of the protocols described for methods B to F (or use a commercially available kit for genomic DNA purification). Use the Decision Guide at the end of experiment 3.3 to assist with selection of optimal method.

Method B Lysozyme Treatment Coupled to Boiling Lysis in STET Buffer

ADDITIONAL MATERIALS

STET buffer (0.1 M NaCl, 10 mM Tris-HCl [pH 8], 1 mM EDTA [pH 8], 5% Triton X-100)

10 mg of lysozyme per ml (freshly prepared in 10 mM Tris-HCl [pH 8])

Weigh out 0.01 g of lysozyme. Add 0.1 ml of 10 mM Tris-HCl buffer. Vortex to mix. Use immediately, and discard any unused lysozyme suspension.

1.8-ml boil-proof microcentrifuge tubes

100°C water bath

METHODS

1. Grow a 5-ml liquid culture of an isolate or prepare a cell suspension from five agar plates (assuming that you recover approximately 1.0 ml per plate) as described in method A, variation 2.

2. Pellet the entire cell culture or suspension volume (use 1.8-ml boil-proof microcentrifuge tubes; you will need to sequentially pellet ~1.5 ml of cells two to four times by centrifugation at 10,000 rpm for 2 minutes each time).

3. Resuspend the pellet in 350 μl of STET buffer, and add 25 μl of freshly prepared lysozyme (10 mg/ml). Mix by vortexing on low speed for 3 seconds. Note that lysozyme does not work efficiently at pH < 8.

4. Immediately transfer the lysozyme-treated cell suspension to a 100°C water bath and boil the samples for exactly 40 seconds.

5. Immediately place the samples on ice for 10 minutes.

6. Centrifuge the lysate at 10,000 rpm for 10 minutes.

   Note: This centrifugation step will pellet the cell debris as well as the chromosome except what is sheared, which likely will be less than 10% of the total DNA in the cells. The fragmented genomic DNA is light enough (like a plasmid) to remain in the supernatant. The amount recovered is sufficient for PCR-based applications, which require only one or a few copies of the target DNA sequence.

7. Transfer the supernatant to a fresh 1.8-ml microcentrifuge tube, and discard the pellet of bacterial debris. Use new tips when collecting the supernatant for separate samples.

8. Keep genomic DNA samples on ice; nucleases present in the crude DNA preparation will degrade the bacterial chromosomes quickly if left at ambient temperature. Your subsequent PCR experiment will not work if you have no template DNA due to carelessness.

9. Be sure the tubes are properly labeled. The DNA samples must be stored in a freezer at −20°C.

ACKNOWLEDGMENT

This protocol was adapted from D. S. Holmes and M. Quigley, *Anal. Biochem.* **114**:193–197, 1981, as described by J. Sambrook, E. F. Fritsch, and T. Maniatis in *Molecular Cloning: a Laboratory Manual,* 2nd ed., Cold Spring Harbor Laboratory Press, Cold Spring Harbor, NY, 1989.

Method C Guanidinium Thiocyanate Extraction

ADDITIONAL MATERIALS

Store the following reagents at ambient temperature.

Tris-EDTA (TE) buffer (pH 8)
Guanidinium thiocyanate buffer (5 M guanidinium thiocyanate, 100 mM EDTA [pH 8], 0.1% [wt/vol] N-lauroylsarcosine [sodium salt])

Store the following reagents at 4°C.

7.5 M ammonium acetate
Chloroform–2-pentanol
2-Propanol (isopropanol)
70% (vol/vol) ethanol (dilute 200-proof ethanol with sterile distilled H_2O)

METHODS

1. Grow a 3-ml liquid culture of an isolate or prepare a cell suspension from an agar plate as described in method A, variation 2.

2. Pipette approximately 1.0 ml of bacterial cells into a 1.8-ml microcentrifuge tube.

3. Centrifuge at 10,000 rpm for 2 minutes.

4. Decant the supernatant into a waste container.

5. Resuspend the pellet by intermittent vortexing in 50 µl of TE buffer.

6. To lyse the cells, add 300 µl of guanidinium thiocyanate buffer. Briefly vortex to mix, and incubate at room temperature for 5 to 10 minutes.

7. Place the tubes on ice.

8. Add 150 µl of cold 7.5 M ammonium acetate, and invert the tube several times to mix.

9. Keep the tube on ice for a further 10 minutes.

10. Dilute the lysate 1:1 by adding 500 µl of cold chloroform–2-pentanol. Mix by gentle vortexing.

11. Centrifuge at high speed (13,000 rpm) for 10 minutes.

12. Transfer the aqueous phase (top layer) to a new microcentrifuge tube. Be careful not to transfer any of the lower organic layer.

13. Add approximately 1.1 volumes of cold 2-propanol (e.g., if you have 500 µl of aqueous phase, add 540 µl of cold 2-propanol). To mix, invert the tube gently but mix the solution thoroughly for 1 minute.

14. Centrifuge at high speed (13,000 rpm) for 5 minutes.

15. Decant the supernatant into a waste container.

Experiment continues

16. Add 200 µl of cold 70% ethanol to the pellet, and invert the tube three or four times to wash the salt off all sides of the tube.

17. Centrifuge at high speed (13,000 rpm) for 5 minutes.

18. Decant the supernatant into a waste container, and repeat steps 15 to 17 two more times.

19. Dry the DNA pellet completely by storing the tube with the cap open in a fume hood at ambient temperature (25°C) until the ethanol wash has evaporated completely and no fluid is visible. (It should take approximately 2 to 5 minutes.)

20. Resuspend the DNA pellet in 50 µl of TE buffer. Mix by pipetting the buffer up and down several times.

21. Be sure the tubes are properly labeled. The DNA samples must be stored in a freezer at −20°C.

ACKNOWLEDGMENT

This protocol was derived from D. G. Pitcher, N. A. Saunders, and R. J. Owens, *Lett. Appl. Microbiol.* **8:** 151–156, 1989.

Method D microLYSIS-PLUS for PCR-Ready DNA

ADDITIONAL MATERIALS

Kit catalog no. 2MLP-100   PCR tubes and caps
1.8-ml microcentrifuge tubes   Thermal cycler

METHODS

1. Grow a 3-ml liquid culture of your isolate or prepare a cell suspension from three agar plates (assuming that you recover approximately 1.0 ml per plate) as described in method A, variation 2.

2. Pellet the entire cell culture or suspension volume (use 1.8-ml microcentrifuge tubes; you will need to sequentially pellet ~1.5 ml of cells two or three times by centrifugation at 10,000 rpm for 2 minutes each time).

3. Resuspend the cell pellet in 20 µl of microLYSIS-PLUS, a very strong lysis agent.

4. Transfer the cell suspension to a PCR tube, then place the tubes in a thermal cycler and run the following profile:

    Step 1: 65°C for 15 minutes

    Step 2: 96°C for 2 minutes

    Step 3: 65°C for 4 minutes

    Step 4: 96°C for 1 minute

    Step 5: 65°C for 1 minute

    Step 6: 96°C for 30 seconds

    Step 7: hold at 20°C

    For cells that are very hard to lyse, step 1 may need to be longer.

5. After cycling, all or part of the microLYSIS-PLUS–DNA mixture can be used directly in PCR as described in experiment 3.2, making up to 40% of the final volume of a single reaction. Although finer titration of the DNA sample ultimately may be required, start by setting up three PCR mixtures in a 50-µl total volume as follows:

    a. 5 µl of DNA + 45 µl of PCR reagents

    b. 10 µl of DNA + 40 µl of PCR reagents

    c. 20 µl of DNA + 30 µl of PCR reagents

6. Be sure the tubes are properly labeled. The DNA samples must be stored at −20°C.

ACKNOWLEDGMENT
This protocol was adapted from that provided for the microLYSIS-PLUS for PCR-ready DNA kit by The Gel Company (gelinfo@gelcompany.com).

Method E Glass Bead Lysis
This protocol is customarily used to lyse yeast cells with tough chitin cell walls and thus is recommended for isolating genomic DNA from bacterial cells that form calcified colonies, which may be refractory to lysis by other methods.

ADDITIONAL MATERIALS
2.0-ml microcentrifuge tubes

Beveled P200 pipette tips

0.1-mm glass beads (acid washed)

Lysis buffer (0.1 M Tris-HCl [pH 8], 50 mM EDTA, 1% sodium dodecyl sulfate [SDS])

5 M NaCl

Phenol-chloroform-isoamyl alcohol

Chloroform

95% (vol/vol) cold ethanol (dilute 200-proof ethanol with sterile distilled H_2O)

70% (vol/vol) cold ethanol (dilute 200-proof ethanol with sterile distilled H_2O)

METHODS

1. Grow a 5-ml liquid culture of an isolate or prepare a cell suspension from five agar plates (assuming you recover approximately 1.0 ml per plate) as described in method A, variation 2.

2. Pellet the entire cell culture or suspension volume (use 2.0-ml microcentrifuge tubes; you will need to sequentially pellet ~2 ml of cells three times by centrifugation at 10,000 rpm for 2 minutes each time).

3. Wash the cells three times with sterile water by resuspending the cell pellet in ~1.5 ml of water and centrifuging at 10,000 rpm for 2 minutes each time. Decant the supernatant, and then repeat.

4. Resuspend the washed pellet in 500 µl of lysis buffer by vortexing.

Experiment continues

5. Add 0.1-mm glass beads, which have a sand-like consistency, to approximately 2 mm below the meniscus. Vortex vigorously for 30 seconds, add 25 μl of 5M NaCl, and vortex for another 30 seconds. Note that a bead beater, rather than a vortexer, may be used to obtain better lysis.

6. Centrifuge at 10,000 rpm for 2 minutes to decrease the foam that forms during the lysis procedure (step 5).

7. Remove the cell lysate by using a P1000 pipetter with the tip placed at the very bottom of the tube. Transfer the lysate to a fresh 1.8-ml microcentrifuge tube.

   Note: At this point, students may stop the procedure, placing their genomic DNA samples on ice to prevent nucleases present in the crude DNA preparation from degrading the bacterial chromosome. The crude lysate may be used as the template DNA for PCR. Be sure the tubes are properly labeled, and that the DNA samples are stored in a freezer at −20°C when not in use. If PCR is not successful with the crude lysates, students may return to the protocol and proceed with purification as outlined in steps 8 to 16.

8. Perform a phenol-chloroform extraction once as follows.

   Caution: Wear gloves because phenol-chloroform is highly corrosive to skin. Perform extractions under a fume hood because phenol-chloroform is highly volatile with noxious fumes.

   a. Add an equal volume of phenol-chloroform to your sample (e.g., if you have 500 μl of supernatant, add 500 μl of phenol-chloroform). *Note:* There are two layers in the bottle—the top layer contains isoamyl alcohol whereas the bottom layer is a solubilized mixture of phenol and chloroform. Make sure you withdraw liquid for extraction from the bottom layer.

   b. Gently mix by inverting the tube for 2 to 3 minutes by hand (layers should be thoroughly mixed).

   c. Centrifuge the tube at high speed (13,000 rpm) for 15 minutes.

   d. Set a P200 pipetter at its maximum volume (200 μl). Then, using a beveled tip (e.g., gel-loading tips work well), collect the top aqueous layer and transfer the extract into a clean 1.8-ml microcentrifuge tube. You will have to aspirate two or three times to collect the entire aqueous layer. Use a fresh tip each time; avoid the interface by tipping the tube at an angle when aspirating the aqueous layer into the pipette tip.

9. Perform a chloroform extraction twice as follows.

   Caution: Wear gloves because chloroform is a probable carcinogen. Perform extractions under a fume hood because chloroform is highly volatile with a sweet but harmful odor.

   a. Add an equal volume of chloroform to your sample (e.g., if you have 500 μl of supernatant, add 500 μl of chloroform).

   b. Gently mix by inverting the tube for 2 to 3 minutes by hand (layers should be thoroughly mixed).

   c. Centrifuge the tube at high speed (13,000 rpm) for 15 minutes.

   d. Using a beveled tip, collect the top aqueous layer and transfer the extract into a clean 1.8-ml microcentrifuge tube as described above (step 8d).

   e. Repeat steps a to d a second time, but transfer the extract to a 2.0-ml microcentrifuge tube; then go on to step 10 in the protocol.

10. To the aqueous extract, add 1.0 ml of cold 95% ethanol and mix gently by inversion 8 to 10 times.

11. Centrifuge at high speed (13,000 rpm) for 5 minutes, and remove the supernatant by gentle aspiration.

12. Add 1.0 ml of 70% cold ethanol to the DNA pellet, inverting the tube three or four times to wash the salt off the pellet and all sides of the tube.

13. Centrifuge at high speed (13,000 rpm) for 5 minutes, and remove the supernatant by gentle aspiration. Be careful not to dislodge the DNA pellet during this step; sometimes it does not remain adhered to the side of the tube.

14. Dry the pellet completely by storing the tube with the cap open in a fume hood at ambient temperature (25°C) until the ethanol wash has evaporated completely and no fluid is visible (this should take approximately 2 to 5 minutes).

15. Resuspend the pellet in 50 μl of TE (vortex briefly, incubate at 37°C for 10 minutes, vortex again, and perform a quick spin to remove condensation from the lid).

16. Be sure the tubes are properly labeled. The DNA samples must be stored in a freezer at −20°C.

ACKNOWLEDGMENT

This protocol was adapted from that provided by Steven Hahn's laboratory (Fred Hutchinson Cancer Research Center, Seattle, WA): http://www.fhcrc.org/science/labs/hahn/Methods/mol_bio_meth/yeast_quick_dna.html.

Method F Single-Colony Microwave Lysis

Use this method only as a last resort. Perform the procedure on at least three or four separate colonies of the same isolate. Using a sterile toothpick, scrape single colonies into the bottom of a PCR tube. Close the cap, and microwave on the high setting for 1 minute. Immediately transfer the tube to ice, and add the PCR reaction mix including *Taq*. Conduct the PCR procedure according to the protocol described in experiment 3.2.

Technical aside regarding DNA preparation

Large-scale methods such as sonication or French press also are options for lysing cells but are not really practical for the teaching-laboratory environment.

MATERIALS

16S rRNA PCR primer 1492R	Ice bucket
16S rRNA PCR primer 27F	DNA template (from Experiment 3.1)
10× PCR buffer	25 mM MgCl₂
10 mM dNTPs	DMSO sterilized via nylon filter
Taq DNA polymerase	Sterile distilled water
PCR tubes and caps	PCR tube rack
1.8-ml microcentrifuge tubes (sterile)	Ethanol-resistant markers
Thermal cycler	

METHODS

Wear gloves when performing all PCR procedures. Always run negative controls (and positive controls when possible).

1. Use the following PCR primers for 16S rRNA gene amplification:

PCR Primers

5′ → 3′ sequence[a]	Primer name	Position	T_m (50 mM KCl)
GGT TAC CTT GTT ACG ACT T	16S_1492R	1492–1510	49°C
AGA GTT TGA TCM TGG CTC AG	16S_27F	8–27	56°C

[a]M = A or C.

2. For each PCR experiment, only 16 template DNA samples will be used at a time.

3. Prepare the PCR master mix by adding the following reagents to a 1.8-ml microcenrtifuge tube.

   Note 1 Keep all PCR reagents and reaction mixtures on ice throughout the setup. Do not discard PCR reagent stocks that are given to you; these will be your working stocks throughout the project. Keep these reagent stocks in a −20°C freezer box when not in use.

   Note 2 To minimize cross-contamination, always use a fresh aliquot of sterile water in the PCR master mix. Each time you perform PCR, use a sterile 5.0-ml pipette (not a pipetter) to aseptically transfer approximately 1.0 ml of water from the stock bottle to a sterile microcentrifuge tube.

   Note 3 Label the sides rather than the caps of PCR tubes with an ethanol-resistant marker. Markings on the caps may rub off during the PCR run.

Experiment continues

PCR Master Mix

			C_F in 50-μl reaction mixture
5.0 μl $\times$ (no. of reactions + 2) =	μl	10$\times$ PCR buffer[a]	1$\times$
2.5 μl $\times$ (no. of reactions + 2) =	μl	DMSO[b]	5% (vol/vol)
2.0 μl $\times$ (no. of reactions + 2) =	μl	25 mM MgCl$_2$	1 mM MgCl$_2$
2.0 μl $\times$ (no. of reactions + 2) =	μl	25 pmol of 1492R primer/μl[c]	1 pmol/μl (1 μM)
2.0 μl $\times$ (no. of reactions + 2) =	μl	25 pmol of 27F primer/μl[c]	1 pmol/μl (1 μM)
31.0 μl $\times$ (no. of reactions + 2) =	μl	Sterile distilled H$_2$O	NA[d]
1.0 μl $\times$ (no. of reactions + 2) =	μl	10 mM dNTPs	0.2 mM each dNTP
0.5 μl $\times$ (no. of reactions + 2) =	μl	5 U of *Taq* polymerase/μl[e]	2.5 U/reaction
46.0 μl $\times$ (no. of reactions + 2) =	μl	**PCR master mix**	

[a] Contains 15 mM MgCl$_2$, so a final concentration of 1.5 mM is obtained in a 50-μl reaction volume if not supplemented with additional MgCl$_2$. Since a 2.5 mM final concentration is desired, 1 mM MgCl$_2$ will be added to the reaction mixture.

[b] Wear gloves when handling DMSO. Use DMSO in the fume hood only. DMSO facilitates denaturation of the DNA template, thereby decreasing the apparent T_m of the primer-template DNA duplex.

[c] Prevent the formation of primer dimers by first incubating the primer stocks at 80°C for 2 minutes and immediately plunging the tubes into ice. Keep the primer stocks cold, and add aliquots to the PCR master mix as needed. The heat treatment should resolve existing secondary structure (e.g., hairpins and nonspecific heteroduplexes), while the cold incubation temporarily prevents nonspecific priming events and secondary structures from forming again before the primers are added to the PCR mix.

[d] NA, not applicable.

[e] The *Taq* DNA polymerase is added to the mixture last, just before the mixtures are placed into the thermal cycler. Due to the viscosity of the enzyme stock solutions, students often struggle to pipette small volumes accurately without extensive practice. Thus, it is advised that the instructor or TA demonstrate how to add the *Taq* to the tubes and ensure that it is properly mixed before continuing with the PCR cycling steps. It is also worthwhile to have students practice pipetting 50% glycerol solutions before trying to work with the *Taq* on their own. Because the template DNA is not highly purified and DMSO inhibits *Taq* activity by ~50%, the amount of *Taq* used for these reactions is twice the amount typically used for standard PCR.

4. Gently vortex the tube containing the PCR master mix several times to mix. You will notice that the *Taq*, which typically is stored in 50% glycerol, initially descends to the bottom of the tube upon expulsion from the pipette tip. Gentle mixing will ensure that the *Taq* is dispersed evenly within the master mix. Return the tube to the ice bucket.

5. Add 4 μl of DNA template to each PCR tube (should be $\leq$1 μg/reaction mixture). For a negative control, add 4 μl of sterile water (same as that used in the PCR master mix) to one PCR tube. Keep the PCR tube rack on ice such that the DNA in tubes remains cold at 4°C. (Remember: nuclease activity and nonspecific priming at ambient temperature.)

6. Add 46 μl of the PCR master mix to each PCR tube. Pipette up and down to mix. Discard leftover master mix.

   Note: Make sure the DNA-PCR master mix solution is at the bottom of each tube before placing tubes in thermal cycler. If drops cling to the sides of the tubes, use a pipette tip to guide the drops down to the bottom.

7. Put the caps on the PCR tubes. Place the tubes into a 96-well thermal cycler (PCR machine), and start the PCR program:

   a. Initial denaturation: 94.0°C for 3.00 minutes

   b. Three-step cycling repeated 35 times:
      Denaturation: 94.0°C for 1.00 minute
      Annealing: 48.0°C for 30 seconds
                 (~5°C below apparent T_m of primers)
      Extension: 72.0°C for 1.00 minute

   c. Final extension: 72.0°C for 7.00 minutes

   d. Storage: 4.0°C for infinite time (∞)

8. When the program has finished, the PCR tubes must be removed from the thermal cycler and stored at −20°C until the next lab period.

BOX 3.1 Troubleshooting PCR

1. Unless a positive control is used, it will not be obvious whether the PCR experiment failed or the genomic DNA preparation method was unsuccessful. Any bacterial DNA lysate that produces a PCR product by the procedure described in Experiment 3.2 can serve as a positive control. This lysate may be obtained either from another student in the class or from a common laboratory strain such as *E. coli.*

2. Try different PCR conditions such as "hot start" PCR (described in Experiment 5.2) or "touchdown" PCR (described by Don et al. [1991]).

3. Try different primer combinations (e.g., 16S rDNA 8F instead of 27F in combination with 1492R, or 1510R instead of 1492R in combination with 27F).

Alternative PCR Primers

$5' \rightarrow 3'$ sequence	Primer name
CGG TTA CCT TGT TAC GAC TT	16S_1510R
AGA GTT TGA TCC TGG CTC AG	16S_8F

EXPERIMENT 3.3 Gel Electrophoresis of Amplicons

MATERIALS

Agarose

125-ml Erlenmeyer flask

Hot mitts

1× Tris-acetate-EDTA (TAE) buffer

6× TAE loading dye

PCR mixtures from experiment 3.2

Spatula

Analytical scale

Power supply

Minigel apparatus

1-kb DNA ladder

Glass petri dish lid

METHODS

1. Prepare a 0.8% agarose gel as follows:

   a. Use a 125-ml Erlenmeyer flask.

   b. Measure out 0.4 g of agarose and add it to the flask.

   c. Add 50 ml of 1× TAE buffer (do not use water).

   d. Heat in a microwave oven for approximately 1 to 2 minutes or until solution just begins to boil. *Caution:* The solution is very hot; do not touch it with bare hands or you may get burned. Use hot mitts to handle the solution. Carefully remove flask and gently swirl the solution. If the agarose has not completely dissolved, return the flask to the microwave for further heating.

   e. Cover the flask with the lid from a glass petri dish to prevent evaporation while the solution cools over the next 10 to 15 minutes. Let the solution cool until it can be handled without hot gloves. Gently swirl the agarose solution every 3 or 4 minutes to ensure homogeneous cooling.

   f. Meanwhile, assemble the minigel casting tray. Put two 13-well combs into the tray, one at the top and another halfway down.

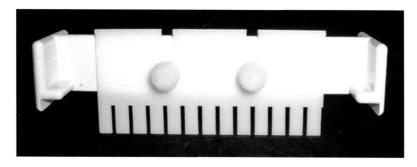

   g. Pour the cooled 0.8% agarose gel solution into the casting tray.

   h. Let the gel solidify (approximately 20 minutes at room temperature).

   i. Gently remove the combs, and place the tray containing the gel into the electrophoresis chamber of the minigel apparatus. Fill the chamber with 1× TAE buffer.

   j. Add 5 μl of ethidium bromide (EtBr) to the buffer chamber. *Note:* **EtBr is a mutagen and suspected carcinogen; wear gloves when handling it.**

Experiment continues

2. Prepare the 1-kb DNA ladder as follows for gel electrophoresis: 4 μl of 6× dye + 18 μl of H_2O buffer + 2 μl of 1-kb ladder stock = 24 μl (total volume).

     This will be enough for two lanes (if running a gel with only one comb, cut the volumes in half).

3. Prepare each PCR sample as follows for gel electrophoresis: 2 μl of 6× dye + 6 μl of H_2O buffer + 4 μl of PCR sample = 12 μl (total volume).

4. Load 12 μl of the 1-kb ladder into a single lane of the top set of wells, and load the remaining 12 μl into a single lane of the bottom set of wells. Then load the 16 PCR samples (12 μl each) into the remaining wells. Subject the samples to electrophoresis for approximately 30 minutes at 95 mA.

5. When electrophoresis is complete, turn off the power supply and then remove the gel from the electrophoresis apparatus. Be sure to wear gloves when manipulating the gel. View the DNA bands using a UV gel-imaging system (a transilluminator). Please consult the teaching assistant (TA) or instructor for instructions on how to use the gel-imaging system. Take a picture of the illuminated gel for your laboratory notebook. Save the gel image as a JPEG file (.jpg extension) for upload to the CURL Online Lab Notebook (http://ugri.lsic.ucla.edu/cgi-bin/loginmimg.cgi).

6. Dispose of your gel, your gloves, and any other products contaminated with EtBr in the appropriate waste container.

7. Examine your gel picture. Your DNA bands corresponding to the 16S rRNA amplicon should align with the 1.5-kb marker on the 1-kb ladder. Make sure you adhere the picture to a page in your lab notebook with tape or glue, and immediately label each lane with the identity of the sample loaded as well as the size of each DNA fragment.

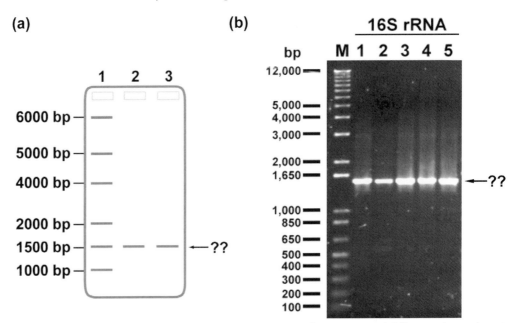

Results of gel electrophoresis. (a) Diagrammatic representation of an agarose gel following electrophoresis. Lanes are labeled as follows: 1, 1-kb ladder; 2, first isolate; 3, second isolate. (b) 1% agarose–1× TAE gel run for 1 hour at 106 V. Lanes are labeled as follows: M, marker (Invitrogen 1 Kb Plus DNA ladder); 1, *B. subtilis*; 2, *P. fluorescens*; 3, *P. acidovorans*; 4, *P. putida*; 5, *P. testosteroni*. Gel picture courtesy of Hwasun Ku, UCLA.

Note: When you read gels, the orientation of the gel should be such that the wells in which samples were loaded are placed at the top and the order can be read from left to right. The larger fragments should be at the top, and the smaller fragments should be near the bottom. You should label the size of each band in the DNA ladder, since this can vary depending on the manufacturer. Your TA or instructor will provide this information. You should also label what sample was loaded in each lane and note the size of each fragment in each lane; this can be deduced by comparing the relative migration of bands to those in the DNA ladder. In the examples in panels a and b above, what is the approximate size of each of the bands indicated by the arrow and question marks?

8. If your electrophoresis results indicate that you have successfully amplified the 16S rRNA gene, you may move on to Experiment 3.4. If no PCR product is obtained, follow the decision guide on the next page, which will steer you through trouble-shooting strategies for this part of the project.

Experiment continues

Decision Guide

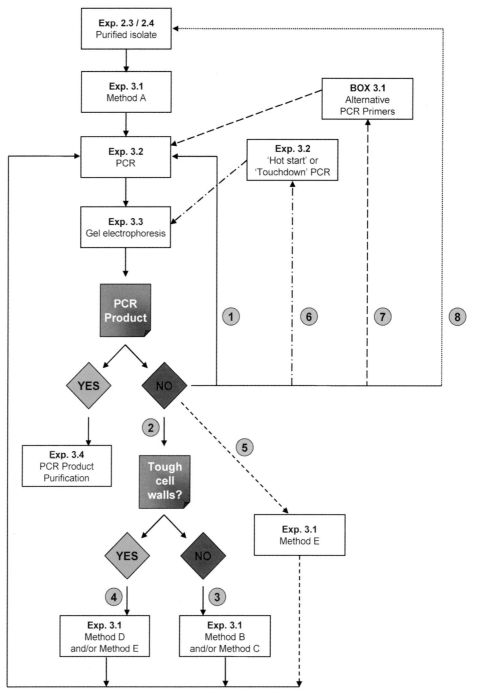

Decision Guide Summary. (1) If no PCR product is obtained after Experiment 3.3, repeat Experiments 3.2 and 3.3; however, be sure to include a positive control. (2) No PCR product at this point suggests the genomic DNA isolation (Experiment 3.1) was not successful. Try an alternative method for preparing genomic DNA, with the choice of method influenced by observations regarding colony morphology. Do the colonies appear calcified, representing cells with potentially tough cell walls to lyse? (3) If no, try method B or C. (4) If yes, try method D or E. (5) If none of these methods yields the desired PCR product, either try method E as a last resort or move on to begin troubleshooting the reaction itself. (6 and 7) Try repeating the PCR step using hot-start or touchdown PCR conditions (6) or alternative primer combinations as described in Box 3.1 (7). (8) If you exhaust all recommended troubleshooting strategies, you should start over with a different isolate.

EXPERIMENT 3.4 Purification of PCR Products

Once amplification of the 16S rRNA gene has been confirmed via electrophoresis, the PCR products should be cleaned using a protocol designed to purify double-stranded DNA fragments produced by PCR. The 16S rDNA fragment, which is within the range of 100 bp to 10 kb, will be purified away from primers, nucleotides, polymerases, and salts present in the PCR mix. The procedure described is for the QIAquick PCR purification kit from Qiagen, which uses spin columns in a microcentrifuge. Numerous other kits are commercially available, including the GenElute PCR cleanup kit (Sigma Aldrich), the Wizard PCR cleanup systems (Promega), the PureLink PCR purification kit (Invitrogen), and the GenCatch PCR cleanup kit (Epoch Biolabs). The manufacturer of each kit supplies instructions.

MATERIALS

QIAquick PCR purification kit P1000 and P200 pipetters
Microcentrifuge Pipette tips

METHODS

1. Add 5 volumes of buffer PB to 1 volume of the PCR sample, and mix by inverting the tube several times. For example, add 250 µl of buffer PB to a 50-µl PCR sample.

2. Place a QIAquick spin column in a 2.0-ml collection tube provided with the kit. Be sure to label the sides of both the column and the collection tube.

3. To bind DNA, apply the sample to the QIAquick column. Do not touch the pipette tip to the column membrane.

4. Centrifuge at high speed (13,000 rpm) for 1 minute.

5. Discard the flowthrough. Place the QIAquick column back into the same tube. The collection tubes are reused to reduce plastic waste.

6. To wash, add 750 µl of buffer PE to the QIAquick column.

    Note: Be sure to use buffer PE to which ethanol has been added (check the markings on the lid). You must use 200-proof ethanol when preparing the buffer, or the DNA will not stay bound to the column during the wash steps.

7. Centrifuge at high speed (13,000 rpm) for 1 minute.

8. Discard the flowthrough, and place the QIAquick column back into the same tube.

9. Centrifuge the column at high speed (13,000 rpm) for an additional 1 minute to remove residual wash buffer, which can inhibit subsequent enzymatic reactions (e.g., DNA sequencing) if not entirely eliminated from the column.

    Note: Residual ethanol from buffer PE is not removed completely unless the flowthrough is discarded before this final centrifugation step.

10. Place the QIAquick column in a clean 1.8-ml microcentrifuge tube. Be sure to label both the cap and side of the tube (in case the cap should break off in the subsequent centrifugation step).

Experiment continues

11. To elute DNA, add 30 μl of buffer EB (10 mM Tris-HCl [pH 8.5]) to the center of the QIAquick membrane. Let the column stand for 1 minute, and centrifuge at high speed (13,000 rpm) for 1 minute. You must use a lid with the centrifuge to prevent the caps of the microcentrifuge tubes from breaking off during the spin.

12. The DNA samples may be stored in the freezer at −20°C.

REFERENCE

This protocol has been adapted from that provided with the QIAquick PCR purification kit as described in the *QIAquick Spin Handbook for PCR Purification Kit, Nucleotide Removal Kit, and Gel Extraction Kit,* Qiagen Corp., November 2006.

EXPERIMENT 3.5 Quantification of Purified PCR Products

Following purification, an aliquot of the DNA first should be run on an agarose gel alongside a DNA ladder to confirm that a purified 16S rDNA product was recovered successfully (method A below). At the same time, this step will provide a qualitative assessment of the total yield after purification. If a DNA band of the expected size is observed, proceed to method B for conduct of a more quantitative determination of the purified product.

Wear gloves throughout DNA analysis.

Method A Qualitative Comparative Analysis

1. Run a gel as described in Experiment 3.3, except prepare the DNA sample as follows. Add 2.4 μl of DNA eluate from Experiment 3.4 to 2 μl of 6× TAE loading dye plus 7.6 μl of dH$_2$O for a total volume of 12 μl. Prepare the 1-kb DNA ladder as described, and load the sample(s) and ladder onto a 0.8% agarose gel.

2. Use the picture you obtain from this gel to make a qualitative comparison of how much DNA you have recovered relative to that visualized in the first picture taken in Experiment 3.3. The intensity of the band in the second picture should be as high as that in the first picture if you recovered the equivalent amount.

Method B Quantitative Spectrophotometric Analysis

MATERIALS

UV/visible spectrophotometer such as a NanoDrop (Thermo Scientific), a SpectraMax Plus384 microplate reader (Molecular Devices), or equivalent

METHODS

1. Obtain optical density (OD) measurements at wavelengths of 260 and 280 nm for an appropriate dilution of your DNA sample. The dilution and/or volume to test will be determined by your TA and instructor in accordance with the specifications and sensitivity of the UV/visible spectrophotometer used for this project.

2. Calculate the DNA concentration by using the OD$_{260}$, and check for purity by using the OD$_{260}$/OD$_{280}$ ratio.

   One absorbance unit at a wavelength (λ) of 260 nm is equal to a DNA concentration of 50 μg/ml. In other words:

   $$\frac{50 \text{ mg/ml}}{1 \text{ Å}} = \frac{x \text{ mg/ml}}{\text{Measured OD}_{260}}$$

   Solve for x, and then multiply by your dilution factor. This calculation will give you the concentration of your purified PCR product in micrograms per milliliter.

3. To check for purity, divide the OD$_{260}$ by the OD$_{280}$. For pure DNA, the OD$_{260}$/OD$_{280}$ should be between 1.7 and 1.8. Ratios lower than 1.7 indicate protein contamination in your sample; ratios higher than 1.8 indicate RNA contamination. A ratio of 2 indicates pure RNA.

To determine the composition and order of bases in the purified PCR products, an in-house or commercial DNA sequencing service may be utilized to perform the DNA sequencing reactions and subsequent gel analysis. Alternatively, if the student laboratory is equipped with a DNA analyzer, students may process the samples themselves as part of the project. It is at the discretion of the instructor to determine the most suitable means by which to accomplish this aspect of the project and provide instructions to students accordingly.

The following primer will be used to sequence the 16S rRNA gene:

519R Sequencing Primer

5′ → 3′ sequence[a]	Primer name	Position	Template
GWA TTA CCG CGG CKG CTG	16S_519R_seq	519–536	16S rDNA

[a]W = A or T and K = G or T.

This primer is complementary to a conserved sequence within the 16S rRNA gene. Notice that this primer is in the reverse orientation, meaning approximately 500 bp within the 5′ region of the 16S rRNA gene (e.g., positions 1 to 519) will be amplified. This portion of the gene is sufficiently variable to allow identification of cultivated bacterial species.

MATERIALS

Purified PCR products from Experiment 3.4   Sequencing primer: 16S_519R

METHODS

PERIOD 1

Prepare and submit samples for DNA sequencing or perform the DNA sequencing reactions and gel analysis as directed by the TA or instructor.

PERIOD 2

Inspect and manually edit DNA sequences to be used for bioinformatics and phylogenetic analysis.

You should assess the quality of the DNA sequence obtained for each of your isolates as follows before any further analysis takes place.

1. For each DNA sequence, first examine the chromatogram file (it should be a .pdf file). A high-quality DNA sequence has reasonably sharp peaks with height that is uniformly above background for all four bases as shown:

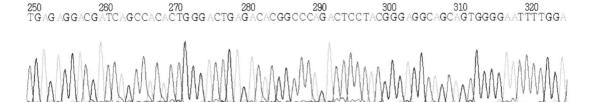

Experiment continues

A sequence of substandard quality also is presented for comparison. Note how difficult it is to resolve any of the base assignments due to overlapping peaks. This sample could have been contaminated with multiple chromosomes from more than one microorganism, the concentration of the DNA subjected to sequencing may have been too low, or the quality of the DNA sample may have been poor.

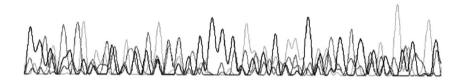

Such results indicate that some troubleshooting should be done to confirm that your isolate is actually pure (Experiments 2.3, 4.1, and 4.2). You may have to make fresh streak plates and repeat the genomic DNA preparations and PCR procedures (Experiments 3.1 to 3.5) for this isolate until a high-quality sequence is obtained. Print the chromatograms, regardless of quality, in color (not black and white or grayscale), and paste or tape printouts of chromatograms in your lab notebook.

2. If visual inspection indicates that your DNA sequence is of a reliable quality to work with further, open the corresponding FASTA file (it should be a plain-text file with .txt extension). The nucleotide bases as reported above the peaks in the chromatogram have been transcribed into an accompanying plain-text file, retaining the base order as called by the DNA sequencing software. The format of a FASTA nucleotide sequence record is as follows, with a greater-than character (>) preceding a short description line that is followed by the DNA sequence in uppercase or lowercase letters:

```
>F08UCLA121ACH30R01_519R
CGGATCGGCTATCTGTGGTACGTCAAACAGCAAGGTATTAACTTACTGCCCTTCCTCCCA
ACTTAAAGTGCTTTACAATCCGAAGACCTTCTTCACACACGCGGCATGGCTGGATCAGGC
TTTCGCCCATTGTCCAATATTCCCCACTGCTGCCTCCCGTAGGAGTCTGGACCGTGTCTC
AGTTCCAGTGTGACTGATCATCCTCTCAGACCAGTTACGGATCGTCGCCTTGGTAGGCCT
TTACCCCACCAACTAGCTAATCCGACCTAGGCTCATCTGATAGCGTGAGGTCCGAAGATC
CCCCACTTTCTCCCTCAGGACGTATGCGGTATTAGCGCCCGTTTCCGGACGTTATCCCCC
ACTACCAGGCAGATTCCTAGGCATTACTCACCCGTCCGCCGCTGAATCCAGGAGCAAGCT
CCCTTCATCCGCTCGACTTGCATGTGTTAGGCCTGCCGCCAGCGTT
```

At both ends of the DNA sequence there likely are N's, each of which represents an ambiguous base call made at a particular position. There also may be a few scattered N's within the internal stretches of the sequence.

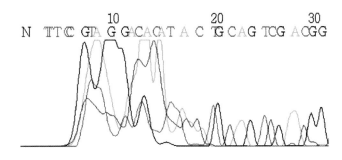

Find the positions of the N's in the chromatogram to see if you can manually resolve the base calls by inspecting the peaks themselves. Pencil the appropriate base call directly onto the chromatogram printout, then modify the DNA sequence FASTA file accordingly. It is unlikely that you will be able to resolve ambiguous or incorrect base calls comprising the first 20 bases or so, as this is immediately downstream of the position within the gene where the 519R primer annealed. These nucleotides may be deleted from the text file (*be sure to save your modified text file with a new name; you always want to retain the original FASTA file unchanged*).

The length of the DNA sequence expected should be approximately 500 bases. It is unlikely that the bases beyond the first 500 or so actually encode the 16S rRNA gene, so these also may be deleted from your FASTA text file. It is probable that you will end up deleting additional bases from the end once you establish where the gene starts, which may be deduced from the multiple sequence alignment constructed in Experiment 7.2.

Be sure to save your modified sequences again in FASTA format.

3. Use any of the following websites to determine the reverse complement of your DNA sequences:

    The Sequence Manipulation Suite http://www.bioinformatics.org/SMS/rev_comp.html

    Baylor College of Medicine (BCM) Human Genome Sequencing Center (HGSC) http://searchlauncher.bcm.tmc.edu/seq-util/Options/revcomp.html

    The Bio-Web Python CGI Scripts for Molecular Biology & Bioinformatics http://www.cellbiol.com/scripts/complement/reverse_complement_sequence.html

Once you obtain the reverse complement of each of your 16S rRNA gene sequences, save your sequences once more in FASTA format for use in the experiments comprising Units 6 and 7.

UNIT 4

Phenotypic Characterization of Bacterial Isolates

SECTION 4.1
MICROBIAL PRODUCTION OF ANTIBIOTICS AND THE EMERGENT RESISTOME

Since the first half of the 20th century, humankind has successfully capitalized on the discovery of antibiotics, utilizing sophisticated chemicals produced by microbes to combat and control infectious disease (Davies, 2007; Tomasz, 2006). Since the 1970s, the United States has been facing the reemergence of infectious disease as a serious health problem (Alekshun and Levy, 2007; Davies, 2007; Levy and Marshall, 2004). The causative agents of infectious disease include bacteria, viruses, protozoans, fungi, and parasitic worms, each with their own mode of transmission and treatment. Of particular concern in the United States are hospital-acquired, or nosocomial, infections, which account for approximately 80,000 deaths annually (Madigan and Martinko, 2006). The hospital environment provides for the selection and maintenance of multiple-drug-resistant bacterial pathogens, which preferentially infect immunocompromised patients at sites such as the bloodstream and the urinary and respiratory tracts. Some of the most common microorganisms responsible for nosocomial infections include *Staphylococcus aureus*, *Escherichia coli*, *Enterococcus* spp., *Pseudomonas aeruginosa*, *Klebsiella pneumoniae*, and *Candida albicans*. Some of these microbes are among the normal flora found in the human body, while others are opportunistic pathogens, which cause disease in hosts with compromised immune systems. Many of these microbes have the potential to be resistant to one or more antibiotics, complicating treatment and increasing the susceptibility of patients and even health care workers themselves to infection.

The origin of antibiotic resistance genes
The appearance and spread of strains resistant to antibiotic treatment are most alarming, as they stand to thwart the initial achievements provided by antimicrobial therapies, which significantly reduced the number of fatal bacterial infections (Alekshun and Levy,

113

2007; Davies, 2007; Levy and Marshall, 2004; Mokdad et al., 2004). Therein lies the dark side of the antibiotic paradigm—the discovery and medicinal application of antibiotics as a treatment for bacterial infections have been paralleled by the emergence of bacteria that possess mechanisms to protect themselves against these toxic substances. Efforts are under way to track the source of drug resistance and uncover the mechanisms responsible. Most of the genes that confer resistance to antibiotics used to treat pathogenic bacteria in clinical settings reside on mobile genetic elements such as transposons or resistance (R) plasmids, with any single R plasmid containing several different antibiotic resistance genes (Alekshun and Levy, 2007). As is discussed in more detail below, such genes often encode enzymes that inactivate the antibiotic, prevent its uptake into the bacterial cell, or actively pump the drug out of the cell. Because the antibiotic resistance genes are not part of the bacterial chromosome, they can be readily transferred from one strain to another by horizontal gene transfer (HGT), making the spread of resistance difficult to control. These genes appear to be orthologs, or genes with functional similarity due to a common ancestral origin, of the resistance elements frequently found clustered with antibiotic biosynthesis operons typical of soil-dwelling producers of antibiotics (D'Costa et al., 2006). For example, sequencing the genome of a vancomycin-resistant clinical isolate of *Enterococcus faecalis* revealed that resistance to vancomycin appeared to be encoded by mutant *vanB* genes within a previously unknown transposable element (Paulsen et al., 2003). VanB is one of three types of DNA ligases, which also include VanA and VanD, involved in the synthesis of peptidoglycan cross-links in the cell wall (Guardabassi et al., 2004). The antibiotic vancomycin inhibits bacterial cell wall synthesis by interacting with the two D-alanine residues at the terminus of the peptidoglycan precursor (D-alanine–D-alanine), preventing these residues from being accessible to the active site of the transpeptidase, which removes one of the two D-alanine residues during a reaction that leads to cross-linking of two peptidoglycan chains (Madigan and Martinko, 2006). Mutations in *vanA* or *vanB* that result in vancomycin resistance in enterococci cause the bacteria to synthesize precursors with D-alanine–D-lactate termini, avoiding interaction with vancomycin (Guardabassi et al., 2004).

It turns out that soil bacterial communities contain genes similar to those encoding vancomycin resistance in the human pathogen *E. faecalis*. In a 2004 study by Guardabassi and colleagues, bacteria from soil collected from various sites in Denmark were isolated using media supplemented with vancomycin (Guardabassi et al., 2004). PCR primers were designed to detect homologues of the *vanA* and *vanB* genes from isolates that grew in the presence of vancomycin. Although all were gram-positive bacteria, two strains belonged to the genus *Paenibacillus* while the other four belonged to a genus of actinomycetes called *Rhodococcus*. Interestingly, most clinically relevant antibiotics in use today are derived from soil-dwelling actinomycetes (Baltz, 2007; D'Costa et al., 2006).

Guardabassi and colleagues found that both *Paenibacillus* strains contained genes with high identity to *vanA* from *Enterococcus faecium* (87% and 92%, respectively), while the *Rhodococcus* strains harbored genes with considerably lower identity (58%) to the *E. faecalis vanA* gene. With regard to *vanB*, both *Paenibacillus* strains encoded genes with moderate identity to the gene derived from the *E. faecalis* transposon discovered during genome sequencing (76% for both) while *Rhodococcus* strains demonstrated only 62% identity to the same gene (Paulsen et al., 2003; Guardabassi et al., 2004). Similar identity levels have been reported for other vancomycin-resistant actinomycetes, including strains that produce the antibiotic itself (Marshall et al., 1998; Hong et al., 2002).

This work suggests that vancomycin resistance elements in enterococci responsible for nocosomial infections are related to those found in soil-dwelling gram-positive bacteria. It is unlikely, however, that the enterococci acquired the genes through direct HGT since only a moderate level of sequence conservation is calculated for *vanA* and *vanB* orthologs and a marked difference in G+C content is observed among the resistance genes in this study (Zirakzadeh and Patel, 2005). Instead, if the *vanA* and *vanB* genes originated on mobile genetic elements in the soil-dwelling *Paenibacillus* or *Rhodococcus* strains, the genes were probably sequentially transferred through several intermediate species, wherein the *vanA* and *vanB* gene sequences diverged, before arriving in the enterococci.

Driven by the selective pressures created by the hospital environment, the vancomycin resistance genes have made their way into yet another opportunistic human pathogen, *Staphylococcus aureus* (Tomasz, 2006; CDC, 2002). It may not be possible to predict the path that mobilized resistance genes will take to make their way into human pathogens. However, understanding resistance mechanisms present in the soil environment, as well as their frequency of occurrence, may provide a means of averting or combating resistance determinants that have the potential to emerge in the clinical setting.

The antibiotic resistome within soil bacterial communities

D'Costa and colleagues assessed the degree to which resistance to the various classes of antibiotics exists within soil bacterial communities (D'Costa et al., 2006). Being coinhabitants of a terrestrial ecosystem, soil-dwelling bacteria should have a battery of antibiotic resistance mechanisms to protect against antibiotics produced either by themselves or by other microbes. Characterization of the collective antibiotic determinants, or resistome, present in the soil may reveal novel resistance mechanisms. For their analysis, D'Costa et al. specifically targeted members of the actinomycete genus *Streptomyces*. Practically speaking, this was a relatively easy bacterial type to isolate, since the presence of *Streptomyces* cells on an agar plate can be detected initially based on the "earthy odor" caused by the production of geosmins (Madigan and Martinko, 2006). *Streptomyces* colonies, which are compact and have a dusty appearance and a characteristic color, can be readily distinguished from other types of bacteria, and *Streptomyces* spores can be readily streaked to obtain pure cultures of the environmental isolates. Recall from Unit 2 that a medium called ISP4, which specifically enriches for this bacterial lineage, is utilized in the "I, Microbiologist" project; therefore, it is likely that students will be working directly with representatives of the *Streptomyces* genus.

Significantly, *Streptomyces* species are renowned for their ability to produce antibiotics (Madigan and Martinko, 2006). More than 500 different antibiotics are produced by *Streptomyces,* and almost 50% of all *Streptomyces* spp. isolated to date have proven to be producers of antibiotics, with several distinct chemicals being produced by a single microorganism. On one hand, by focusing their study on this group, D'Costa and colleagues bias their analysis of the soil resistome to include only the bacteria that are the most likely to actually harbor mechanisms of resistance, if for no other reason than to avoid suicide. On the other hand, by constructing the study around this group of known antibiotic producers, their analysis is sure to be successful, with a question and an experimental system that will produce a manageable amount of data with predictable outcomes. Either way, significant insights stand to be gained from this traditional, albeit subjective, approach to studying the antibiotic resistome.

Experimental strategy used by D'Costa and colleagues to study the soil resistome

An outline of the overall procedure for the D'Costa et al. (2006) study is shown in Fig. 4.1. The authors cultivated spore-forming bacteria from soil samples collected from a variety of locations, including forest, rural agricultural sites, and urban areas. Cells from candidate colonies morphologically resembling that expected for *Streptomyces* were examined microscopically to confirm the presence of filaments with spore-bearing structures. Aspirants that met these initial criteria were streak purified to homogeneity and then identified based on PCR amplification and sequencing of the 16S ribosomal RNA (rRNA) genes derived from the genomic DNA of purified isolates. The authors identified 480 strains as *Streptomyces*, which they later screened against 21 different antibiotics, which included natural products and their semisynthetic derivatives, as well as completely synthetic compounds. A number of the antibiotics examined had been commercially available for several decades, while others had been approved for clinical use only recently. The drugs tested comprised all major bacterial targets, including the cell wall, the cytoplasmic membrane, the machinery required for nucleic acid and protein synthesis, and enzymes involved in folic acid metabolism (Fig. 4.2).

FIGURE 4.1 Overview of the experimental strategy employed by D'Costa et al. (2006).

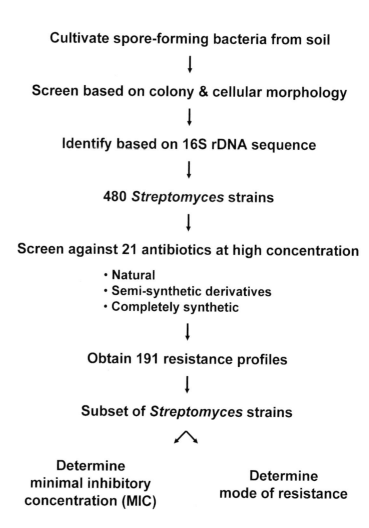

Cultivate spore-forming bacteria from soil

↓

Screen based on colony & cellular morphology

↓

Identify based on 16S rDNA sequence

↓

480 *Streptomyces* strains

↓

Screen against 21 antibiotics at high concentration

- Natural
- Semi-synthetic derivatives
- Completely synthetic

↓

Obtain 191 resistance profiles

↓

Subset of *Streptomyces* strains

Determine minimal inhibitory concentration (MIC)   **Determine mode of resistance**

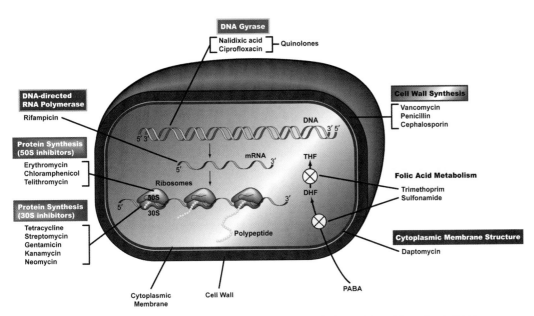

FIGURE 4.2 Antibiotic targets in bacterial cells. Major targets include nucleic acid synthesis (DNA gyrase, RNA polymerase), protein synthesis (50S, 30S, and tRNA), cell wall synthesis, cytoplasmic membrane structure, and folic acid metabolism. Illustration by Cori Sanders (iroc designs).

Overview of antibiotic cellular targets and mechanisms of action

Antibiotics can be classified on the basis of their mode of action, which is related to their chemical structure, as well as their spectrum of microbial activity. The sulfa drugs were among the first fully synthetic chemical compounds used as antimicrobial agents. As structural analogs of growth factors, these compounds interfere with the biosynthesis of nucleic acids and essential proteins. Sulfa drugs have been in clinical use since the 1930s, and resistance to a single drug is now rather common; a combinatorial treatment is therefore administered, resulting in the blockage of two consecutive steps in a biosynthesis pathway. For example, sulfamethoxazole with trimethoprim is a drug combination which blocks biosynthesis of nucleic acids that require the vitamin folic acid as a precursor (Fig. 4.3). Sulfamethoxazole inhibits synthesis of dihydrofolic acid (DHF) by competing with *para*-aminobenzoic acid (*p*-aminobenzoic acid, or PABA), a constituent of folic acid. Trimethoprim blocks production of tetrahydrofolic acid (THF) from DHF by binding to and inhibiting the activity of the enzyme DHF reductase.

β-Lactam antibiotics impede synthesis of the peptidoglycan layer in bacterial cell walls. They do this by binding transpeptidase, preventing its ability to catalyze the peptide cross-linking reaction between glycan chains required to maintain the integrity and strength of the cell wall. Penicillin, a β-lactam product of the fungus *Penicillium chrysogenum*, was first discovered in 1929 by British scientist Alexander Fleming. Although it was the first clinically effective antibiotic, its widespread use did not occur until the early 1940s during World War II, when it served as a revolutionary treatment of grampositive bacterial infections such as that caused by *Streptococcus pneumoniae*. Cephalosporin, a β-lactam antibiotic that was first discovered as the product of another fungus called *Cephalosporium acremonium*, tends to be more resistant to enzymes that destroy the β-lactam ring, a major cause of the development of penicillin resistance, as discussed below. Later, semisynthetic derivatives of penicillin (i.e., ampicillin) and cephalosporin (i.e., cefuroxime) that incorporated structural changes allowing the drugs to be trans-

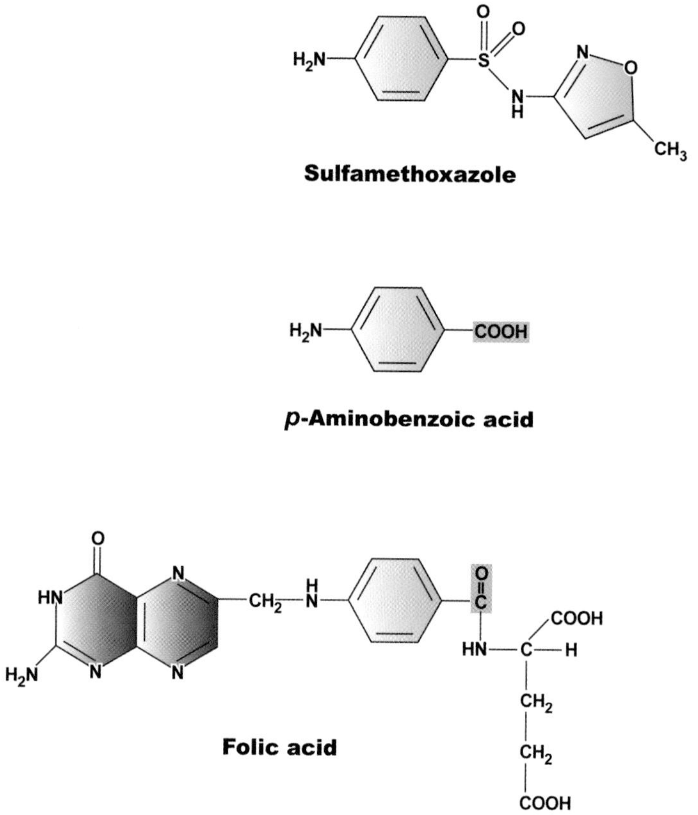

Sulfamethoxazole

p-Aminobenzoic acid

Folic acid

FIGURE 4.3 Sulfa drugs. The natural precursor p-aminobenzoic acid (PABA) is transported into cells where it is eventually converted to folic acid. Sulfamethoxazole is a structural analog and competitive antagonist of PABA, interfering with the synthesis of folic acid. Illustration by Cori Sanders (iroc designs).

ported across the outer membrane of some gram-negative cells broadened the spectrum of activity for this class of antibiotics (Fig. 4.4).

Daptomycin, which is produced by actinomycetes, is a cyclic lipopeptide that was recently approved by the Food and Drug Administration (FDA) as a narrow-spectrum antibiotic for the treatment of infections by gram-positive bacteria (Fig. 4.5). This hydrophobic molecule inserts into bacterial cell membranes and induces the rapid depolarization of the electrostatic potential generated by electron transport chains, thereby inhibiting cellular processes such as ATP synthesis.

Quinolones interact with DNA gyrase, preventing the DNA supercoiling required for condensation and packaging of the bacterial chromosome. Considered broad-spectrum antibiotics, these completely synthetic compounds are effective at treating both gram-positive and gram-negative bacteria. Examples include nalidixic acid and its fluorinated derivative ciprofloxacin, which is used to treat urinary tract infections and anthrax caused by penicillin-resistant *Bacillus anthracis* (Fig. 4.6).

Rifampin (also called rifampicin) is a semisynthetic derivative of a compound made by *Amycolatopsis mediterranei*. This molecule specifically targets the β-subunit of the DNA-dependent RNA polymerase in bacterial cells, thereby inhibiting RNA synthesis (Fig. 4.7). This drug is particularly important in the treatment of mycobacterial infections such as tuberculosis and leprosy.

Several antibiotics inhibit protein synthesis by interacting with either the 50S subunit or the 30S subunit of the ribosome at different steps during translation. Because protein synthesis is a cellular process universally conserved in the *Bacteria,* these drugs have been administered extensively in the clinical setting as broad-spectrum antimicrobials. For instance, erythromycin, which is produced by *Streptomyces erythreus,* and its semisyn-

(a)

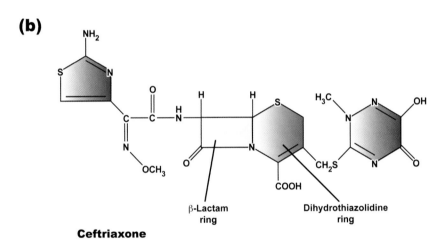

Natural Penicillin

N-acyl group | 6-Aminopenicillanic acid

β-Lactam ring | Thiazolidine ring

Ampicillin

N-acyl group | 6-Aminopenicillanic acid

β-Lactam ring | Thiazolidine ring

(b)

Ceftriaxone

β-Lactam ring | Dihydrothiazolidine ring

FIGURE 4.4 β-Lactam antibiotics. (a) Natural penicillin is composed of 6-aminopenicillanic acid, which contains the β-lactam ring (yellow) and a five-member thiazolidine ring (red), plus an N-acyl group (left of dashed line). Semisynthetic derivatives of penicillin G, such as ampicillin, result from substitutions in the N-acyl group. (b) Example of semisynthetic cephalosporin, which retains the β-lactam ring (yellow) but replaces the thiazolidine ring with a six-member dihydrothiazine ring (blue). Shown is ceftriaxone, which is used to treat *Neisseria gonorrhoeae* infections. Illustration by Cori Sanders (iroc designs).

thetic derivative, telithromycin, are used to treat patients allergic to β-lactam antibiotics (Fig. 4.8a). They interfere with the activity of the 50S subunit of the ribosome during translation. Tetracycline (Fig. 4.8b), which is produced by several species of *Streptomyces*, binds aminoacyl-tRNA and disrupts functionality of the 30S ribosomal subunit, thereby inhibiting protein synthesis. Both natural and semisynthetic derivatives of tetracycline are widely used in veterinary medicine. The biosynthesis of tetracycline and its derivatives involves hundreds of genes and dozens of intermediate steps—a complex chemistry not possible to recapitulate in the laboratory. The aminoglycoside streptomycin (Fig. 4.8c), produced by *Streptomyces griseus*, also interferes with protein synthesis at the 30S subunit of the ribosome; however, the mechanism of inhibition is completely different from that

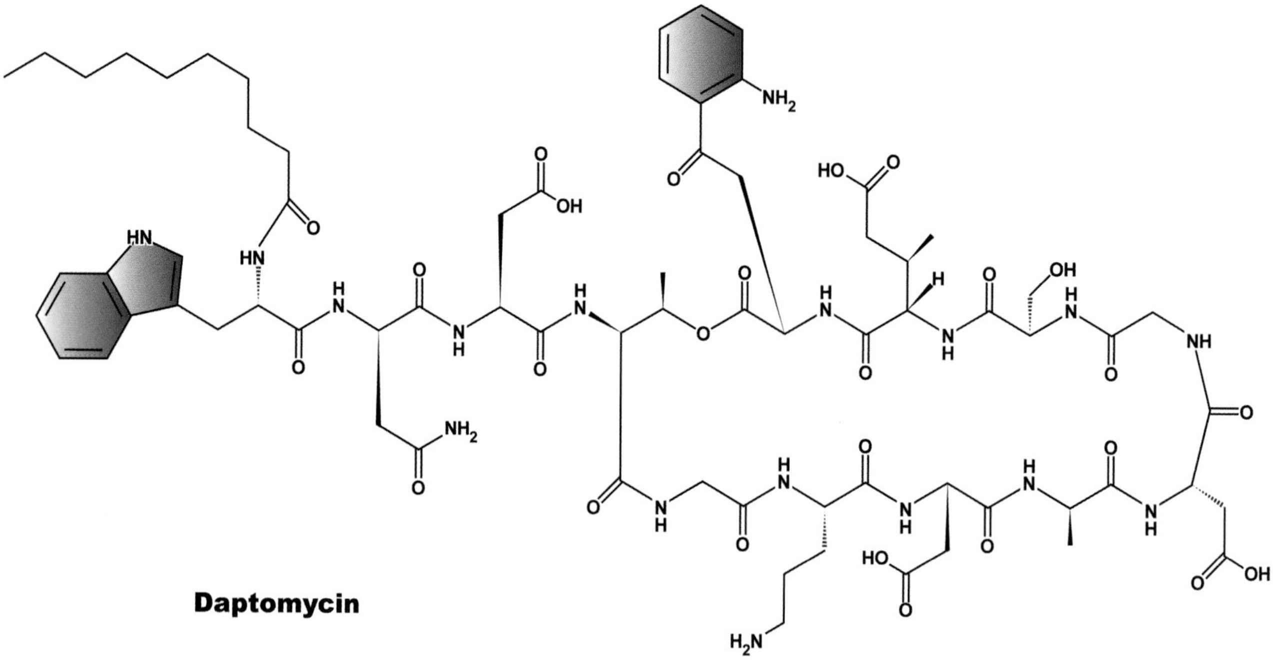

Daptomycin

FIGURE 4.5 Daptomycin, a hydrophobic cyclic lipopeptide that inhibits ATP synthesis by disrupting the electrostatic potential of the bacterial cell membrane. Illustration by Cori Sanders (iroc designs).

FIGURE 4.6 Quinolone compounds, which include fluoroquinolone derivatives of nalidixic acid such as ciprofloxacin. Illustration by Cori Sanders (iroc designs).

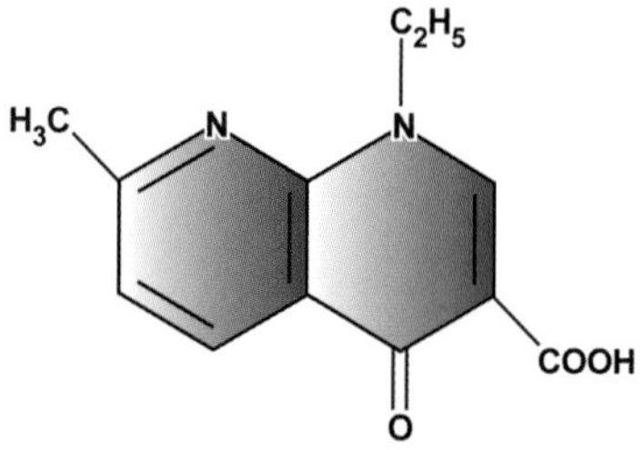

Nalidixic acid

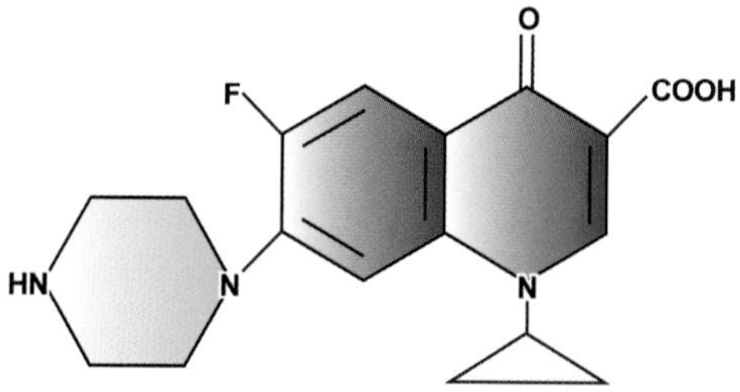

Ciprofloxacin

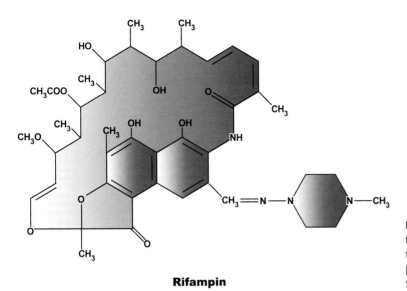

FIGURE 4.7 Antibiotics that affect transcription. Rifampin and its derivatives are macrocyclic lactones that interfere with RNA synthesis by binding to the β subunit of the holoenzyme. Illustration by Cori Sanders (iroc designs).

Rifampin

of tetracycline. This antibiotic binds to the 16S rRNA of the ribosome and prevents initiation of translation.

Resistance profiles of Streptomyces strains cultivated by D'Costa and colleagues span all classes of antibiotics

Notably, the screen performed by D'Costa et al. (2006) was conducted with antibiotics that spanned all of the aforementioned classes, but the assays were performed with drug concentrations at 20 µg/ml, which is considerably higher than what might be administered to patients or what might exist in the soil environment (Goh et al., 2002; Hamscher et al., 2002; Halling-Sørensen et al., 1998). Thus, the study excluded microorganisms that might exhibit resistance when exposed to drugs at low to intermediate levels, as could potentially be encountered in a hospital setting. Although the number of resistance genes among *Streptomyces* spp. in the soil environment might be expected to be higher under such circumstances, it is not known whether one would capture a more diverse range of resistance mechanisms than are encountered when the high antibiotic concentrations are used.

All cells exhibit some low level of resistance to any given antibiotic; this occurrence is termed the intrinsic resistance (Alekshun and Levy, 2007). The level of intrinsic resistance is due to a number of cellular functions such as the presence of efflux pumps and the reduced permeability of the cell membrane (Tamae et al., 2008). This observation is underscored by the work of Davies and coworkers, who have studied the effects of subinhibitory concentrations of a number of different antibiotics on transcriptional pathways (Goh et al., 2002; Yim et al., 2006, 2007). That such a sizeable number of transcriptional changes can be brought about by antimicrobial compounds at low concentrations has led to the view that antibiotics are primarily small-molecule intercellular signaling agents rather than toxins, since the latter function is typically detected at concentrations too high to occur under natural conditions (Davies et al., 2006; Yim et al., 2007).

Cultures of the environmental isolates were prepared from spores in 96-well microtiter plates and were used to inoculate a second set of microtiter plates containing media supplemented with antibiotics at a final concentration of 20 µg/ml. The cultures were

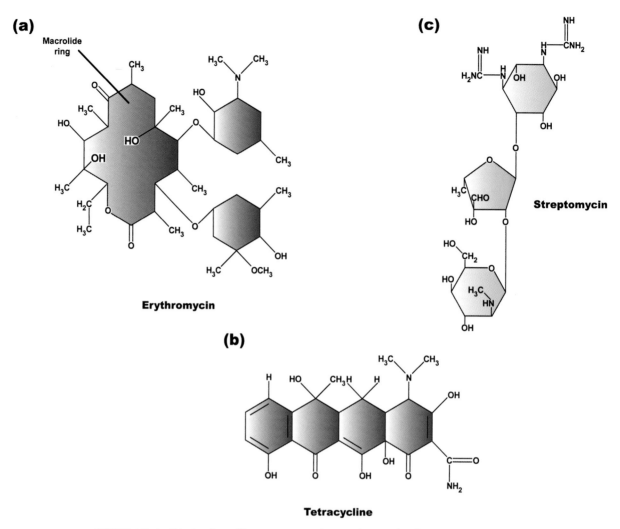

FIGURE 4.8 Antibiotics that affect protein synthesis in bacterial cells. (a) Erythromycin, which is composed of a large lactone (macrolide) ring connected to sugar molecules, disrupts the initiation of translation by interfering with the activity of the large ribosomal subunit (50S). (b) Tetracycline, which is composed of a naphthacene ring system substituted at numerous positions, depending on the derivative, prevents translation by interfering with the activity of the small ribosomal subunit (30S). (c) Streptomycin, which is made up of amino sugars linked by glycosidic bonds, inhibits translation by perturbing the activity of the 30S subunit. Illustration by Cori Sanders (iroc designs).

incubated at 30°C for several days and scored for resistance, defined as growth in the presence of the antibiotics. As shown in Fig. 4.9a, the screen revealed 191 different antibiotic resistance profiles, with every strain being resistant to multiple drugs in distinct combinations. As indicated in Figure 4.9b, the majority of the strains were resistant to six to eight different antibiotics. Together, these results suggest that there is a considerable amount of genetic diversity among the *Streptomyces* strains, giving rise to an assorted range of multidrug resistance capabilities.

Importantly, the microorganisms tested in the screen must have the cellular target for the antibiotic to function effectively: an intrinsic resistance to a drug exists if the bacterium lacks the target altogether. For example, bacteria with the ability to use exogenous sources of folic acid for nucleic acid synthesis need not synthesize their own, thus rendering sulfa drugs targeting folic acid metabolism completely ineffective. Figure 4.10 summarizes the percentage of isolates exhibiting resistance to the drugs tested according

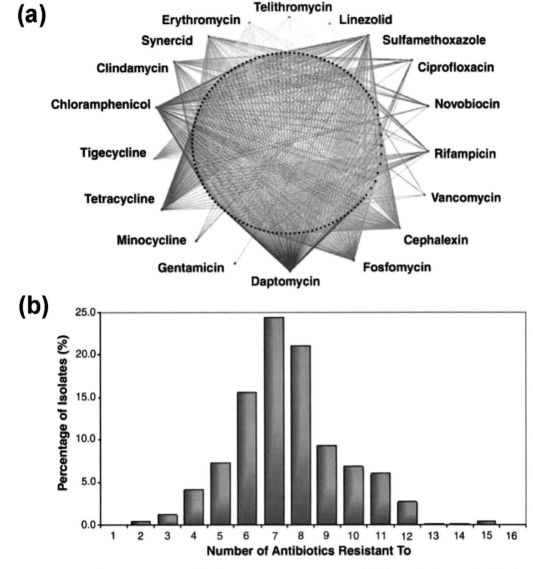

FIGURE 4.9 Antibiotic resistance profiles for 480 *Streptomyces* strains. (a) Schematic diagram showing the density and diversity of resistance profiles. The central circle of 191 black dots represents different resistance profiles, where a line connecting the profile to the antibiotic indicates resistance. (b) Resistance spectrum of the 480 isolates. Strains were individually screened on solid medium containing individual antibiotics at 20 μg/ml (final concentration). Resistance was defined as reproducible growth in the presence of the antibiotic. Reprinted from D'Costa et al. (2006) with permission.

to the cellular target for each antibiotic. In general, although isolates demonstrated resistance to all classes of antibiotics, some drugs appeared to elicit a lower frequency of resistance than others. For instance, 60% of the isolates exhibited resistance to tetracycline, whereas none (0%) of the isolates exhibited resistance to several of the aminoglycosides including streptomycin, neomycin, and gentamicin. Both drug classes interfere with the 30S subunit of the ribosome, adversely affecting protein synthesis; however, the two groups differ in their spectrum of microbial activity (Madigan and Martinko, 2006). For example, at one time tetracycline saw widespread medicinal use in humans as a broad-spectrum antibiotic. In contrast, the aminoglycosides have proven clinically useful primarily against gram-negative pathogens. The results of the D'Costa et al. (2006) study

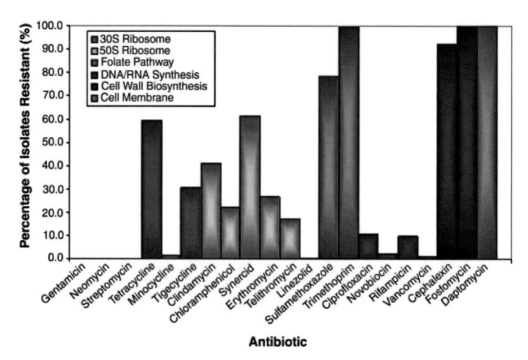

FIGURE 4.10 Levels of resistance to representatives from each class of antibiotic. Reprinted from D'Costa et al. (2006) with permission.

may suggest that there is no obvious selective pressure for gram-positive *Streptomyces* strains to develop or maintain resistance to aminoglycosides. This result is somewhat surprising since these antibiotics are actually produced by species of *Streptomyces*. That no isolates were resistant to any of the aminoglycosides may imply that the antibiotic concentration used for the screen was simply too high, out of range to capture even moderately resistant representatives.

For a subset of the *Streptomyces* strains, the extent of resistance was quantified by determining the minimal inhibitory concentration (MIC). As shown in Fig. 4.11, the antibiotic susceptibility assay performed to detect the MIC of a particular drug against an individual microorganism can be done by using liquid or solid culture medium. The MIC is typically reported in micrograms of the antibiotic tested per milliliter. For the D'Costa et al. (2006) study, the assay was performed with liquid cultures in 96-well plates rather than test tubes, using twofold increments of antibiotic spanning a range of 1 to 256 μg/ml. The lowest concentration at which no growth was obtained was scored as the MIC.

Resistance mechanisms revealed by digging in the dirt

D'Costa et al. (2006) investigated the mechanism of antibiotic resistance for a subset of the *Streptomyces* strains. Table 4.1 lists the various modes by which bacteria exhibit intrinsic resistance or develop drug resistance based on the induced expression or activation of innate gene products or those acquired by mutation or HGT. In the D'Costa et al. (2006) study, several potentially novel resistance mechanisms were explored, with specific interest in those that could be attributed to enzymatic deactivation of the antibiotic or alteration of the bacterial cell target on which the antibiotic exerts its toxic effects. There are many well-characterized examples in which microorganisms possess enzymes that convert an antibiotic to an inactive form. For example, bacteria may express

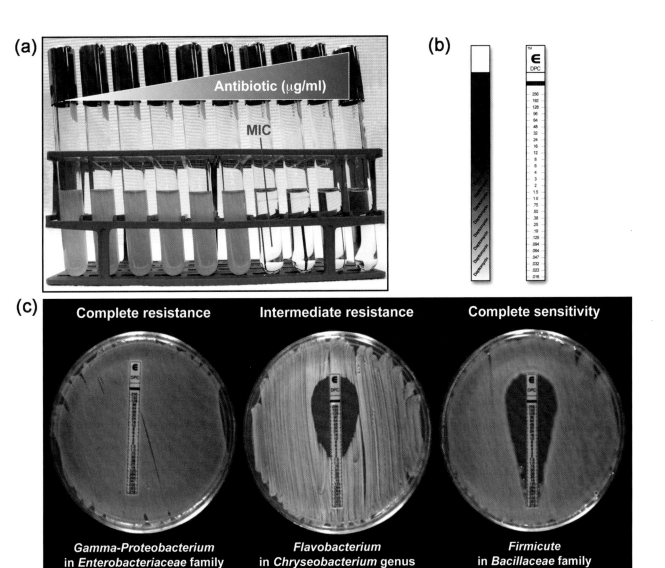

FIGURE 4.11 Antibiotic susceptibility assays used to determine the minimal inhibitory concentration (MIC). (a) A series of tubes containing media with increasing concentrations of a particular antibiotic are prepared and inoculated with the bacterial strain; they are then allowed to grow under appropriate incubation conditions. Growth, which is measured as the amount of turbidity in each tube, occurs only in tubes with antibiotic concentrations lower than the MIC. Photograph taken by Cori Sanders (iroc designs). (b) Each strip, called an Etest (AB Biodisk), is configured as shown in the diagram, starting with the lowest concentration of antibiotic (in micrograms per milliliter) at the bottom end of the strip and the highest concentration approaching the top end of the strip. Illustration by Cori Sanders (iroc designs). (c) As shown, the surface of an agar plate is overlaid with a liquid culture of the test organism, and then the Etest strips are placed on the agar surface. During incubation, the antibiotic diffuses into the agar, inhibiting growth of the test organism at all concentrations of the antibiotic greater than or equal to the MIC. All three organisms were isolated from the soil and identified by analysis of their 16S rRNA genes following the "I, Microbiologist" protocols. The first two are gram-negative isolates, which demonstrate different levels of resistance to daptomycin (MICs of 2 and < 0.01 μg/ml, respectively). Both microbial isolates also exhibit resistance to erythromycin (data not shown). The third is a gram-positive isolate with a profile consistent with the prescribed use of the drug as a narrow-spectrum antibiotic against infections with gram-positive bacteria. Photograph provided by To Hang (Shela) Lee, UCLA class participant in winter 2006.

Table 4.1 Mechanisms of bacterial resistance to antibiotics

Resistance mechanism	Examples
Reduced permeability	Outer membrane of gram-negative cells is impermeable to penicillin G
Efflux	Transport proteins that pump toxic compounds out of cell
Development of resistant biochemical pathway	Bypass need to synthesize folic acid; instead acquire exogenously
Inactivation of antibiotic	Express modifying enzymes such as β-lactamase, methylase, acetylase, phosphorylase, adenylase, glucosylase
Alteration of cellular target	Accumulate mutations in RNA polymerase for rifampin, ribosome for erythromycin or streptomycin, and DNA gyrase for quinolones

β-lactamase, an enzyme that cleaves the β-lactam ring of penicillins and cephalosporins (Fig. 4.12). Others may express enzymes that alter the antibiotic by adding functional groups to specific positions on the molecules, rendering them inactive. Such common modifications include acetylation, methylation, phosphorylation, and adenylation.

There were five examples of enzymatic inactivation of antibiotics in the D'Costa et al. (2006) study. As shown in Table 4.2, a subset of those strains exhibiting resistance to 11 different antibiotics were screened for inactivation. Liquid cultures of each *Streptomyces* strain were grown in the presence of 20 μg of antibiotic per ml, and the cell suspensions were centrifuged to pellet the cells and allow collection of the supernatant, which would

FIGURE 4.12 Enzymatic deactivation of antibiotics streptomycin and penicillin. Illustration by Cori Sanders (iroc designs).

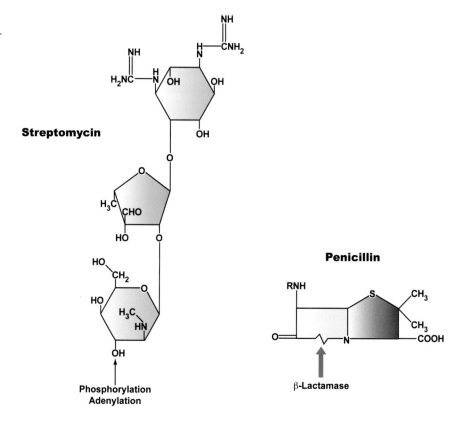

Table 4.2 Antibiotic inactivation of the soil library comprised of 480 isolates[a]

Antibiotic	Resistant	No. of strains Screened for inactivation	Confirmed (% of those screened)
Cephalexin	442	16	3 (18.8)
Ciprofloxacin	52	52	0 (0)
Clindamycin	107	46	0 (0)
Daptomycin	480	80	64 (80)
Erythromycin	128	128	9 (7)
Novobiocin	12	12	0 (0)
Rifampin	49	49	20 (40.8)
Synercid	294	71	13 (18.3)
Telithromycin	83	83	4 (4.8)
Trimethoprim	478	80	0 (0)
Vancomycin	5	5	0 (0)

[a]Excerpt from Table 1 in D'Costa et al. (2006), reprinted with permission.

presumably contain the enzymes causing deactivation if secreted from the cell. A culture of a susceptible indicator strain was overlaid on the surface of an agar plate, and paper disks containing the antibiotic were placed on the agar surface. The supernatants were spotted onto the antibiotic disks. After incubation, lack of a zone of growth inhibition around a disk indicated *Streptomyces* strains capable of inactivating the antibiotic as the mechanism of resistance.

As denoted in Table 4.2, 80% of the tested strains exhibiting resistance to daptomycin did so via enzymatic inactivation. This antibiotic was recently introduced to treat skin and soft tissue infections caused by gram-positive, multidrug-resistant pathogens. The high frequency of daptomycin resistance observed in this study was surprising, as it represents only the second documented case of resistance by this mechanism, although additional, as yet uncharacterized modes of resistance in the soil clearly must exist to account for the remaining 20% of resistant strains (Debono et al., 1988). Another significant observation was that approximately 40% of the tested strains displaying resistance to rifampin, a drug used to treat mycobacterial tuberculosis, acted by inactivation of the antibiotic. This finding was not anticipated because resistance in clinical isolates typically has been ascribed to alterations in the cellular target, namely, point mutations in the gene encoding the β subunit of RNA polymerase. A third case of enzymatic inactivation was observed for telithromycin-resistant isolates, in which almost 5% of the tested strains acted by a mechanism distinct from that previously observed for inactivation of the natural predecessor, erythromycin (Noguchi et al., 1995; Cundliffe, 1992). Telithromycin was approved by the FDA in 2004 for treatment of respiratory tract infections; it has been shown that at least one *Streptomyces* strain (JA#7) modifies telithromycin to what appears to be a larger, monoglucosylated product as indicated by mass spectrometry analysis (Fig. 4.13). This finding was important because this resistance mechanism has never been seen in clinical isolates exhibiting resistance to telithromycin.

There were two examples of cellular target alteration in the D'Costa et al. (2006) study. Not surprisingly, five *Streptomyces* strains were highly resistant to vancomycin, and four of the five strains contained the mutant *vanA* gene cluster encoding vancomycin resistance in the human pathogen *E. faecalis* and the soil-dwelling *Paenibacillus* and

(a)

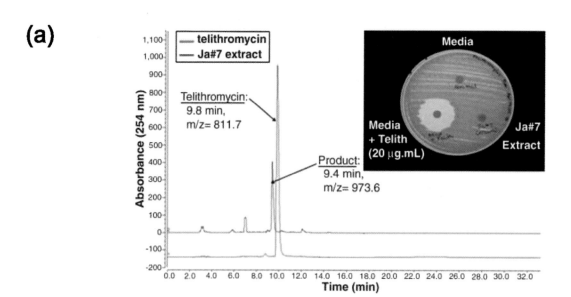

(b)

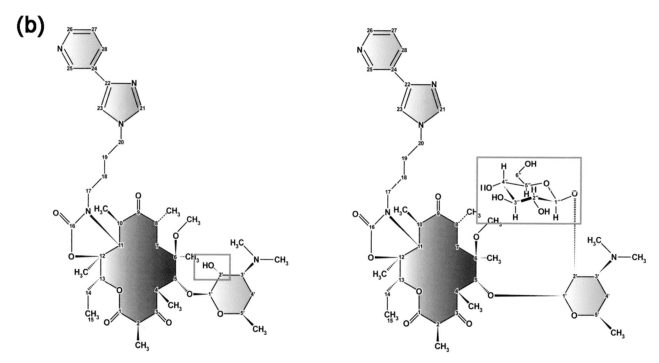

FIGURE 4.13 Modification of telithromycin by *Streptromyces* strain JA#7. (a) The supernatant from a culture of JA#7, grown in the presence of 20 μg of telithromycin/ml, was analyzed by high-pressure liquid chromatography (HPLC) followed by mass spectrometry. The glucosylation of telithromycin was accompanied by a shift in retention time (9.8 min to 9.4 min), an increase in the mass-to-charge ratio ($m/z = 811.7$ to 973.6 Da), and the gain of resistance by the indicator strain, *M. luteus* (inset). Reprinted from D'Costa et al. (2006) with permission. (b) Comparison of the structures of telithromycin (R = OH, orange box) versus the inactivated product (R = glycosyl group, orange box). Illustration by Cori Sanders (iroc designs) based on image from D'Costa et al. (2006).

Rhodococcus strains (Paulsen et al., 2003; Guardabassi et al., 2004). Resistant bacteria escape the damaging effects of vancomycin on cell wall synthesis by synthesizing an alternative peptidoglycan precursor molecule, which is not recognized by the antibiotic, thus allowing cells to survive in the presence of vancomycin at up to 128 to 256 μg/ml. Of the 480 strains screened in this study, 52 demonstrated resistance to the fluoroquinolone ciprofloxacin. Resistance in clinical isolates has been attributed chiefly to the accumulation of point mutations within a region of the gene encoding DNA gyrase *(gyrA)*, thereby eliminating any potential for ciprofloxacin to adversely affect enzyme activity. A 266-bp region of the N-terminal region of *gyrA* was PCR amplified and sequenced from 38 of these strains. As shown in Fig. 4.14, 9 of the 38 strains contained point mutations in this region of the gene, including locations commonly associated with clinical ciprofloxacin resistance (e.g., S83 and D87) and novel locations (e.g., S97, M100, and S110). The MIC varied depending on the position(s) of the mutation(s), with a range of 16 to 128 μg/ml. These point mutations appear to be a function of natural sequence variation in soil bacteria, suggesting that an intrinsic resistance is present in the population. Alternatively, the semisynthetic or completely synthetic antibiotics used in agriculture as well as human and veterinary medicine might find their way into the soil, where they provide the environmental selective pressure needed for resistance mechanisms to develop.

In conclusion, a vast, untapped reservoir of antibiotic resistance has materialized in soil microbial communities. Humankind could potentially return to the preantibiotic days of infection control if these environmental strains are able to transfer their genes to clinically relevant bacteria. The D'Costa et al. (2006) study in fact did reveal novel mechanisms of antibiotic resistance that could emerge in clinical environments, implying that efforts to explore and catalog the soil resistome should continue and expand to include microbes other than *Streptomyces*. In addition, performing such analyses with a range of antibiotic concentrations may reveal an even higher density of resistance than formerly appreciated.

FIGURE 4.14 Protein sequence alignment of the quinolone resistance-determining region (QRDR) within the N-terminal domain of the *gyrA* gene for ciprofloxacin-resistant strains. Mutations are highlighted in light orange boxes, and those with a white background represent amino acids not displaying any similarity to the wild-type sequences (*S. coelicolor* and *S. avermitilis*). Black sites are completely conserved among all strains sequenced, blue sites display 80 to 99% identity, and green sites demonstrate 60 to 80% identity. Sites labeled with a star are novel with respect to the mutations observed in the corresponding strain. The MIC of ciprofloxacin for each strain is indicated to the left of the strain name. Reprinted from D'Costa et al. (2006) with permission.

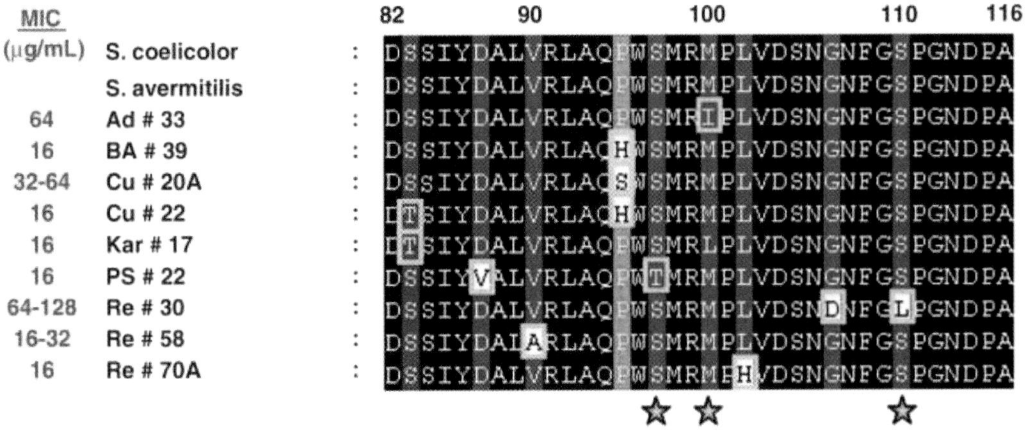

REFERENCES

Alekshun, M. N., and S. B. Levy. 2007. Molecular mechanisms of antibacterial multidrug resistance. *Cell* **128:**1037–1050.

Baltz, R. H. 2007. Antimicrobials from actinomycetes: back to the future. *Microbe* **2:**125–131.

Centers for Disease Control and Prevention (CDC). 2002. *Staphylococcus aureus* resistant to vancomycin—United States. *Morb. Mortal. Wkly. Rep.* **51:**565–567.

Cundliffe, E. 1992. Glycosylation of macrolide antibiotics in extracts of *Streptomyces lividans. Antimicrob. Agents Chemother.* **36:**348–352.

Davies, J. 2007. Microbes have the last word. *EMBO Rep.* **8:**616–621.

Davies, J., G. B. Spiegelman, and G. Yim. 2006. The world of subinhibitory antibiotic concentrations. *Curr. Opin. Microbiol.* **9:**445–453.

D'Costa, V. M., K. M. McGrann, D. W. Hughes, and G. D. Wright. 2006. Sampling the antibiotic resistome. *Science* **311:**374–377.

Debono, M., B. J. Abbott, M. Molloy, D. S. Fukuda, A. H. Hunt, V. M. Daupert, F. T. Counter, J. L. Ott, C. B. Carrell, L. C. Howard, L. D. Boeck, and R. L. Hamill. 1988. Enzymatic and chemical modifications of lipopeptide antibiotic A21978C: the synthesis and evaluation of Daptomycin (LY146032). *J. Antibiot.* (Tokyo) **41:**1093–1105.

Goh, E., G. Yim, W. Tsui, J. McClure, M. G. Surette, and J. Davies. 2002. Transcriptional modulation of bacterial gene expression by subinhibitory concentrations of antibiotics. *Proc. Natl. Acad. Sci. USA* **99:**17025–17030.

Guardabassi, L., H. Christensen, H. Hasman, and A. Dalsgaard. 2004. Members of the genera *Paenibacillus* and *Rhodococcus* harbor genes homologous to enterococcal glycopeptide resistance genes *vanA* and *vanB. Antimicrob. Agents Chemother.* **48:**4915–4918.

Halling-Sørensen, B., S. Nors Nielsen, P. F. Lanzky, F. Ingerslev, H. C. Holten Lützhøft, and S. E. Jørgensen. 1998. Occurrence, fate, and effects of pharmaceutical substances in the environment—a review. *Chemosphere* **36:**357–393.

Hamscher, G., S. Sczesny, H. Höper, and H. Nau. 2002. Determination of persistent tetracycline residues in soil fertilized with liquid manure by high-performance liquid chromatography with electrospray ionization tandem mass spectrometry. *Anal. Chem.* **74:**1509–1518.

Hong, H.-J., M. S. B. Paget, and M. J. Buttner. 2002. A signal transduction system in *Streptomyces coelicolor* that activates the expression of a putative cell wall glycan operon in response to vancomycin and other cell wall-specific antibiotics. *Mol. Microbiol.* **44:**1199–1211.

Levy, S. B., and B. Marshall. 2004. Antibacterial resistance worldwide: causes, challenges and responses. *Nat. Med.* **10:**S122–S129.

Madigan, M. T., and J. M. Martinko. 2006. *Brock Biology of Microorganisms,* 11th ed. Pearson Prentice Hall, Pearson Education, Inc., Upper Saddle River, NJ.

Marshall, C. G., I. A. Lessard, I. Park, and G. D. Wright. 1998. Glycopeptide antibiotic resistance genes in glycopeptide-producing organisms. *Antimicrob. Agents Chemother.* **42:**2215–2220.

Mokdad, A. H., J. S. Marks, D. F. Stroup, and J. L. Gerberding. 2004. Actual causes of death in the United States, 2000. *JAMA* **291:**1238–1245.

Noguchi, N., A. Emura, H. Matsuyama, K. O'Hara, M. Sasatsu, and M. Kono. 1995. Nucleotide sequence and characterization of erythromycin resistance determinant that encodes macrolide 2′-phosphotransferase I in *Escherichia coli. Antimicrob. Agents Chemother.* **39:**2359–2363.

Paulsen, I. T., L. Banerjei, G. S. A. Myers, K. E. Nelson, R. Seshadri, T. D. Read, D. E. Fouts, J. A. Eisen, S. R. Gill, J. F. Heidelberg, H. Tettelin, R. J. Dodson, L. Umayam, L. Brinkac, M. Beanan, S. Daugherty, R. T. DeBoy, S. Durkin, J. Kolonay, R. Madupu, W. Nelson, J. Vamathevan, B. Tran, J. Upton, T. Hansen, J. Shetty, H. Khouri, T. Utterback, D. Radune, K. A. Ketchum, B. A. Dougherty, and C. M. Fraser. 2003. Role of mobile DNA in the evolution of vancomycin-resistant *Enterococcus faecalis. Science* **299:**2071–2074.

Tamae, C., A. Liu, K. Kim, D. Sitz, J. Hong, E. Becket, A. Bui, P. Solaimani, K. P. Tran, H. Yang, and J. H. Miller. 2008. Determination of antibiotic hypersensitivity among 4,000 single-gene-knockout mutants of *Escherichia coli. J. Bacteriol.* **190:**5981–5988.

Tomasz, A. 2006. Weapons of microbial drug resistance abound in soil flora. *Science* **311:**342–343.

Yim, G., F. De la Cruz, G. B. Spiegelman, and J. Davies. 2006. Transcription modulation of *Salmonella enterica* serovar Typhimurium promoters by sub-MIC levels of rifampin. *J. Bacteriol.* **188:**7988–7991.

Yim, G., H. H. Wang, and J. Davies. 2007. Antibiotics as signaling molecules. *Philos. Trans. R. Soc. Lond.* Ser. B **362:**1195–1200.

Zirakzadeh, A., and R. Patel. 2005. Epidemiology and mechanisms of glycopeptide resistance in enterococci. *Curr. Opin. Infect. Dis.* **18:**507–512.

READING ASSESSMENT

1. Describe the mechanism(s) used by bacteria to spread antibiotic resistance genes in a clinical setting. Why is this difficult to control?

2. Identify the cellular target for each of the following antibiotics. Provide one example of a resistance mechanism that a bacterium could develop for each antibiotic listed.

    a. Rifampin

    b. Erythromycin

    c. Daptomycin

    d. Trimethoprim

    e. Ciprofloxacin

3. Which of the following strains from Experiment 4.4 are expected to reveal environmental isolates producing antibiotics similar to vancomycin? Hint: This strain displays an increased sensitivity to vancomycin in comparison to the wild-type strain.

    a. *E. coli fis tolC*

    b. *E. coli smpA surA*

    What is the most likely cellular target for antibiotics produced by isolates that exhibit a zone of growth inhibition in the presence of this strain?

4. As discussed in the text, D'Costa et al. (2006) restricted their analysis to only *Streptomyces* strains. What advantages did this strategy offer? What were the disadvantages? Would you expect the outcome of the study to have been different had the authors expanded their screen to include other types of bacteria? If so, how? If not, why not?

5. As discussed in the text, the initial screen performed by D'Costa et al. (2006) for antibiotic resistance among the *Streptomyces* strains was done at relatively high

drug concentrations. Would you expect the outcome of the study to have been different if the screen had been done with subinhibitory drug concentrations? If so, how? If not, why not?

6. Given the observation in the D'Costa et al. (2006) study that 60% of the isolates exhibited resistance to tetracycline and 0% of the isolates exhibited resistance to aminoglycosides (refer to Fig. 4.10), provide two reasons why the two classes appear to differ in their spectrum of microbial activity. Keep in mind that both classes are naturally produced by *Streptomyces* species and both target the 30S ribsome. (*Hint:* Is there any selective pressure for *Streptomyces* strains to develop or maintain resistance to tetracyclines or aminoglycosides?)

7. What considerations should be used by a physician when deciding whether to prescribe broad- or narrow-spectrum antibiotics?

MICROBIAL CONTRIBUTIONS TO THE PRODUCTION OF BIOMASS AND BIOFUELS

Making the case to pursue development of alternative, renewable energy sources

The quest to discover renewable and sustainable sources of energy has led scientists to look at the metabolic handiwork of microorganisms for a solution to declining oil supplies and increased air pollution. The burning of fossil fuels such as coal and petroleum causes carbon dioxide (CO_2) to be released into the atmosphere. This by-product of the technological progress brought about by the industrial revolution has contributed to the ecological crisis known as global warming (Hansen et al., 2000). Global energy consumption is projected to increase by 50% between 2005 and 2030, with a concomitant increase in CO_2 emissions by approximately 51% if dependence on oil and coal as exclusive energy sources continues (Energy Information Administration, 2008). While the world still relies on petroleum as a primary liquid-energy source, given its importance in the transportation and industrial sectors, production of this limited resource is projected to peak within the next 25 to 50 years—within our lifetime (Buckley and Wall, 2006).

The future of energy production need not be as bleak or crippling as it sounds if the world would consider biofuels as a partial means of meeting increasing demands for energy and concurrently decreasing greenhouse gas emissions. Biofuels include substances such as ethanol, biodiesel, and hydrogen, which all represent forms of chemical energy that can be converted to electricity to heat our homes or to kinetic energy to fuel our cars. Because they have evolved a unique and diverse repertoire of metabolic capabilities, microorganisms possess the machinery to generate these application-ready forms of chemical energy from raw organic biomass or solar power (Antoni et al., 2007). The energy transformations that involve microbes are more eco-friendly, or "green," than fossil fuel technologies, in that hazardous materials are neither used nor produced (Buckley and Wall, 2006). In addition, the conversion of biomass to biofuels can be a carbon-negative process, where the total amount of CO_2 in the atmosphere actually decreases (Tilman et al., 2006).

The technology required to generate bioethanol is the most advanced in comparison to other bioenergy categories. Bioethanol can be made from some of the most abundant sources of biomass, called feedstocks, which include mixed prairie grasses; monoculture crops such as corn, sugarcane, willow, hybrid poplar, and switchgrass; and waste materials such as corn stalks and wood chips (Tilman et al., 2006). As a liquid fuel, bioethanol is fairly easy to store and is compatible with existing infrastructure, although transport requires special handling due to its water-adsorbing properties (Buckley and Wall, 2006). The theoretical yield of bioethanol as a fuel source is about 0.5 g per g of raw biomass, a recovery of approximately 90% of the potential energy. In practice, current technologies recuperate around 60% of the available energy.

Corn as a feedstock for bioenergy production

In the United States, corn is a customary substrate for manufacturing bioethanol, in part because its use is subsidized by the federal government but also because the process is technically straightforward. Corn is composed of starch, a complex polysaccharide consisting of α-1,4-linked glucose molecules, with up to 500 sugar units per chain (Fig. 4.15). Microorganisms express enzymes that facilitate the conversion of corn to bioethanol at two stages in the overall process. Amylases hydrolyze the α-1,4-linkage between the glucose units in an unbranched sugar chain, releasing α-D-glucose and the disaccharide maltose, which is further catabolized by maltase into two molecules of glucose (Staley et al., 2007). Through the cellular reactions of glycolysis and alcoholic fermentation, this simple sugar is sequentially converted to ethanol with the release of CO_2. Microorganisms commonly used for this process include the yeast *Saccharomyces cerevisiae* and bacteria such as *Zymomonas mobilis* and recombinant strains of *E. coli* and *Klebsiella oxytoca* (Buckley and Wall, 2006).

Prairie grasses as feedstocks for bioenergy production

Domesticated, perennial plants such as switchgrass also have been investigated as potential feedstocks for bioethanol production. Switchgrass is a herbaceous (nonwoody) perennial. The cell walls of herbaceous plants consist of cellulose, hemicellulose (a hetero-

FIGURE 4.15 Structure and stepwise catabolism of starch to glucose, which undergoes alcoholic fermentation to produce ethanol and carbon dioxide. Illustration by Cori Sanders (iroc designs).

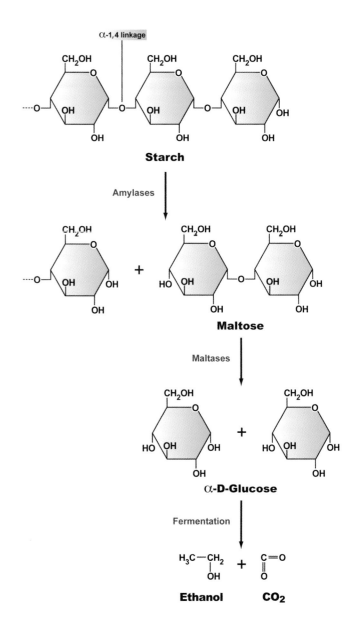

geneous polysaccharide composed of xylans and glucose [xyloglucan]), and pectin (another heterogeneous polysaccharide composed of rhamnose and α-(1-4)-linked D-galacturonic acids [rhamnogalacturon]). Other sugars also may be incorporated into pectin.

As shown in Fig. 4.16, cellulose consists of glucose molecules connected by β-1,4-linkages, which cause the sugar chains to twist 180° relative to one another (Staley et al., 2007). This arrangement allows the polymer to form long fibers, with up to 14,000 sugar units per chain. Cellulose is degraded to glucose through the activities of several cellulolytic enzymes. Endo-β-1,4-glucanases hydrolyze interior linkages within the long sugar chains, releasing smaller fragments that are cleaved by exo-β-1,4-glucanases into disaccharides called cellobiose. A third type of enzyme called α,β-glucosidase hydrolyzes cellobiose into individual glucose units, which serve as the substrate for glycolysis and fermentation to ethanol plus CO_2. Neither hemicellulose nor pectin has the tensile strength of cellulose, and both are readily hydrolyzed to sugars that enter a variety of

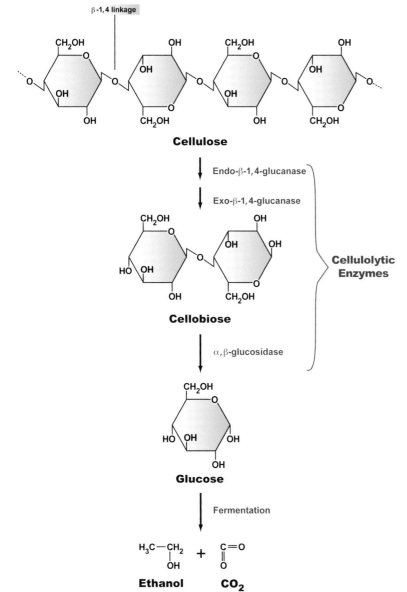

FIGURE 4.16 Structure and stepwise catabolism of cellulose to glucose, which undergoes alcoholic fermentation to produce ethanol and carbon dioxide. Illustration by Cori Sanders (iroc designs).

catabolic pathways in microbial cells. Xylan is cleaved by xylanase into the pentose sugar D-xylose. A number of bacteria and fungi possess the hydrolytic enzymes required for degradation of these plant products, including microorganisms that are cultivated on the VXylA media as part of the "I, Microbiologist" project.

The cell walls of switchgrass also contain lignocellulose, a complex polymer made up of cellulose and lignin. As shown in Fig. 4.17, lignin is a very large, complex, heterogeneous polymer of aromatic molecules formed from a rather haphazard polymerization of alcohol precursors that becomes embedded within the cellulosic cell wall (Staley et al., 2007). Presenting itself as a somewhat indestructible compound, lignin is not metabolized as a source of carbon or energy by any known prokaryotic organism. Lignin is degraded naturally only by fungi via a reaction that involves its indiscriminate oxidation to low-molecular-weight phenolic products. Taken together, although there are

FIGURE 4.17 One of the various possible structures of lignin, a complex aromatic substance that provides structural support for woody plants and trees. Illustration by Cori Sanders (iroc designs) based on an image from Staley et al. (2007).

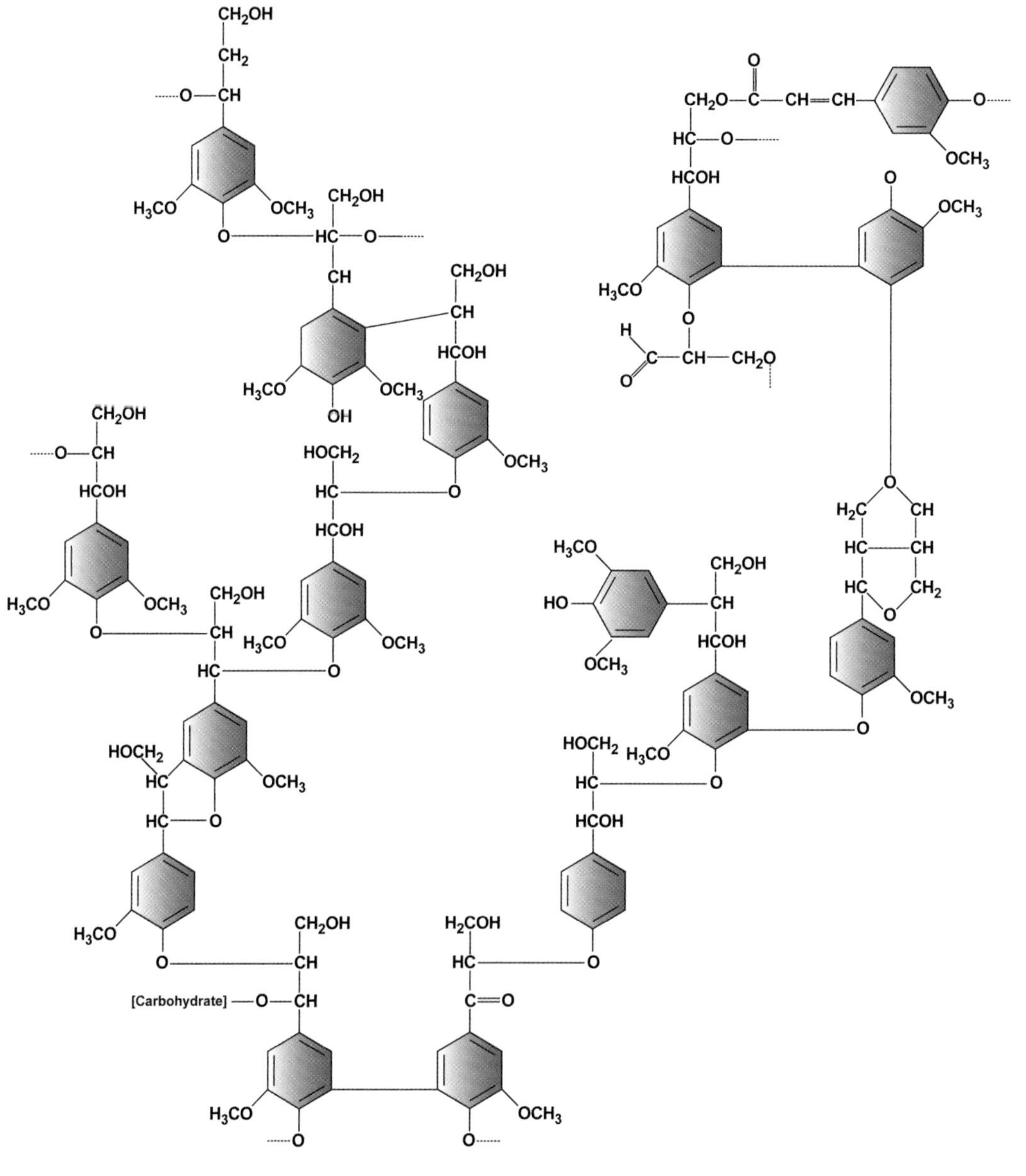

fungal and bacterial species to convert cellulose, hemicellulose, and lignocellulose to sugars and fermentative microbes to complete the transformation to ethanol, the process is much more difficult and potentially expensive in comparison to that for corn.

Woody plants as bioenergy crops

Hybrid poplar and willow trees, which are nonherbaceous (woody) plants, are being cultivated not only for their promise as bioenergy feedstocks, but also for their bioremediation potential. Specifically, the secondary xylem of nonherbaceous plants (Fig. 4.18) consists of hemicellulose, lignocellulose, and pectin, which are subject to the same sorts of catabolic transformations as the plant products of herbaceous plants. Furthermore, these relatively fast-growing woody plants remove CO_2 from the atmosphere, stabilize soil by reducing erosion and runoff, and restore nutrients to depleted agricultural lands (see BFIN in Web Resources).

Production of biofuels from biomass

All constituents of the feedstock must be degraded during the production of biofuels, a process that involves several steps, some of which are mediated by microbes while others are dependent on harsh chemical treatments. Due to the recalcitrant nature of lignin to biodegradation, these phenolic molecules are removed from the feedstocks prior to the hydrolysis stage by treatment with acidic chemicals or heat, permitting increased accessibility to cellulose and hemicellulose by microbial enzymes (Fig. 4.19). By screening bacterial isolates for cellulase activity (Teather and Wood, 1982), participants in the "I, Microbiologist" project are actively contributing to research involving microbial energy transformations that take place during the hydrolysis stage in the biomass conversion process. The final stage in which the sugars derived from cellulose and hemicellulose are fermented to ethanol in bioreactors takes place in multiple steps. Because no single organism, or "superbug," provides the solution for biodegradation of all feedstocks, the search continues for microbes that express cellulases and other hydrolytic enzymes with a broader substrate range or the engineering of fermentative organisms with more efficient catalytic activity and increased tolerance to product concentrations in bioreactors.

Microbial contributions to production of biomass

Not only can microorganisms be involved in the direct conversion of bioenergy crops to biofuels, but also they can assist in the cultivation process required to generate the plants used as biomass. There is an abundance of microorganisms that thrive in the rhizosphere and promote plant growth by a variety of mechanisms. As biocontrol agents,

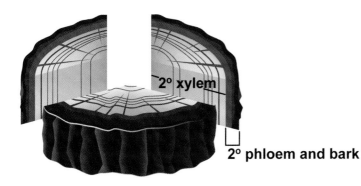

FIGURE 4.18 Composition of nonherbaceous (woody) plants. The woody part of the stem, or 2° xylem, is composed of hemicellulose, lignocellulose (lignin and cellulose mixed together), and pectin. Illustration by Cori Sanders (iroc designs).

2° xylem

2° phloem and bark

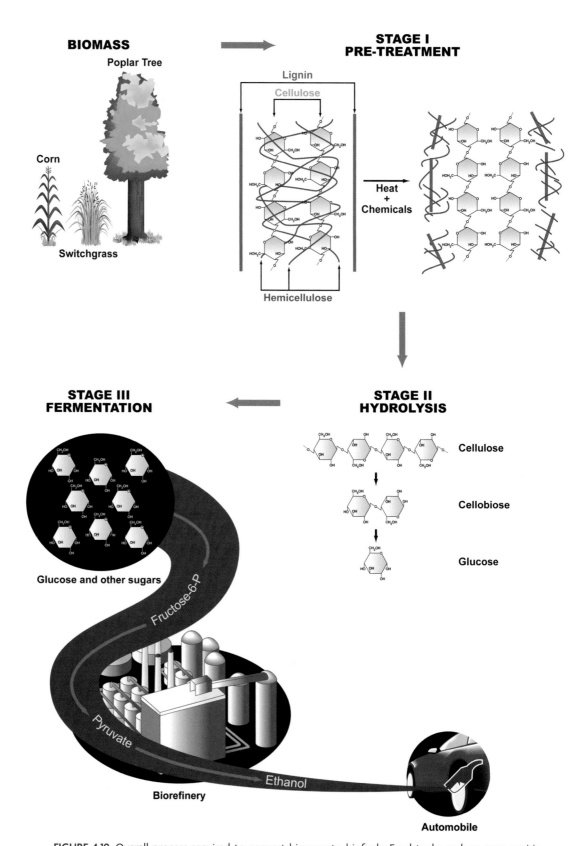

FIGURE 4.19 Overall process required to convert biomass to biofuels. Feedstocks such as corn, prairie grasses like switchgrass, or trees such as poplar or willow are first treated with chemicals and/or heat (stage I) to prepare cellulose for hydrolysis by microbial enzymes (stage II). The monosaccharide product glucose then undergoes fermentation in massive bioreactors (stage III), resulting in the production of ethanol, a renewable fuel source for our vehicles. Illustration by Cori Sanders (iroc designs) based on an image from www.doegenomestolife.org/biofuels/.

some bacteria and fungi are involved in management of plant diseases caused by soil pathogens (Whipps, 2001). For instance, the soft rot potato pathogen *Erwinia carotovora* subsp. *atroseptica* can be managed by an antibiotic called 2,4-diacetylphloroglucino produced by *Pseudomonas fluorescens*. As biofertilizers, soil microorganisms can provide nutrients to the plants in a number of ways (Vessey, 2003). As introduced in Unit 2, some bacteria, including *Rhizobium* and the associative nitrogen fixers, provide nitrogen in the form of ammonia directly to plants based on their ability to transform atmospheric nitrogen (N_2) into a form that can be assimilated by plants. Other microorganisms stimulate plant growth by increasing the availability of essential nutrients such as phosphate and iron, much of which exists in soil in insoluble forms. Some bacteria secrete phosphatases and organic acids that convert phosphate to a soluble form, while others may facilitate absorption of soluble iron by exploiting the scavenging abilities of bacterial siderophore activities. There also is evidence to suggest that bacterially mediated effects on plant growth may be attributed to the production of phytohormones, which are involved in the initiation of root formation, as well as cell division and enlargement. Thus, biofertilizers may promote changes in root size and morphology, with the production of longer and more highly branched roots increasing the surface area available to the plant, thereby facilitating nutrient absorption. Taken together, inoculation of the rhizosphere of bioenergy crops with plant growth-promoting bacteria could be considered a means of maintaining soil health and nutrient status.

Which is the optimal feedstock for bioenergy production?

The optimal feedstock for production of bioethanol may not be corn or switchgrass, which both require fertile soil supplemented with fertilizers, pesticides, and ample amounts of water. In fact, fossil fuels are used to produce at least 80 tons of nitrogen-based fertilizers via the Haber-Bosch process, consuming 1% of the world's annual energy supply (Smith, 2002; Lee and Hirsch, 2006; Heffer and Prud'Homme, 2008). Despite having an overall yield of only 10 to 20%, the Haber-Bosch process is the source of approximately half of all fixed nitrogen applied to agricultural food production and supplies roughly 40% of the world's dietary protein supply (Smil, 2001; Fixen and West, 2002).

$$3H_2 + N_2 \xrightarrow{\substack{\uparrow \text{ heat (400-500°C)} \\ \uparrow \text{ pressure (250 atm)} \\ \uparrow \text{ energy (requires catalyst)}}} 2NH_3$$

Thus, the use of feedstocks that require application of fertilizers produced via nonrenewable energy sources is counterproductive and not sustainable in the long term with the current amount of energy input. Plants that associate with bacteria capable of biological nitrogen fixation could be considered as alternative bioenergy crops, thereby diminishing the need for costly chemical fertilizers (Tilman et al., 2001).

The expanded use of monoculture bioenergy plants has had ecological effects such as reducing biodiversity in natural terrestrial environments due to destructive agricultural practices. Conservation of native ecosystems is recognized as an important contributor to efforts aimed at reducing the negative effects of global climate change, controlling the spread of pestilent diseases caused by insect vectors, and inhibiting the persistence of invasive plant species. In addition, although conversion of corn to biofuel is less expensive with fewer technical hurdles than the use of switchgrass, its production causes a net increase in greenhouse gas (GHG) emissions (Tilman et al., 2006). GHG emissions are expressed as CO_2 equivalents measured across the entire life cycle of the plant. This quantity considers how much CO_2 is removed from or released into the atmosphere during plant growth, as well as how much fossil fuel-based CO_2 is released during biomass generation and biofuel production. These latter two components include the energy required to run the farm equipment used for planting and harvesting feedstocks, the energy input during production and application of fertilizers and pesticides, and the fuel costs associated with transport of the feedstocks to biorefineries. Thus, any given biofuel is considered carbon positive or negative, leading to a net increase or decrease, respectively, in GHG emissions, or is regarded as carbon neutral, causing no net change in atmospheric CO_2 levels. Of note, although corn as a biofuel source is carbon positive, it still has 12% lower GHG emissions than the gasoline or diesel it replaces (Hill et al., 2006).

Dual function of mixed prairie grasses as a bioenergy source and a carbon sink

High-diversity mixtures of prairie grasses have been investigated recently as a bioenergy crop with potential to be converted to biofuels without having to compete for fertile land used for food production, reduce terrestrial habitat diversity, or increase carbon emissions (Tilman et al., 2006). These native perennial plants can be grown on nutrient-depleted, marginal soils with little to no application of water or fertilizers. Tilman and coworkers specifically explored the biofuel potential of 18 different species of plants, which included woody and herbaceous legumes, C_4 and C_3 grasses, herbaceous forbs, and woody oak species. They planted 152 plots of land with combinations of 1, 2, 4, 8, or 16 plant species; the plots with 16 species were referred to as the low-input high-diversity (LIHD) mixtures of native grassland perennial plants. None of the plots were fertilized, and they were irrigated only while plants germinated and established themselves in their respective plots. Over a 10-year period, these scientists made a number of annual measurements aimed to determine bioenergy yields from aboveground biomass and the capacity to sequester carbon in plant roots and soil. The plots were burned each spring to remove aboveground biomass from the previous year. The authors hypothesized that the inclusion of legumes in LIHD mixtures would eliminate the need for nitrogen fertilization of the bioenergy crops, which would benefit from the activities of rhizobia and nitrogen-fixing diazotrophs in the soil. The authors also postulated that LIHD biofuels would be carbon negative, exhibiting a net reduction in GHG emissions.

Aboveground, living plant matter was harvested annually from each plot in early August; this was followed by root mass sampling in mid-August. The harvest of aboveground plant matter was obtained by clipping and subsequent dry-weight determinations of four locations per plot. The gross amount of usable energy obtained from aboveground biomass was subsequently calculated based on the amount of energy released upon combustion by one of three mechanisms: cofiring with coal to produce electricity, conversion to bioethanol, or gasification and transformation to synfuel and electricity. As shown in Fig. 4.20, there was a rise in bioenergy production as the number of plant species per plot increased, with the LIHD plots generating an average of 238% more bioenergy than any monoculture plot in the last 3 years of the study. Thus, there appears to be a higher bioenergy yield from feedstocks characterized by high plant diversity.

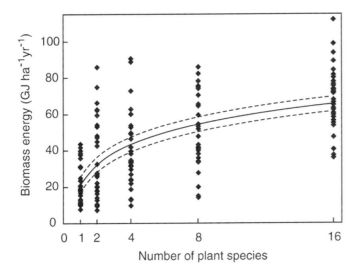

FIGURE 4.20 Effect of plant biodiversity on biomass energy yield. The solid curved line reflects log fit, and the dashed curved lines give 95% confidence intervals for this fit. Units are gigajoules per hectare per year. Reprinted from Tilman et al. (2006) with permission.

The belowground matter comprising plant roots was acquired by collecting 5- by 30-cm soil cores, which were sieved to separate roots from soil particles. The roots were gently rinsed with water and then weighed such that root mass per unit area could be calculated. The authors estimated the amount of carbon present in roots, which can be determined by oxidizing the roots with potassium dichromate in sulfuric acid; the carbon content is proportional to the amount of reduced dichromate (Staley et al., 2007). After several measurements, it was concluded that 40% of root biomass is carbon (Tilman et al., 2006). As shown in Fig. 4.21, the amount of CO_2 sequestered in roots is also a function of plant biodiversity. By the end of the 10-year study, the roots of the plants in the LIHD plots captured 160% more CO_2 did than those in the monoculture plots.

For a subset of the plots, the release of carbon from the plants into the soil, a process facilitated by microbial decomposition of dead plant matter, was measured for three different soil depths at four sites per plot. Soil samples were collected before the plots were planted and again at the end of the 10-year study. The net change in soil organic carbon (ΔC) during the elapsed time was reported as the net rate of CO_2 sequestration in soil. As shown in Fig. 4.22, soil carbon storage displayed a similar pattern to root

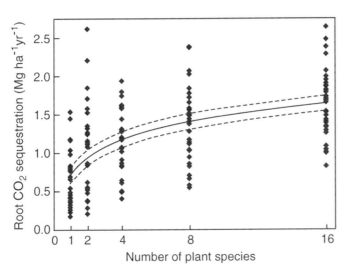

FIGURE 4.21 Effect of plant biodiversity on CO_2 sequestration in plant roots. The solid curved line reflects log fit, and the dashed curved lines give 95% confidence intervals for this fit. Units are megagrams per hectare per year. Reprinted from Tilman et al. (2006) with permission.

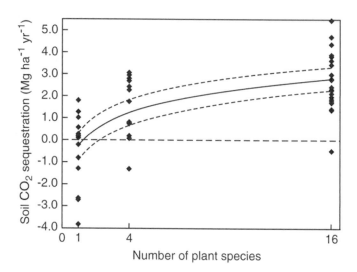

FIGURE 4.22 Effect of plant biodiversity on CO_2 sequestration in soil. The solid curved line reflects log fit, and the dashed curved lines give 95% confidence intervals for this fit. Units are megagrams per hectare per year. Reprinted from Tilman et al. (2006) with permission.

carbon storage. In total for the observed decade, the LIHD plots sequestered 103% more CO_2 in plant roots and soil than did the monoculture plots, making a case for the mixed prairie grasslands to function as a sink for greenhouse gases produced by burning fossil fuels.

The amount of CO_2 sequestered by the LIHD mixture was addressed in the context of GHG savings, which, as discussed above, considers the entire life cycle of the feedstock as calculated from the amount of CO_2 removed from or released into the atmosphere during biomass cultivation, in addition to the amount of fossil fuel-based CO_2 released during transport of biomass as well as during production and combustion of biofuels. Expressed as CO_2 equivalents, the reduction in GHG emissions by LIHD biofuels when used instead of petroleum-based fuels was compared to the reductions by bioethanol derived from corn-based feedstocks and by biodiesel obtained from soybean when substituted for gasoline or diesel fuel. As shown in the top panel of Fig. 4.23, depending on

FIGURE 4.23 Environmental effects of bioenergy resources. (Top) GHG reduction for biofuels relative to emissions for fossil fuels for which they substitute. (Bottom) Fertilizer application rates. Reprinted from Tilman et al. (2006) with permission.

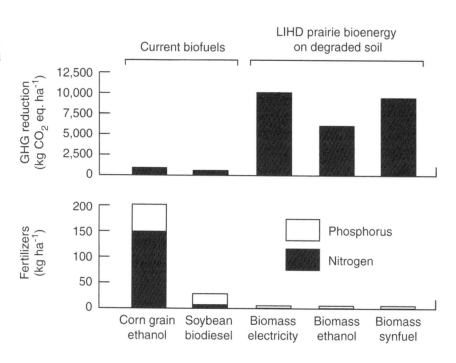

the mechanism by which energy was obtained from the LIHD biomass, the GHG reductions are between 6- and 20-fold greater for the LIHD biofuels than for the corn-based bioethanol or soybean biodiesel. Thus, these data argue that LIHD biomass may be considered carbon negative because its growth and ultimate combustion as a biofuel lead to a net decrease in GHG emissions.

Recall that no nitrogen-based fertilizers were applied to the LIHD plots. Instead, nitrogen was supplied by the legumes included in the prairie grass mixtures (Tilman et al., 2001). Interestingly, the total soil nitrogen concentration increased almost 25% in the LIHD plot over the course of the 10-year study, whereas the soil nitrogen concentration in monoculture plots remained unchanged. Therefore, biological nitrogen fixation is replenishing the supply of nitrogen to the plants, an observation that demonstrates the utility of microbes in supporting biomass generation for biofuel production. This area of research will clearly benefit from continued efforts to diversify the medley of nitrogen-fixing bacteria cultivated from the soil, with the goal of exploiting them as biofertilizers. As an added benefit, the enzymes responsible for nitrogen fixation (i.e., nitrogenases) can also be used to produce hydrogen, another biofuel currently under investigation (Buckley and Wall, 2006).

It was estimated that to sustain the LIHD plots, low annual inputs of phosphorus-based fertilizers would be required to replace harvested biomass, which contains 0.2% phosphorus (lower panel of Fig. 4.23). In contrast, currently used varieties of corn and soybeans require larger amounts of fertilizer application, with corn needing massive quantities of both nitrogen- and phosphorus-based fertilizers to flourish. These data indicate that the use of LIHD mixtures as feedstocks does not necessitate the conversion of fertile farmland to support biofuel production. In fact, on a global scale, using only abandoned and deteriorated farmland, LIHD biomass is poised to provide enough energy to support 13% of that needed for transportation and 19% of that needed for electricity. That this feedstock also may be used as a CO_2 sink argues for its immediate consideration as a renewable and sustainable energy source—one that capitalizes on the metabolic repertoire available only to microbes.

KEY TERMS

Biocontrol agents Soil microorganisms that mediate development of plant diseases caused by microbial pathogens via interactions and activities such as production of antibiotics, competition for nutrients and colonization sites, and degradation of toxins.

Biofertilizers Soil microorganisms, including microbes that are nitrogen fixers, that increase the availability and uptake of mineral nutrients for plants.

Biofuels Chemical energy sources derived from the potential energy in biomass; examples include bioethanol, biodiesel, hydrogen, methane, and butanol.

C₃ plant A plant that uses the Calvin cycle to fix CO_2 from the atmosphere into biomass; the first organic product of carbon fixation is a three-carbon compound. Examples of C_3 plants are rice, wheat, and soybeans.

C₄ plant A plant that uses a more efficient form of CO_2 fixation than C_3 plants, reflecting an evolutionary adaptation to an atmosphere containing more oxygen (O_2). The Calvin cycle is preceded by an enzymatic step in which CO_2 is first incorporated into a three-carbon compound to form a four-carbon product in one cell type (mesophyll cells). The four-carbon compounds are exported to a different cell type (bundle sheath cells), where CO_2 is released and reassimilated into biomass via the Calvin cycle. Spatial separation prevents the counterproductive process called photorespiration, which is carried out by RuBisCO, the enzyme that binds CO_2 and catalyzes the first step in the Calvin cycle. This arrangement prevents RuBisCO from contact with O_2, which it can bind when CO_2 concentrations are low, causing CO_2 to be released rather than consumed, with no ATP produced. Examples of C_4 plants are sugarcane, corn, and switchgrass.

Feedstocks Biomass supplies for biofuel production.

Forbs Herbaceous flowering plants such as clovers (which are legumes), sunflowers, and milkweeds.

Global warming A specific example of climate change, describing an increase in the average temperature of Earth's atmosphere and oceans as a result of higher greenhouse gas emissions influenced by human activities combined with natural phenomena such as solar irradiance.

Hectare A unit of area equal to 10,000 m²; commonly used for land measurements. One hectare is equal to approximately 2.47 acres. Abbreviated ha.

Herbaceous plant A nonwoody plant with aboveground stems and leaves, all of which die at the end of a growing season. For perennials such as switchgrass, new growth arises from the roots, belowground stems, or crown tissue at the soil surface.

Legumes Plants that form symbiotic associations with nitrogen-fixing bacteria known as rhizobia.

Nonherbaceous plant A woody plant with aboveground stems and leaves. Stems remain alive during winter and grow shoots for new leaves the following year. Examples of biofuel crops include trees such as the willow and poplar.

Perennial A plant that lives longer than 2 years.

Rhizobium A bacterium that establishes a nitrogen-fixing, endosymbiotic relationship with leguminous plants.

Synfuel Synthetic fuel; liquid fuel obtained from coal, natural gas, or biomass.

REFERENCES

Antoni, D., V. V. Zverlov, and W. H. Schwartz. 2007. Biofuels from microbes. *Appl. Microbiol. Biotechnol.* **77**:23–35.

Buckley, M., and J. Wall. 2006. *Microbial Energy Conversion.* American Academy of Microbiology, Washington, DC. www.asm.org.

Energy Information Administration. 2008. *International Energy Outlook.* www.eia.doe.gov/iea.

Fixen, P. E., and F. B. West. 2002. Nitrogen fertilizers: meeting contemporary challenges. *Ambio* **31**:169–176.

Hansen, J., M. Sato, R. Ruedy, A. Lacis, and V. Oinas. 2000. Global warming in the twenty-first century: an alternative scenario. *Proc. Natl. Acad. Sci. USA* **97**:9875–9880.

Heffer, P., and M. Prud'Homme. 2008. Outlook for world fertilizer demand, supply, and supply/demand balance. *Turk. J. Agric. For.* **32**:159–164.

Hill, J., E. Nelson, D. Tilman, S. Polasky, and D. Tiffany. 2006. Environmental, economic, and energetic costs and benefits of biodiesel and ethanol biofuels. *Proc. Natl. Acad. Sci. USA* **103**:11206–11210.

Lee, A., and A. M. Hirsch. 2006. Signals and responses: choreographing the complex interaction between legumes and α and β-rhizobia. *Plant Signaling Behavior* **1**:161–168.

Smil, V. 2001. *Enriching the Earth—Fritz Haber, Carl Bosch, and the Transformation of World Food Production,* p. 81, 145, and 157. MIT Press, Cambridge, MA.

Smith, B. E. 2002. Nitrogenase reveals its inner secrets. *Science* **297**:1654–1655.

Staley, J. T., R. P. Gunsalus, S. Lory, and J. J. Perry. 2007. *Microbial Life,* 2nd ed. Sinauer Associates, Inc., Sunderland, MA.

Teather, R. M., and P. J. Wood. 1982. Use of Congo Red-polysaccharide interactions in enumeration and characterization of cellulolytic bacteria from the bovine rumen. *Appl. Environ. Microbiol.* **43**:777–780.

Tilman, D., P. B. Reich, J. Knops, D. Wedin, T. Mielke, and C. Lehman. 2001. Diversity and productivity in a long-term grassland experiment. *Science* **294**:843–845.

Tilman, D., J. Hill, and C. Lehman. 2006. Carbon-negative biofuels from low-input high-diversity grassland biomass. *Science* **314**:1598–1600.

Vessey, J. K. 2003. Plant growth promoting rhizobacteria as biofertilizers. *Plant Soil* **255**:571–586.

Whipps, J. M. 2001. Microbial interactions and biocontrol in the rhizosphere. *J. Exp. Bot.* **52**:487–511.

Web Resources

Bioenergy Feedstock Information Network (BFIN) http://bioenergy.ornl.gov/main.aspx

European Fertilizer Manufacturers Association (EFMA) http://cms.efma.org/EPUB/easnet.dll/execreq/page?eas:dat_im=000BCE&eas:template_im=000BC2

U.S. Department of Energy, Energy Efficiency and Renewable Energy Biomass Program Information Resources for Students http://www1.eere.energy.gov/biomass/abcs_biofuels.html

READING ASSESSMENT

1. Describe the energy crisis that the world potentially faces beginning in the year 2033.

2. What is a biofuel? Why is the development of this technology significant?

3. In the United States, why is use of corn as a biofuel feedstock currently more appealing than switchgrass or poplar trees? What advantages do the latter two options offer in the longer term over corn?

4. Which biofuel, if any, would you recommend to government leaders or private investors as an appropriate candidate for the country to invest its monetary resources to develop as a green energy source? Why?

5. What contributions can microbes make to the production of feedstocks?

6. What aspects of microbial metabolism have been exploited in the biofuel production process? Provide one example.

7. As part of the "I, Microbiologist" project, what is the significance of identifying microbes that produce cellulase?

8. Using the experiments and tools available for the "I, Microbiologist" project, design an experiment that should allow you to detect cellulase activity in a natural environment. Provide a brief explanation of why you would sample a particular source or take specific steps. How could such choices affect the experimental outcomes?

9. As discussed in the text, Tilman and coworkers made several correlative observations with respect to the effect of plant biodiversity. In comparison to monoculture plots, plots planted with the greatest number of plant species also showed an increased ability to (circle all correct answers):
   a. produce bioenergy from above-ground biomass.
   b. reduce fossil fuel consumption at biorefineries.
   c. consume fertilizer produced by the Haber-Bosch process.
   d. sequester carbon dioxide in the plant roots.
   e. sequester carbon dioxide in the plant leaves.

10. What observations did Tilman and colleagues make with respect to the nutrient changes in the soil of mixed prairie grass feedstock plots at the end of the 10-year study? Were these observations seen with monoculture plots? Why or why not?

UNIT 4

EXPERIMENTAL OVERVIEW

In Experiment 2.3, students finished purifying up to 48 different soil bacterial isolates. Efforts now can be made to describe the metabolic potential of these microorganisms, using any or all of the assays in Unit 4. Instructors also may consider incorporating experiments of their own design to phenotypically characterize the bacteria isolated by students.

In Experiments 4.2 and 4.3, students will make wet mounts and Gram stains, respectively, of their isolates for microscopic examination (Experiment 4.1). All 48 of the bacterial isolates may be screened for antibiotic production by using the susceptible gram-positive indicator *Micrococcus luteus* or the gram-negative *E. coli* strains (Experiment 4.4). Isolates displaying interesting phenotypes may be considered good candidates for PCR and sequencing of their 16S rRNA genes (Unit 3). This same subset of isolates then may be subjected to the antibiotic resistance screen (Experiment 4.5) and the cellulase activity screen (Experiment 4.6).

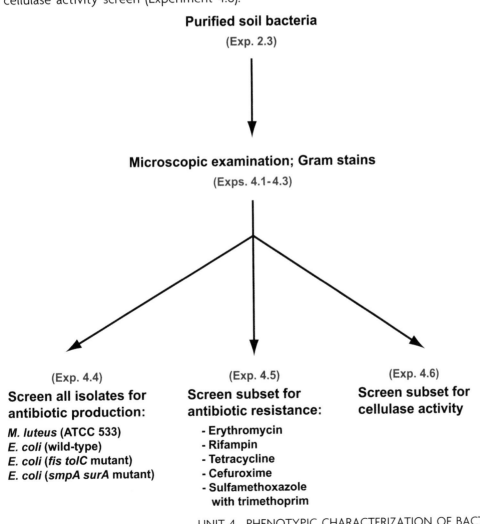

Purified soil bacteria

(Exp. 2.3)

Microscopic examination; Gram stains

(Exps. 4.1-4.3)

(Exp. 4.4)
Screen all isolates for antibiotic production:

M. luteus **(ATCC 533)**
E. coli **(wild-type)**
E. coli **(*fis tolC* mutant)**
E. coli **(*smpA surA* mutant)**

(Exp. 4.5)
Screen subset for antibiotic resistance:

- Erythromycin
- Rifampin
- Tetracycline
- Cefuroxime
- Sulfamethoxazole
  with trimethoprim

(Exp. 4.6)
Screen subset for cellulase activity

EXPERIMENT 4.1 Microscopic Examination and Characterization of Isolates

MATERIALS

Olympus CX41 microscope   Infinity 2-2C digital microscope camera
Computer   Lens paper
Lens cleaner   Preprepared slides of thread
Immersion oil

METHODS

Once purified isolates have been obtained (Experiment 2.3), a digital microscope may be used to take pictures of individual cells, documenting cellular morphology. The procedure described is for the Infinity 2-2C digital microscope camera, which is attached to an Olympus CX41 high-grade microscope. Due to the simplicity of their operation, this microscope-and-camera combination is well suited for the undergraduate educational environment, representing a cost-efficient training instrument that delivers quality optics, performance, and versatility. Key microscope specifications for observing environmental isolates include phase-contrast optics, a total magnification of at least ×1,000, and the ability to easily interface with a digital camera. Other binocular phase-contrast microscopes are commercially available and specifically developed for undergraduate training; these include the Motic BA310, Nikon Eclipse E100, Leica CM E, and National Optical 162-PH. Instructions for each microscope and digital camera specifications are supplied by the manufacturer. Irrespective of the supplier, it is imperative that students handle this delicate equipment with the utmost care. Instructions for handling, setup, and operation of the Olympus CX41 microscope and Infinity digital microscope camera are provided below and can be broadly applied to alternative microscope and camera systems.

Handling the microscope and camera setup

1. Use both hands when removing the microscope from the cabinet in which it is stored, one hand on the bottom and the other supporting the body of the microscope. There is a handle on the back side of the body to assist with microscope maneuvers. The microscope should be returned with the body facing the back of the cabinet. Do not hold the microscope with one hand, because the eyepiece lenses may fall out.

2. Place the microscope on the bench so its body is facing toward you. Turn the revolving binocular tube toward you.

3. Affix the digital camera to the microscope adaptor.

4. Unwind and plug in the power cords.

5. Plug the USB cord for the camera into the computer port.

Key parts of the microscope

The numbers listed below refer to the numbers shown on the diagram of the Olympus CX41 microscope (see page 153).

1. Main switch
   0 = Off
   1 = On

Experiment continues

2. Light intensity control knob

   a. The numbers around the knob designate the reference voltage values. To increase the illumination (e.g., make it brighter), turn the knob clockwise. To lower the illumination (e.g., make it darker), turn the knob counterclockwise.

3. Stage and specimen holder

   a. Stage. The stage is a plane surface on which slides are placed for observation.

   b. Specimen holder. The specimen holder assembly allows the slide to be held in a precise position while being moved from side to side and/or back to front with the stage motion controls. A slide may be placed in the holder by opening the spring-loaded curved finger, placing the slide into the holder from the front, and gently releasing the curved finger. *Note:* **Releasing the curved finger with great force will damage the glass slide.**

4. *x*-axis and *y*-axis knobs

   a. These knobs control the motion of the slide holder on the stage. The upper knob is the *y*-axis knob and moves the specimen forward and back. The lower knob, the *x*-axis knob, moves the specimen left and right.

5. Revolving nosepiece with three objective lenses

   a. All of the three objective lenses are parfocal, which means that if one lens is in proper focus, another may be moved into position by rotating the nosepiece and be in approximate focus. However, when you change from one magnification to another, you will probably need to adjust the light levels and the fine focusing.

   b. Low-power objective (10× magnification). Total magnification obtained with this lens is 10× by 10× eyepiece = 100×. This objective may be used for phase-contrast or bright-field observations.

   c. High dry objective (40× magnification). The total magnification obtained with this lens is 40× by 10× = 400×. This objective may be used for phase-contrast or bright-field observations.

   d. Oil immersion objective (100× magnification). The total magnification obtained with this lens is 100× by 10× = 1,000×. This objective may be used for phase-contrast or bright-field observations, but it must be used with immersion oil.

6. Coarse- and fine-adjustment knobs with prefocusing lever

   a. These controls raise and lower the stage, thereby changing the distance between the specimen and front element of the objective lens. The larger knob is for coarse adjustment, and the smaller knob is for fine adjustment. The fine-adjustment knob has a limited range, so it is necessary to bring the specimen into focus initially by using the low-power objective with the coarse adjustment and then using the fine adjustment to sharpen the image adequately. Once the image is in focus, you may increase the magnification by turning the high dry objective and later the oil immersion objective into place; however, only fine adjustments should be made with these latter lenses since they are parfocal.

b. The prefocusing lever (found on the left side of microscope) ensures that the objective lens does not come in contact with the specimen and simplifies focusing. After focusing the specimen with the coarse-adjustment knob, turn the prefocusing lever clockwise to lock. The upper limit of coarse adjustment is now set in a locked position. Focusing using the fine-adjustment knob is unaffected. After using the coarse-adjustment knob to lower the stage for changing specimens, refocusing is easily achieved by raising the stage to the prefocusing position.

7. Binocular tubes with eyepiece lenses

    a. This assembly holds the eyepiece (ocular) lenses, which magnify the image of a specimen 10-fold (note the number 10×). It is designed to allow adjustment of the distance between the eyepieces and the focus of one of them.

    b. Adjusting the interpupillary distance. While looking through the eyepieces, adjust for binocular vision so the fields of view for each eye coincide completely. Note your interpupillary distance (indicated by the index dot) so that it can be quickly duplicated in the future.

    c. Using the eye shades. If you are not wearing glasses, extend the folded eye shades out to prevent extraneous light from entering between the eyepieces and the eyes. If you wear glasses, keep the eye shades in the folded-down position. This will prevent scratching of the eyepieces or your glasses.

8. Diopter adjustment knob

    a. Adjusting the diopter. Look through the right eyepiece with your right eye, rotate the coarse- and fine-adjustment knobs until the specimen is in focus, and then look through the left eyepiece with your left eye. Turn the diopter adjustment ring to focus the specimen.

9. Field iris diaphragm ring

    a. The field iris diaphragm allows for the adjustment of light intensity as needed. When closed, the field iris diaphragm can protect the specimen against unnecessary heating.

    b. For better contrast, the field iris diaphragm should be opened slightly larger than the field of view. The iris may have to be adjusted for increased light intensity during changes to a higher magnification.

10. Condenser height adjustment knob

    a. This knob allows the condenser lens to be placed at the proper height to allow the best illumination of the specimen. The proper height for the condenser lens can be determined by performing the following series of steps for Kohler illumination:

    i. Close down the field iris diaphragm (step 9).

    ii. Adjust the condenser to the proper height so that the halo around the spot of light is "between" red and blue.

    iii. Open the field iris diaphragm; however, do not open the iris completely. Better contrast is achieved if the field iris diaphragm is open just beyond the field of view. *Note:* **The iris may have to be adjusted when changing to a higher magnification.**

Experiment continues

11. Auxiliary lens centering knob

    a. The auxiliary lens centering knob allows the light path passing through the condenser to be centered in order to maximize the amount of light passing into the objective lens.

    b. The field iris diaphragm can be centered by performing the following series of steps:

    i. With the 10× objective engaged and the specimen in focus, turn the field iris diaphragm ring (step 9) counterclockwise to stop down the diaphragm near its minimum size.

    ii. Bring the image of the diaphragm into focus by adjusting the condenser height adjustment knob (step 10).

    iii. Rotate the two centering knobs until the diaphragm image is in the center of the field of view.

    iv. Open the field iris diaphragm.

12. Aperture iris diaphragm knob

    a. The aperture iris diaphragm (condenser) regulates the light path passing through the condenser.

    b. The aperture iris diaphragm determines the numerical aperture (NA) of the illumination system. Matching the NA of the illumination system with that of the objective provides better image resolution, contrast, and depth of field. Setting the condenser aperture iris diaphragm to between 70% and 80% of the NA of the objective lens is usually recommended. If the aperture iris diaphragm is set too small, an image "ghost" may be observed.

Microscope settings

Wet mounts
  1. Coverslip
  2. Phase optics

Stained preparations
  1. No coverslip
  2. Bright-field optics

Summary of Microscope Settings for Bright-Field and Phase Optics

Part or characteristic	Bright-field optics			Phase optics		
	Low	High dry	Oil	Low	High dry	Oil
Objective lens	10×	40×	100×	10×	40×	100×
Condenser wheel	0	0	0	1	2	3
Magnification (including eyepiece)	100×	400×	1,000×	100×	400×	1,000×
NA	0.25	0.65	1.25	—[a]	—	—
Resolution (μm)	1.34	0.52	0.27	1.34	0.52	0.27
Depth of focus (μm)	28	3.04	0.69	28	3.04	0.69
Field of view (mm)	2.0	0.5	0.2	2.0	0.5	0.2
Working distance (mm)	10.5	0.56	0.13	10.5	0.56	0.13

[a]—, no data.

The Olympus CX41 microscope

1 - ON / OFF Switch
2 - Light Intensity Control
3 - Stage and Specimen Holder
4 - X-Axis and Y-Axis Knobs
   (moves slide left/right or forward/backward)
5 - Nose Pieces with Objectives
   (Low 10X, High 40X, Oil 100X)
6 - Coarse/Fine Adjustment Knobs
   (raises/lowers stage)
   and Pre-Focus Lever Lock

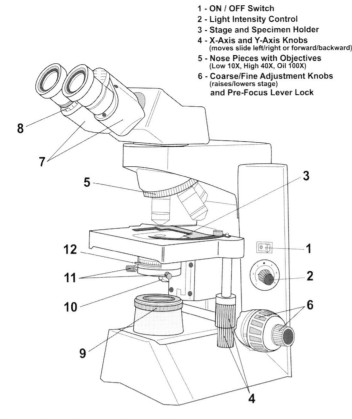

7 - Binocular Tubes with Eyepiece Lenses (10X)
   (note interpupillary distance)
8 - Diopter Adjustment Knob
   (focus right eyepiece lens with #6 and left with diopter)
9 - Field Iris Diaphragm
   (adjusts light intensity)
10 - Condenser Adjustment Knob
   (height/Kohler illumination)
11 - Auxiliary Lens Centering Knob
   (centers #9)
12 - Condenser/Aperture Iris Diaphragm
   (NA = 70-80% of objective lens)

How to prepare the microscope to make observations

The series of steps described on this page should be done before every use of the microscope.

1. Always clean the lenses before starting your microscopic observation. Use lens paper to clean lenses. *Never use Kleenex, Kimwipes, or paper towels.* Wipe the surface once. Do not use the same area of the lens paper a second time because this grinds the surface with dirt. Use one area of the paper with each stroke, and then change to another area. The ocular lenses can be scratched by careless cleaning.

2. Obtain one of the preprepared slides of threads. With the aid of the mechanical stage jaw, clamp the slide and move it so the threads will be in focus under low power.

3. Adjust the interpupillary distance. See step 7b under "Key parts of the microscope" above.

4. Adjust the diopter as described in step 8 above.

Experiment continues

5. Center the field iris diaphragm. With the 10× objective engaged and the condenser ring set to 0, focus the specimen by raising the stage to the highest position with the coarse-adjustment knob and then lower the stage while looking through the eyepieces until the threads come into focus. Now turn the field iris diaphragm ring (step 9 above) counterclockwise to stop down the diaphragm near its minimum size. Bring the image of the diaphragm into focus with the condenser height adjustment knob (step 10 above). Rotate the two centering knobs until the diaphragm image is in the center of the field of view. Open the field iris diaphragm.

6. Adjust the height of the condenser. The proper height for the condenser lens can be determined by performing the series of steps for Kohler illumination described in step 10a above.

The following steps should be done when performing observations using bright-field or phase optics at all magnifications.

Note: Remember that the lenses are parfocal. Always begin looking at a specimen at the lowest magnification and work your way up to higher magnifications. This procedure not only allows optimal visualization of specimens but also prevents inadvertent scratching of the lenses.

Bright-field microscopy

1. Low-power objective (10×)

    a. The working distance for the low-power objective is about 10.5 mm (distance between the slide and the lens).

    b. Turn the condenser wheel to position 0 for bright-field.

    c. Focus the specimen using the coarse-adjustment knob (step 6 above) and lock the prefocus lever.

    d. Adjust the light with the light intensity control knob (step 2 above), so it is comfortable to your eyes.

    e. Move the slide right, left, forward, and backward until the specimen is in the center of the field.

    f. Focus sharply using the fine-adjustment knob, and note the working distance so you can zero in quickly in the future. Do not remove the slide. Continue to step 2.

2. High dry lens (40×)

    a. Turn the nosepiece so the high dry lens is in position.

    b. It should be in approximate focus, since these lenses are parfocal. More light may be needed going from low to high power.

    c. Use the fine-adjustment knob to get sharp focus. Do not remove the slide. Continue to step 3.

3. Oil immersion lens (100×)

    a. Straddle the nosepiece between high dry and oil immersion objectives or swing the objective lenses to the opposite side, and put only one small drop of immersion oil on top of the coverslip.

    b. Rotate the nosepiece so that the 100× lens is in position.

c. It should be in approximate focus, since these lenses are parfocal. More light may be needed going from lower to higher power.

d. Use the fine-adjustment knob to get sharp focus.

e. Lower the stage, and remove the slide.

f. Clean the oil from the slide and 100× lens with lens paper.

Important Note: *Never use oil with 10× or 40× objectives.* If you put oil on the slide to use the 100× oil immersion lens and then decide to go back to lower objectives, you must clean the oil off the slide first. Please use the utmost care with regard to this detail. Oil in the lower objectives is difficult to clean out and can ruin the objectives.

Phase-contrast microscopy

1. Low-power objective (10×)

   a. With the aid of the mechanical stage jaw, clamp the slide and move it so the edge of the coverslip will be in focus under low power.

   b. The working distance under low power is about 10.5 mm (distance between the slide and the lens).

   c. Turn the condenser wheel to position 1 for phase contrast.

   d. Raise the stage to the prefocus position using the coarse-adjustment knob.

   e. Adjust the light with the light intensity knob (step 2 above) so it is comfortable to your eyes.

   f. Focus on the edge of the coverslip or an air bubble. Then move the specimen nearer to the center of the coverslip. Focus sharply using the fine-adjustment knob, and note the working distance so you can zero in quickly in the future. Take notes on your observations. Do not remove the slide. Continue to step 2.

2. High dry lens (40×)

   a. Turn the nosepiece so the high dry lens is in position.

   b. Turn the condenser wheel to position 2 for phase contrast.

   c. It should be in approximate focus since these lenses are parfocal. More light may be needed going from low to high power.

   d. Use fine adjustment to get sharp focus. Take notes on your observations. Do not remove the slide. Continue to step 3.

3. Oil immersion lens (100×)

   a. Straddle the nosepiece between the high dry and oil immersion objectives, or swing the objective lenses to the opposite side and put only one small drop of immersion oil on top of the coverslip. The oil has about the same refractive index as the glass in the slide and in the microscope.

   b. Gently turn the nosepiece so the 100× oil immersion lens clicks into position and is immersed in the oil drop.

   c. Turn the condenser wheel to position 3 for phase contrast.

   d. Increase the light as needed by adjusting the light intensity knob and the aperture iris diaphragm. Focus gently using the fine-adjustment knob.

Experiment continues

e. Aside: Do not remove the slide, but change the condenser wheel to position 0 to repeat the examination of the wet mount preparation using the oil immersion lens with bright-field optics instead of phase contrast. Note the difference between bright-field and phase-contract optics.

Important Note: Never use oil with 10× or 40× objectives. If you put oil on the slide to use the 100× oil immersion lens and then decide to go back to lower objectives, you must clean the oil off the slide first. Please use the utmost care with regard to this detail. Oil in the lower objectives is difficult to clean out and can ruin the objectives.

Digital camera operation

1. Once the microscope has been set up properly as described under "How to prepare the microscope to make observations" above, place the slide containing the specimen in the holder on the stage and adjust the microscope settings as needed to view the specimen under bright-field or phase optics with the 100× oil immersion lens using the steps described above under "Bright-field microscopy" and "Phase-contrast microscopy," respectively.

2. Open "Infinity Analyze" on the desktop of the computer to which the microscope and camera are attached.

3. Notice that the focus for the camera is slightly different from that for the eyepiece. You may not notice this on settings lower than 100×, and it is difficult to discern moving cells. Find something stationary to focus on initially, and turn the fine-adjustment knob slightly while looking at the video capture.

4. Note that there are many image adjustment functions on the left-hand side of the screen, including Exposure, Gain, options for different light sources, a section called Capture Options, and scrolling down to the bottom of this toolbar, settings for Saturation, Hue, Brightness, and Contrast. To familiarize yourself with these functions, try using a preprepared slide of threads.

    a. This microscope uses a halogen bulb, so always select Halogen as your light source.

    b. Under Capture Options, set both Averaging and Subsampling to 1.

    c. Start by adjusting Exposure until the threads are visible and as close to what you see in the microscope as possible. However, the image will not appear optimized until you have modified other settings as well.

    d. Experiment with the other settings. Some values, like Brightness, Contrast, and Gamma, are intimately associated such that if you adjust one, you will need to adjust the other(s).

    e. Note that too much gain can bleach lighter colors, like the yellow thread.

    f. Images high in color or contrast, like the threads, need a high saturation value. For the threads, you will probably want this at the maximum value.

    g. If your purple thread appears too blue, adjust the Hue setting. This probably will change only slightly.

    Note 1 Your image may never appear as bright and sharp as when you look through the eyepiece, due to the limited resolution of the camera, but when you have achieved the desired color balance producing a reasonable image of the threads, you will be ready to work with the wet mounts of your specimens.

    Note 2 Different objectives change the lighting, so that most of your image adjustment values will have to be altered whenever you change magnification.

5. Adjust Exposure until your image appears roughly as bright as through the eyepiece. Using the Auto Exposure function may save you some time, but you also will want to fine tune this setting manually.

6. If your wet mount is essentially colorless through the eyepiece, you can turn down Saturation until the background roughly matches up with the grey you can see in the eyepiece. This may be at or close to 0. You probably will not need to touch Hue.

7. Try modifying Gamma and Gain first, then Brightness and Contrast, until you are satisfied with the image.

8. When you are ready to save your image, scroll up to near the top of your settings toolbar. Press Capture, and then select File → Save As. Save the image as file type JPEG in the appropriate location on the computer.

9. Remember to clean the oil off the 100× objective when you are finished!

Web Resources
Below are websites of several microscope manufacturers.
Olympus CX41 http://www.olympusamerica.com/seg_section/product.asp?product=1027&p=96
Motic BA310 http://www.motic.com/productDetail.aspx?r=NA&lang=en&cid=30&pid=1543
Nikon Eclipse E100 http://www.nikoninstruments.com/e100/
Leica CM E from Cole-Parmer http://www.coleparmer.com/catalog/product_view.asp?sku=4940220
National Optical model 162-PH http://www.microscopesfromnational.com/product/162-PH

Recommended resource for microscope slides, lens paper, and coverslips:
Carolina Biological Supply Company http://www.carolina.com/category/life+science/microscope+slides.do

MATERIALS

Wire inoculating loop   Tap water (in 18-mm tube)
Clean, grease-free microscope slides   Glass coverslips
Kimwipes   Bunsen burner

METHODS

How to prepare a wet mount for phase-contrast microscopy:

From a broth culture

1. Sterilize the inoculating loop by passing it through the flame of a Bunsen burner, and use it to aseptically obtain a small loopful of broth culture from a test tube by following the six steps depicted in the diagram below.

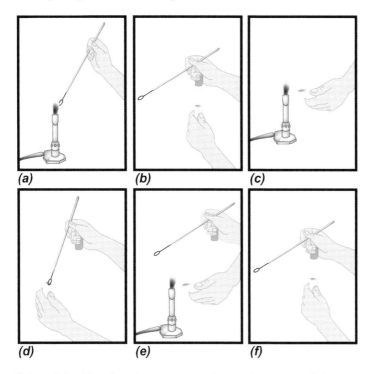

(a)   *(b)*   *(c)*

(d)   *(e)*   *(f)*

2. Place a small loopful of broth culture onto a clean microscope slide.

3. Gently place a coverslip on top of the drop. The coverslip should not be floating. Use a Kimwipe to remove excess water if necessary.

Experiment continues

From an agar plate

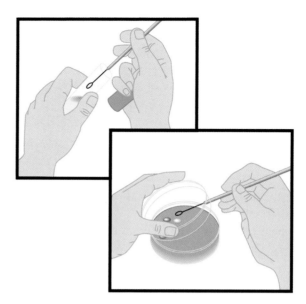

1. With a sterile inoculating loop, obtain a small loopful of tap water from a test tube. It is not necessary to use sterile water. Place a water drop on the clean microscope slide. Two or three wet mounts can be made on a single slide.

2. Resterilize the inoculating loop by passing it through the flame of a Bunsen burner as shown in panel a of the figure on page 159. Allow the loop to cool briefly in air before proceeding with the next step.

3. Remove a small portion of a colony from the surface with the tip of your cool inoculating loop. You might touch the tip of the loop to the agar first to confirm that the loop is cool (*Hint:* You will hear a "sizzle" sound if the loop is still hot). With a circular motion, emulsify the sample in the drop of water on the slide.

4. Gently place a coverslip on top of the drop. The coverslip should not be floating. Use a Kimwipe to remove excess water if necessary.

MATERIALS

Negative control strain: *E. coli* B (nonmotile strain, UCLA laboratory strain 1003)
Positive control strain: *S. epidermidis* (UCLA laboratory strain 1218)

Wire inoculating loop	Tap water (in 18-mm tube)
Clean, grease-free microscope slides	Glass coverslips
Bunsen burner	Crystal violet (primary stain)
Gram's iodine	95% ethanol
Safranin (counterstain)	Paper towels
Absorbent pads	

METHODS

Note: Set out all reagents, including negative and positive controls strains, near a sink. Be sure to cover the counter and floor near the sink with absorbent pads.

Part I Preparation of a Smear for Staining

1. Use only clean, grease-free microscope slides.

   If staining a broth culture, immerse a sterile wire loop in broth to collect a drop, and spread the drop into a thin film covering an area about the size of a nickel. Allow the film to air dry at ambient temperature. Then fix the smear by passing the slide rapidly through the hot part of a Bunsen burner flame, smear side up, two or three times. *Do not overheat it.* The slide should never be so hot that you cannot touch it.

   If staining a specimen from solid medium, use a sterile wire loop to first place a small drop of water on the slide. Tap water is generally adequate and need not be sterile. Pick up a small amount of a single colony with the edge of the sterile loop, and suspend the cells by rubbing the loop in a water drop. The final suspension should be very faintly turbid. Spread the suspension out, air dry, and heat fix as described above.

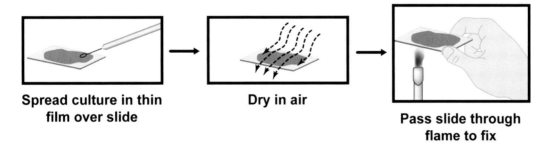

Spread culture in thin film over slide → **Dry in air** → **Pass slide through flame to fix**

Note: The fixed smear should be thin enough to read the print on this page through it.

   Hints: It may be useful when staining smears for the first few times to ring the area of the smear on the underside of the slide with a glass marking pen or wax pencil before staining. Because of the time needed to properly dry the sample, it also may be helpful to set up several smears at one time.

   Controls: It is best to include smears of a known gram-negative organism and a known gram-positive organism on each slide to ensure that proper staining is achieved. The recommended setup is diagrammed below.

Experiment continues

G +	Unknown	G −

Positive control strain (G +): *S. epidermidis*

Negative control strain (G −): *E. coli* B

2. After the slide cools, follow the staining procedure provided in Part II.

Part II Gram Stain (Hucker's Modification)

The Gram stain procedure is a widely used differential staining technique in diagnostic bacteriology. It is often one of the first procedures done when characterizing a new bacterium. Bacteria are divided into two major groups based on their reaction to the Gram stain. Those that retain the primary dye and stain purple are gram positive, while those that lose the primary dye and stain pink are gram negative. The mechanism of differentiation is based on the structure of the cell wall, which differs markedly in gram-positive and gram-negative bacteria as diagrammed below.

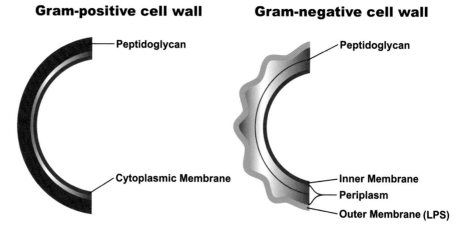

Gram-positive cell wall
- Peptidoglycan
- Cytoplasmic Membrane

Gram-negative cell wall
- Peptidoglycan
- Inner Membrane
- Periplasm
- Outer Membrane (LPS)

1. As shown in the diagram below, flood the heat-fixed smear with crystal violet solution for approximately 1 minute, although the exact time is not critical.

2. Rinse the smear in gently flowing tap water until no additional dye washes out. Shake off excess water. All cells should be stained purple after this step.

Gram⁺   **Gram⁻**

   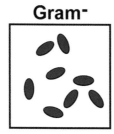

Primary Stain

3. Remove any remaining water by washing it off with iodine solution, and then cover the smear for 1 minute with iodine solution.

4. Rinse off the iodine with water; shake off excess water. All cells should remain purple after this step.

Gram⁺   **Gram⁻**

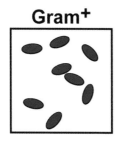   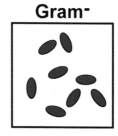

Fix Primary Stain

5. Cover the smear with a few drops of 95% ethanol, and tilt the slide back and forth a few times. Drain off the alcohol. Repeat several times with fresh alcohol until no more purple color is seen coming out of the smear. The decolorization time depends on the thickness of the smear. *This is the critical step in the procedure.*

6. Immediately rinse off remaining alcohol with tap water. Gram-positive cells should be purple, while gram-negative cells should now be colorless.

Gram⁺   **Gram⁻**

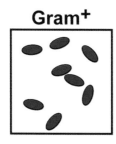   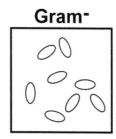

Alcohol Rinse

7. Shake off excess water, and remove any additional water by washing off with the safranin solution. Cover the smear with safranin for 1 minute.

Experiment continues

8. Wash off excess safranin with water, and shake off excess water. Gram-positive cells should appear purple, and gram-negative cells should appear pink.

Gram⁺ **Gram⁻**

 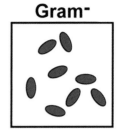

Secondary Stain

9. Dry by blotting (do not rub) between sheets of paper towel. Pass once rapidly through the flame. Examine with the oil immersion lens.

10. Some potential problems that may occur with the Gram stain:
    - Old cultures of gram-positive organisms may stain gram negative.
    - If the smear is too thick, the gram-negative cells may not decolorize properly.
    - Samples may become over-decolorized and give a false Gram reaction.
    - Gram stain appears greenish when viewed under phase optics.
    - Some organisms have a cell wall with additional layers of a composition that is refractory to Gram stain, giving ambiguous or inconsistent results.

REFERENCE
Hucker, G. J. 1921. A new modification and application of the Gram stain. *J. Bacteriol.* **6**:395–397.

EXPERIMENT 4.4 Testing for Antibiotic Production

Soil environments are rich reservoirs for antibiotic-producing bacterial strains. Bacteria can be tested for their ability to produce zones of growth inhibition against susceptible gram-positive or gram-negative indicator strains. Those environmental isolates that secrete antibiotics into the surrounding media cause a clearing in a growing lawn of the indicator strain. This is a good assay for screening all 48 initial isolates, choosing those with interesting antibiotic production profiles for further analysis.

Micrococcus luteus is a very susceptible gram-positive indicator strain and is good for an initial screen of soil isolates. The isolates also may be subjected to additional screens against three gram-negative indicator strains: wild-type *E. coli* and two double-mutant derivatives, *fis tolC* and *smpA surA*. The *fis tolC E. coli* mutant exhibits greater sensitivity to ciprofloxacin than does the wild-type strain, while the *smpA surA E. coli* mutant shows increased sensitivity to vancomycin compared to the wild-type strain (Tamae et al., 2008). Ciprofloxacin, a broad-spectrum antibiotic, is a quinolone derivative that functions as an inhibitor of DNA gyrase and is considered the drug of choice for treating anthrax caused by penicillin-resistant *Bacillus anthracis*. Vancomycin inhibits cell wall synthesis at a step distinct from penicillin and its derivatives. Considered a narrow-spectrum antibiotic, vancomycin typically is not effective against gram-negative bacterial strains and is used as the drug of last resort against penicillin-resistant *Staphylococcus aureus*, an opportunistic gram-positive pathogen. Students might hypothesize that isolates which produce zones of growth inhibition against the *E. coli* mutants but not the wild-type strain (or even *M. luteus*) may be a source of novel antibiotic compounds.

MATERIALS

Strains:

M. luteus ATCC 533 (incubate at 30°C for 24 h)

Wild-type *E. coli* (UCLA laboratory strain 1246; incubate at 37°C for 24 h)

E. coli fis tolC mutant (*fis*/Kan^r, *tolC*/Tet^r, UCLA laboratory strain 1247; incubate at 37°C for up to 48 h)

E. coli smpA surA mutant (*smpA*/Kan^r, *surA*/Cam^r, UCLA laboratory strain 1248; incubate at 37°C for up to 48 h)

Luria Bertani (LB) plates for *M. luteus* and wild-type *E. coli*

LB^Tet plates for *E. coli fis tolC* (tetracycline at 15 μg/ml [final concentration])

LB^Cam plates for *E. coli smpA surA* (chloramphenicol at 20 μg/ml [final concentration])

Tetracycline stock (1.5 mg/ml) (store 0.5-ml aliquots at −20°C)

Chloramphenicol stock (20 mg/ml) (store 0.5-ml aliquots at −20°C)

LB broth

13-mm glass test tubes

Sterile glass beads (~12 per 13-mm tube)   Sterile flat toothpicks

Template: 50-square grid

METHODS

Use aseptic technique throughout procedures. Wear gloves as needed.

Experiment continues

PERIOD 1
Prepare *M. luteus* and *E. coli* Cultures

1. Prepare tubes containing LB broth. For mutant *E. coli* strains, supplement the broth with appropriate antibiotic such that a final concentration of either 15 μg of tetracycline per ml or 20 μg of chloramphenicol per ml is achieved. One may do this by adding 5 μl of an antibiotic stock solution to 5 ml of LB broth, which results in a 1,000-fold dilution of the stocks.

   Note: If preparing multiple tubes at a time, scale up the dilution volumes accordingly. For instance, you could add 50 μl of the antibiotic to 50 ml of LB broth, mix by swirling the bottle, and aseptically transfer 5 ml of the LB broth containing antibiotic to each test tube.

2. Using a sterile toothpick, inoculate 5 ml of LB broth (with or without antibiotic) with a single colony of *M. luteus* or wild-type or mutant *E. coli*. Thoroughly vortex the culture tube, and incubate for 24 hours in a shaker or rotator at 30°C for *M. luteus* and at 37°C for the *E. coli* strains.

PERIOD 2
Picking and Patching Colonies

1. Label enough LB, LB^Tet, or LB^Cam plates for each indicator strain with your name and the date. Use the numbered-grid petri plate template below to number squares on the bottom of each plate (*not* the lid), placing an orientation mark on the bottom of the plate for reference.

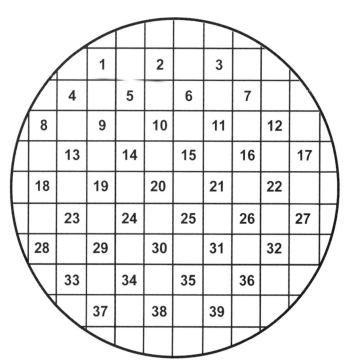

2. Obtain overnight cultures of *M. luteus* and *E. coli* (wild type and two mutants). Do not be alarmed if there appears to be a large, visible *M. luteus* cell aggregate that grows in the overnight culture; the liquid medium will still be saturated with *M. luteus*.

3. Using the glass bead shaking technique (refer to Experiment 2.1 for details), spread 150 µl of the *M. luteus* or *E. coli* overnight culture onto the appropriate plate. Be sure to obtain an even lawn of each culture over the entire plate.

4. Using sterile toothpicks and the numbered grid template above, pick single colonies from your streak plates and patch onto appropriate plates (10 to 12 patches total per plate). Be sure to patch only every other square in order to leave room for evidence of clearings. To be certain that the *M. luteus* or *E. coli* lawns do not overtake the colonies formed by isolates before they have a chance to exhibit growth inhibition, be sure to grid a fairly large amount of each colony onto the plates (approximately this size: O). Also, do not stab your toothpick into the agar.

   Note 1 You may want to group the slower-growing isolates on plates separate from those that grow faster under normal cultivation conditions.

   Note 2 For isolates that must be propagated using liquid cultivation, grow up a turbid culture (you may need to centrifuge the cells and resuspend the pellet in a smaller volume such as 200 µl of broth) and spot 10 to 20 µl of the culture onto the indicator lawn. Allow to dry thoroughly before flipping the plate over for the incubation step.

5. Incubate *M. luteus* plates at 30°C and *E. coli* plates at 37°C for 24 hours. Note that the mutant *E. coli* strains must be incubated longer (up to 48 hours) as they do not grow as fast as the wild-type strain.

PERIOD 3
Plate Inspection

1. Remove the plates from the incubator, and check for evidence of clearings. Measure the diameter (not the radius) of each clearing (in centimeters), and record the data in your laboratory notebook along with a description of the patched colony morphology. Take a picture of the plates using the laboratory camera for inclusion in your notebook.

2. Repeat the assay for isolates that exhibit zones of growth inhibition to confirm the size of the clearing. Take a picture of the plates using the laboratory camera for inclusion in your notebook. Save the image file as a JPEG (.jpg) for upload to the CURL Online Lab Notebook (http://ugri.lsic.ucla.edu/cgi-bin/loginmimg.cgi).

   Technical Note: If your isolates do not grow on LB broth, you may try to perform the assay on medium from which the isolate was originally obtained (e.g., ISP4, N₂-BAP A, RDM, or R2A). Please note that the indicator strains do not grow on VXylA.

PERIOD 4
Data Analysis

Plot the relative antibiotic production (diameter of clearing, in centimeters) for each isolate. Set the minimum at zero, arbitrarily assigned to isolates that appear to produce no zone of growth inhibition against the indicator strain. An example of a plot is shown below.

Experiment continues

EXPERIMENT 4.4

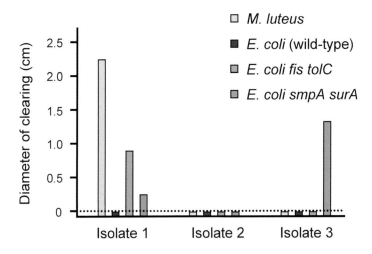

REFERENCE

Tamae, C., A. Liu, K. Kim, D. Sitz, J. Hong, E. Becket, A. Bui, P. Solaimani, K. P. Tran, H. Yang, and J. H. Miller. 2008. Determination of antibiotic hypersensitivity among 4,000 single-gene-knockout mutants of *Escherichia coli*. *J. Bacteriol.* **190:**5981–5988.

MATERIALS

18-mm test tubes	R2A broth
Sterile cotton swabs	Glucose+MSB
Antibiotic discs or gradient strips	N$_2$-BAP broth
Mueller-Hinton agar plates	VXylA broth
RDM broth	

METHODS

Use aseptic technique throughout procedures. Wear gloves as needed.

PERIOD 1
Inoculate Overnight Cultures

1. For each of your isolates, prepare overnight cultures by inoculating tubes containing 5 ml of the appropriate medium with a single colony. To minimize costs, this experiment need not be performed on all 48 isolates but instead may be restricted to the 16 isolates for which the 16S rRNA gene is sequenced.

PERIOD 2
Place Antibiotic Discs on Culture-Painted Plates

1. Dip a sterile cotton swab into the culture tube, thoroughly moistening the cotton swab with the cell suspension.

2. Roll the swab against the sides of the tube to squeeze out excess liquid, although you do not want to let the swab go dry.

3. Use the swab to paint the entire surface of a Mueller-Hinton agar plate as shown below:

Experiment continues

4. Let the plate dry for 15 minutes with the lid slightly ajar until no liquid can be seen.

5. Place antibiotic discs on the plate using the sample template below. Discs should be at least 10 mm from the edge of the plate and 30 mm from the center of the adjacent disc.

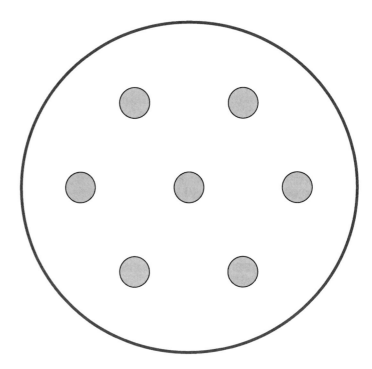

Note: Use no more than seven discs per plate. Some students find it easier to read the results if, instead of using a single plate with seven discs, multiple plates containing three or four discs per plate are used.

A broad collection of antibiotic discs and gradient strips are commercially available. The following five disc types are recommended as a starting point for the "I, Microbiologist" project, as they represent several different antibiotic classes and target diverse cell structures or processes (Fig. 4.2). Students should record the concentration of antibiotic contained in each disc since it varies by manufacturer.

Antibiotic	Abbreviation	Drug target
Erythromycin	E15	50S ribosome
Tetracycline	TE30	30S ribosome
Rifampin	RA5	Transcription
Cefuroxime	CXM30	Cell wall synthesis
Sulfamethoxazole with trimethoprim	SXT	Folic acid synthesis

6. Flip the plate upside down, and incubate for 24 to 48 hours at 30 or 37°C, depending on the optimal conditions for a particular isolate.

    Technical Note: If your isolates do not grow on Mueller-Hinton agar, you may try to perform the assay with the medium from which the isolate was originally obtained (e.g., R2A, RDM, ISP4, N_2-BAP, or VXylA).

PERIOD 3
Data Analysis

1. Remove the plates from the incubator, and check for zones of growth inhibition around the discs. Measure the diameter (not the radius) of each clearing (in centimeters), and record the data in your laboratory notebook. Take a picture of the plates with the laboratory camera for inclusion in your notebook. Save the image file as JPEG (.jpg) for upload to the CURL Online Lab Notebook (http://ugri.lsic.ucla.edu/cgi-bin/loginmimg.cgi).

2. Plot the relative antibiotic resistance (inverse of the diameter of clearing in reciprocal centimeters) for each isolate. To represent isolates that are completely resistant to a given antibiotic (e.g., no clearing visible around the disc), set the maximum at 2.0 cm^{-1}. Because the diameter of a single disc is 0.5 cm, the inverse of this value (1/0.5 cm) is 2.0 cm^{-1}. The inverse diameter of clearing for other isolates is calculated in the same way, using the equation $d_i = 1/d$, where d_i is the inverse diameter (in reciprocal centimeters) and d is the diameter (in centimeters) as measured on the plate.

    An example plot is shown below:

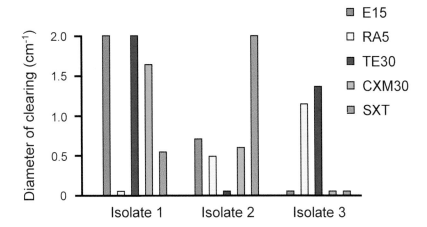

REFERENCES

Bauer, A. L., W. M. M. Kirby, J. C. Sherris, and M. Turck. 1966. Antibiotic susceptibility testing by a standardized single disc method. *Am. J. Clin. Pathol.* **45:**493–496.

Mueller, J. H., and J. Hinton. 1941. A protein-free medium for primary isolation of gonococcus and meningococcus. *Proc. Soc. Exp. Biol. Med.* **48:**330–333.

MATERIALS

Control strain: *E. coli* PCT203 (Amp[r], UCLA laboratory strain 1239)

RDM plates

R2A plates

ISP medium 4[hex, ben] plates

Congo red solution (1 mg/ml)

1 M NaCl

Water bath (55°C)

N$_2$-BAP plates

VXylA plates

0.1% (wt/vol) carboxymethyl cellulose (CMC) plates (hard agar)

0.1% (wt/vol) CMC overlay medium (soft agar)

1 M HCl

Heat block (45 to 48°C)

Note 1 The Congo red, HCl, and NaCl solutions must be made fresh and should not be stored for longer than 3 months.

Note 2 The use of well-dried plates is essential for this experiment to work properly; otherwise, zones of digestion will be too wide due to secondary growth.

Note 3 Students should try method A first. If an isolate does not grow on CMC hard agar, method B must be used instead. Be sure to alert your instructor or TA in advance so that soft agar is melted for method B. To melt soft agar, place tubes in a steamer for 30 minutes and then transfer them to a 55°C water bath. Only as much soft agar as is needed for the assay should be melted, as it cannot be reused.

Note 4 For simplicity, this experiment need not be performed on all 48 isolates but instead may be restricted to the 16 isolates for which the 16S rRNA gene is sequenced.

METHODS

Use aseptic technique throughout procedures. Wear gloves as needed.

PERIOD 1
Inoculate Plates
Method A CMC Hard Agar

1. Using sterile toothpicks, pick single colonies from your streak plates and paint each isolate onto half of a CMC hard agar plate (two isolates per plate as shown in the diagram below). Do not stab your toothpick into the agar. For isolates that must be propagated using liquid cultivation, grow up a turbid culture (you may need to centrifuge the cells and resuspend the pellet in a smaller volume such as 200 µl of broth) and spot approximately 50 µl of culture onto half of a plate. Allow the plate to dry thoroughly before flipping it over for the incubation step. Be sure to paint cells from the positive control strain onto one of your plates. This strain contains a plasmid with genes encoding cellulase (β-1,4-glucanase), a catabolic enzyme that hydrolyzes cellulose, which is found in the leaves and stems of plants.

Experiment continues

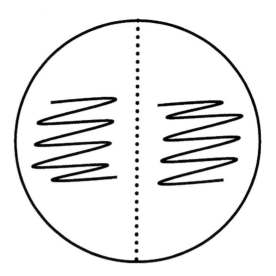

2. Incubate the plates at 30°C for 24 to 48 hours as needed (skip period 2 and go on to period 3 for the next series of steps for method A).

Method B Overlay with CMC Soft Agar

1. Use the medium type that corresponds to the medium on which each isolate is being propagated. Label plates with your name and the date. Patch only one isolate per set of duplicate plates. Include a set of LB^Amp plates upon which the positive control strain is patched (it need not be plated in duplicate). Using sterile toothpicks, pick single colonies from your streak plates and patch the cells onto the surface of plates in duplicate so that you have two identical patterns of the same organisms on two plates. Do not stab your toothpick into the agar. For isolates that must be propagated using liquid cultivation, grow up a turbid culture (you may need to centrifuge the cells and resuspend the pellet in a smaller volume such as 200 μl of broth) and then spot 10 to 20 μl of culture onto the plates. Allow the plates to dry thoroughly before flipping them over for the incubation step.

2. Incubate the plates at 30°C for 24 to 48 hours as needed for each isolate.

PERIOD 2
CMC Overlay (Method B Only)

1. Soft agar for overlays will be in the 55°C water baths—you should transfer the amount you need to the heat block (set at 45 to 48°C) on your laboratory bench. You should let the soft agar tubes equilibrate in the heating block for 10 to 15 minutes after transfer from water baths before pouring the overlays, otherwise you may kill the bacteria you have patched onto the plates. However, if the agar is too cool the overlays will be lumpy.

2. Carefully pour CMC overlay medium onto the surface of each agar plate on which bacterial cells have grown. Avoid pouring the soft agar directly onto the patches of cell growth; instead aim for the agar surface itself. While rocking the plate gently, spread the melted soft agar over the entire surface of the plate before it has time to solidify. Avoid splashing it onto the sides of the petri dish. Place the plate on a level surface, and allow it to stand until the agar has solidified before moving it further.

3. Invert the plates, and incubate them at 30°C for 24 hours.

PERIOD 3
Staining (Methods A and B)

Method A only: Before proceeding with the staining steps, use a sterile toothpick to gently scrape the cells from the surface of the CMC hard agar. However, if cells did not grow on this medium, go back to period 1 and try method B instead.

Method B only: Proceed with only one of the two sets of duplicate plates. If zones of clearing are not visible after staining, incubate the second set of plates for an additional 24 to 48 hours and repeat the staining procedure.

Staining procedure for Methods A and B

1. Flood the agar plates with 1-mg/ml Congo red solution, and incubate at room temperature for 15 minutes.

   Caution: Wear gloves when handling Congo red, which is a toxic benzidine-based dye. It is suspected of metabolizing into benzidine, a known human carcinogen. Dispose of contaminated liquid and solid waste into proper containers.

2. Pour off the Congo red solution into a waste collection container. (Do not pour down the sink drain.)

3. Flood the plates with 1 M NaCl, and incubate at room temperature for 15 minutes.

4. You may be able to see the clearing zones reflecting hydrolysis of CMC by secreted cellulase enzymes after treatment with NaCl. However, to stabilize the visualized zones of hydrolysis for at least 2 weeks, briefly flood the plates with 1 M HCl. This reagent causes the dye to change from red to blue and inhibits further enzyme activity.

5. Pour off the HCl into a waste collection container. (Do not pour down the sink drain.)

6. If you cannot see the clearing zones after staining, leave the plates out at ambient temperature to let them dry. The zones of clearing may become more apparent with time.

   Note: You should also place the second plate back in the incubator for a while. Repeat the staining procedure (steps 1 to 6) with the second plate. If no zones of clearing are visible, what should you conclude about your isolate with respect to its ability to metabolize cellulose?

7. Take a picture of the plates using the laboratory camera for inclusion in your notebook. Save the image file as a JPEG (.jpg) for upload to the CURL Online Lab Notebook (http://ugri.lsic.ucla.edu/cgi-bin/loginmimg.cgi).

ACKNOWLEDGMENT
The procedure was modified from one obtained from Ann Hirsch, University of California, Los Angeles, who adopted the protocol from the reference below.

REFERENCE
Teather, R. M., and P. J. Wood. 1982. Use of Congo red-polysaccharide interactions in enumeration and characterization of cellulolytic bacteria from the bovine rumen. *Appl. Environ. Microbiol.* **43:**777–780.

UNIT 5

Cultivation-Independent Community Analysis of Soil Microbiomes

SECTION 5.1
ACCESSING "UNCULTIVATABLE" MICROBIOMES BY USING METAGENOMIC APPROACHES

Metagenomics is an emerging approach to the scientific exploration of microbial life in the biosphere. The methodology was developed to overcome the limitations of cultivation-based analyses (e.g., the "microbial bottleneck" discussed in Unit 2) in that it provides a means to characterize environmental communities of microorganisms at the molecular level without having to culture them first.

The prefix "meta" has a Greek origin and is used in the English language to describe a concept (i.e., metagenomics) that is a parallel extension of another concept (i.e., genomics). Genomic analysis is the study of a collection of genes derived from a single microorganism, whether they are on the chromosome or on associated plasmids. This methodology relies upon the availability of pure cultures of the organism from which segments of the genome can be isolated and cloned for DNA sequencing and bioinformatics analysis. Metagenomic analysis, as an analogy to genomic analysis, is the study of a collection of genes, in this case isolated from environmental samples rather than a single organism. This exciting new field applies the techniques and general strategies used in genomics to analyze entire microbial communities composed of only a few to thousands of members (see, e.g., Baker and Banfield [2003] and Venter et al. [2004]), bypassing the need to cultivate and purify individual microorganisms.

As depicted in Fig. 5.1, two general strategies are employed to study microbes within a community that are uncultured (i.e., never grown in the laboratory in isolation from other microorganisms) and unculturable (i.e., not able to grow in the laboratory without other microoganisms) (Eisen, 2007). The left side of the flow diagram depicts the steps involved in a community sampling approach (i.e., gene survey or phylotype survey) targeting 16S ribosomal DNA (rDNA), or the DNA sequences encoding 16S ribosomal

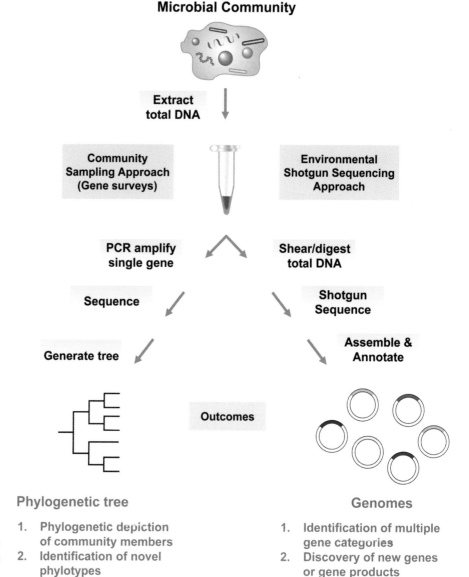

Microbial Community

Extract total DNA

Community Sampling Approach (Gene surveys)

Environmental Shotgun Sequencing Approach

PCR amplify single gene

Shear/digest total DNA

Sequence

Shotgun Sequence

Generate tree

Assemble & Annotate

Outcomes

Phylogenetic tree

Genomes

Phylogenetic tree

1. Phylogenetic depiction of community members
2. Identification of novel phylotypes

Genomes

1. Identification of multiple gene categories
2. Discovery of new genes or gene products
3. Phylogenetic anchors link genes to phylotypes

FIGURE 5.1 Overview comparing two general approaches used to characterize microbial communities directly from environmental samples. Neither method requires cultivation of microbial isolates. Illustration by Cori Sanders (iroc designs).

RNA (rRNA) genes. This approach is comparable to the cultivation-independent portion of the "I, Microbiologist" research project. The right side of the flow diagram outlines an alternative approach called environmental shotgun sequencing (ESS), which is what researchers may choose to utilize when the sequencing process itself is not a limiting or cost-prohibitive step. Both approaches, which are described in more detail below, are sequence based, in that they focus on the detection of phylotypes, novel genes, and metabolic potential. With some modifications, metagenomic strategies also may be function based, providing a way to investigate the activity of gene products (i.e., messenger RNA [mRNA] and proteins) that community members express (e.g., metabolites, enzymes, and antibiotics). These complementary metagenomic methodologies not only will increase our knowledge about the composition and metabolic potential of microbial communities but also may reveal novel insights into how microbes impact local and global ecosystems, which could have practical applications in human health, agriculture, environmental remediation, biotechnology, and even biodefense and forensic science (Eisen, 2007; National Research Council, 2007).

Gene surveys: a sequence-based community sampling approach

As shown in Fig. 5.2, when one takes an environmental sample, in our case the soil, and simultaneously extracts genomic DNA from microorganisms present in that sample, including bacteria, fungi, and other eukaryotic cells, one ends up with a purified "metagenomic" DNA sample that constitutes the soil microbiome—the genetic material derived from the microorganisms present in the environment. However, when one considers the abundance of microbial species (greater than 10^3 or 10^4) of unknown distribution within the population of 10^7 to 10^9 cells, as well as the complex structure and highly variable composition of soil, which contains traces of organic contaminants such as humic acid and heavy metals that interfere with procedures requiring high-quality DNA (e.g., polymerase chain reaction [PCR]), recovery from the soil of genomic DNA that is free of impurities and truly representative of the entire community presents an ongoing challenge (Daniel, 2005). Although a number of strategies have been developed, each with its pros and cons, the resulting consensus is that no single method is optimal for all soil samples. Instead, to minimize bias attributed to the DNA extraction method, the selected approach should be based on soil characteristics, including the type of contamination, as well as consideration of the techniques to be used during ensuing steps (Martin-Laurent et al., 2001; Fortin et al., 2004; Lakay et al., 2007).

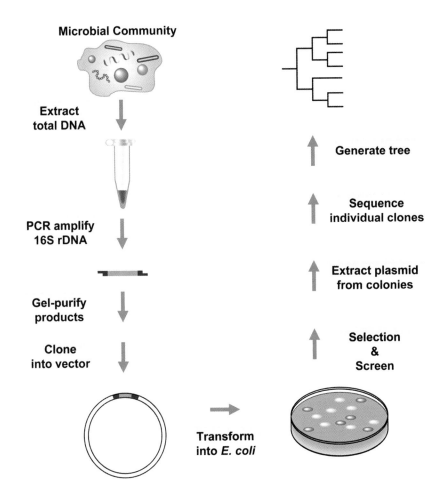

FIGURE 5.2 Overview of the community sampling approach used in the metagenomic studies of soil samples in the "I, Microbiologist" project. Illustration by Cori Sanders (iroc designs).

DNA extraction strategies can be divided into two general categories, both of which involve methods that produce a crude lysate that must be purified further. Cells may be separated from the soil matrix prior to cell lysis, or they may be lysed directly while still within the soil matrix and then the nucleic acids are separated from the soil particles and other cellular debris (Daniel [2005] and references within; Fortin et al., 2004; Lakay et al., 2007). Although more DNA is recovered using the direct-lysis approach, the techniques required to achieve lysis (e.g., treatment with detergent followed by bead beating) are quite harsh and often result in DNA shearing; the recovered metagenomic DNA contains nucleic acids from not only the lysed bacterial cells but also extracellular, archaeal, and eukaryotic DNA. In contrast, DNA obtained by cell separation methods (e.g., cation-exchange chromatography followed by differential centrifugation) tends to be entirely prokaryotic and of a larger average size, suitable for a number of downstream applications. Despite being less efficient in terms of the amount of DNA recovered, the latter methods seem to produce DNA that is less contaminated with humic acids. For the "I, Microbiologist" project, we will use a commercial soil DNA extraction kit that permits isolation of metagenomic DNA directly from soil, although prior to cell lysis the soil sample is treated briefly with a solution that facilitates precipitation of humic acid and other soil-based impurities (see Experiment 5.1).

PCR is used to amplify the 16S rRNA gene simultaneously from the chromosomes of community members represented in the metagenomic DNA sample. As discussed in Unit 7, Carl Woese and Norman Pace pioneered the use of this gene as an evolutionary chronometer in their reconstructions of the universal tree of life (Pace, 1997; Woese et al., 1990). To target the bacterial community specifically, PCR is performed using a set of primers designed to anneal to segments within the 5′ and 3′ ends of the 16S rRNA gene that are thought to be universally conserved in *Bacteria,* thereby excluding from the analysis *Archaea* and *Eucarya,* whose genomic DNA undoubtedly will be present in the community-derived sample. There is variability in the amount of PCR product obtained, depending upon the amount of soil used during the metagenomic DNA isolation step or the PCR conditions employed. Recall that there are probably more than 10,000 bacterial species in a single soil sample, so there will be at least as many genomes represented in the sample, for which the relative abundance will not be known. Thus, there may be only one or a few copies of any single DNA target in metagenomic DNA samples. As discussed in Unit 3, this situation may lead to a considerable number of nonspecific priming events, resulting in the formation of secondary products during the first PCR cycle. These accumulate during subsequent PCR cycles and can produce an astronomically high background in later cloning steps. Other PCR artifacts include chimeras and heteroduplex molecules, which become a concern when performing PCR under lower-stringency conditions (Acinas et al., 2005).

To minimize the influence of PCR artifacts and unwanted PCR products on downstream applications, it is essential to purify the PCR products by a gel extraction technique prior to cloning the genes into a vector (see Experiment 5.4). This method uses agarose gel electrophoresis (discussed in Unit 3) to separate the desired PCR products from the unwanted secondary products based primarily on size. The band containing DNA fragments of a size consistent with that expected for 16S rDNA (i.e., ~1,500 bp) is excised from the gel and extracted using commercially available kits. Unwanted PCR products of the same size as the 16S rRNA genes will not be excluded by this gel extraction technique, but the procedure results in a manageable amount of background during subsequent analysis of clones. The purified PCR product is ready to be ligated into a suitable cloning vector. It should be noted that microorganisms differ in the number of rRNA operons *(rrn)* within their respective genomes, with a range of from 1 to 15 copies per genome (Klappenbach et al., 2000). Morcover, the number of *rrn* operons is positively correlated with the rate of response to growth substrates, or the ability to form colonies rapidly under favorable growth conditions. Thus, one potential bias that can arise in the metagenomic PCR products from the same habitat is an overrepresentation of 16S rRNA genes derived from rapidly growing bacteria and an underrepresentation of genes resulting from slowly growing bacteria.

Although a number of alternative cloning systems or restriction-based strategies could be considered, the "I, Microbiologist" project has been designed to expedite this particular step and therefore a simple and efficient kit has been selected for this procedure (the TOPO cloning kit; see Experiment 5.6). Provided with this kit is a linearized plasmid called pCR2.1-TOPO, which has single 3′ deoxythymidine (T) overhangs with topoisomerase covalently bound to the 3′ phosphate (Fig. 5.3a). *Bacteria,* many *Archaea,* and some viruses contain topoisomerases, which either introduce or remove supercoiling in

FIGURE 5.3 Schematic overview of the mechanism by which metagenomic PCR products are cloned into pCR2.1-TOPO. (a) Linearized vector with topoisomerase covalently linked to single, overhanging thymidine. (b) Once the *Taq*-amplified PCR products are mixed with the TOPO vector, the 5′-OH on each end of the PCR product initiates a nucleophilic attack on the phospho-tyrosyl bond between the vector and topoisomerase, releasing the enzyme while simultaneously forming a new phosphodiester bond between the vector and PCR product. (c) Ligation step—sealing the DNA strands results in a circularized vector containing the metagenomic PCR product as the DNA insert. This recombinant vector, which is not a stable DNA duplex due to the nick that still resides on the strand opposite the newly formed phosphodiester linkage, is immediately transformed into *E. coli* competent cells where the nick will be sealed and the plasmid will be replicated and passed on to daughter cells during binary fission. Illustration by Cori Sanders (iroc designs).

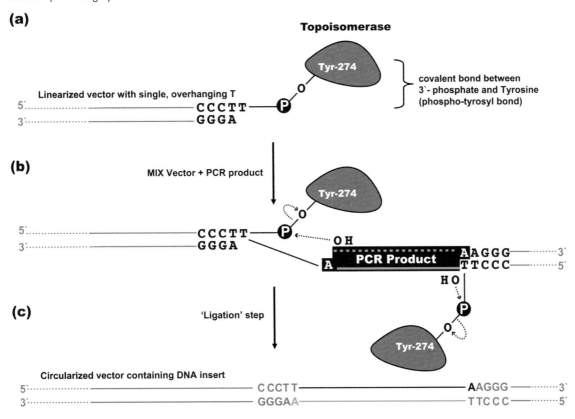

circular DNA molecules, depending on the type of enzyme (type I remove supercoils while type II introduce negative supercoils) (Madigan and Martinko, 2006). For example, topoisomerase I from vaccinia virus binds to duplex DNA at specific sites and then cleaves the phosphodiester backbone following the sequence 5'-CCCTT in one strand (Shuman, 1991). As shown in Fig. 5.3b, the energy from the broken phosphodiester bond is conserved due to formation of a covalent linkage between the 3' phosphate of the cleaved strand and a tyrosine residue at position 274 (Tyr-274) of topoisomerase. This reaction is reversible, in that the phospho-tyrosyl bond between the DNA and enzyme later can be attacked by the 5' hydroxyl of the original cleaved strand, releasing topoisomerase (Shuman, 1994). The TOPO TA cloning kit exploits this reaction to efficiently clone PCR products.

Taq polymerase has a non-template-dependent terminal transferase activity that adds a single deoxyadenosine (A) to the 3' ends of PCR products; this activity is required for the TOPO cloning reaction to work efficiently, as each 3' A forms a complementary base pair with the 3' T on the vector (Zhou and Gomez-Sanchez, 2000) (Fig. 5.3b). It also should be noted that *Taq* does not possess a 3'-to-5' exonuclease proofreading activity, which otherwise would allow it to backtrack along the DNA duplex and correct erroneous incorporation of nucleotides opposite the DNA template strand during replication (Tindall and Kunkel, 1988; Lawyer et al., 1993). Consequently, since this enzyme has lower replication fidelity than other commercially available enzymes, up to 16% of the PCR products generated following amplification of a 1-kb DNA target sequence may be mutated in comparison to only 2.6% for higher-fidelity enzymes such as *Pfu* polymerase (Stratagene, 2007). Thus, it is recommended that both the plus and minus strands of a PCR product be sequenced to verify base calls and that more than one clone for any single gene be sequenced to deduce a consensus for ambiguous base calls (Gyllensten and Allen, 1993; Lawrence et al., 1993). Alternatively, the PCR products may be generated with a high-fidelity enzyme that lacks the terminal transferase activity and then incubated for a short time postamplification with *Taq* and a pool of dATPs to allow addition of the 3' A to each end of the PCR products (Invitrogen Corporation, 2006).

Once the purified PCR products are inserted into pCR2.1-TOPO, the recombinant vectors are transformed into chemically competent *Escherichia coli* cells. The DNA insertion site on pCR2.1-TOPO is located within the *lacZ* gene (Fig. 5.4), which when expressed produces β-galactosidase, allowing cells to metabolize 5-bromo-4-chloro-3-indolyl-β-D-galactopyranoside (X-Gal), a substrate analog of lactose. Cleavage of X-Gal yields an insoluble blue product, so that growth of cells harboring a functional *lacZ* gene results in blue-pigmented colonies on agar media containing X-Gal. Ligation of PCR products into pCR2.1-TOPO disrupts *lacZ* gene expression, resulting in unpigmented, recombinant colonies that are unable to synthesize β-galactosidase. The pCR2.1-TOPO vector also contains a kanamycin resistance gene (Kn[r]) and an ampicillin resistance gene (Ap[r]), which are expressed by all transformed cells when plated on medium containing kanamycin or ampicillin, respectively. Therefore, following the transformation step, cells are plated onto agar medium containing X-Gal and either kanamycin or ampicillin, and recombinant colonies (which should appear white) are chosen for further analysis. For the "I, Microbiologist" project, kanamycin is used as a selective agent to avoid issues with the formation of satellite colonies typical of ampicillin-based selections. Although resistance to both drugs occurs via enzymatic deactivation, phosphorylation of kanamycin occurs inside the cell whereas β-lactamase is secreted outside the cell, inactivating

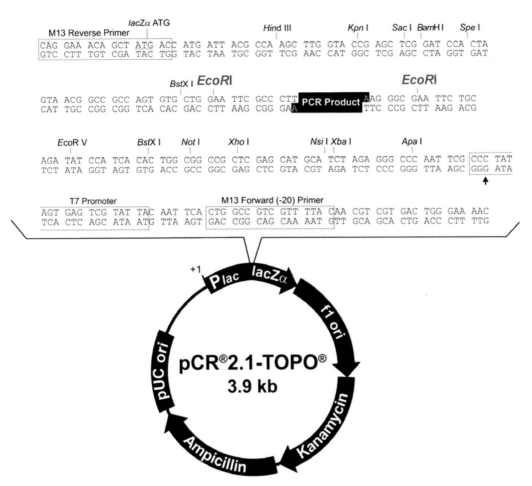

FIGURE 5.4 Plasmid map for pCR2.1-TOPO. Note that the insertion site is within the *lacZ* gene and there are two selectable markers on the plasmid, ampicillin and kanamycin resistance genes. Shown in exaggerated detail is the region within the *lacZ* gene in which the PCR product is inserted, flanked by sites where DNA sequencing primers anneal and restriction endonucleases cleave the DNA duplex. Note that the EcoRI sites (highlighted in larger red type) are bordering the insertion site for the PCR product (black box). Illustration from the *TOPO TA Cloning® User Manual* (Invitrogen Corporation, 2006).

ampicillin in the surrounding medium and facilitating the local growth of sensitive cells (Shaw et al., 1993; Smith and Baker, 2002).

Plasmids are extracted from recombinant colonies and purified in preparation for a secondary screen designed to confirm that metagenomic 16S rDNA was successfully cloned into pCR2.1-TOPO. Candidate plasmids are subjected to restriction analysis with an endonuclease that specifically cleaves two sites next to the region of the vector where the PCR product was inserted, such that cleavage by this enzyme results in release of the DNA insert from the vector fragment. The upper half of Fig. 5.4 shows an exaggerated map of pCR2.1-TOPO detailing the locations of a number of restriction sites bordering the insertion site for the PCR product. For "I, Microbiologist," the restriction endonuclease EcoRI will be used to excise the insert DNA from the vector. To screen for the presence of a DNA band of the expected size (~1.5 kb), the digested plasmids are subjected to agarose gel electrophoresis, alongside appropriate controls (for review of electrophoresis, see Section 3.1). Because there may be internal EcoRI restriction sites in addition to the external sites present in the vector sequence on both sides of the PCR product insert, plasmid digestion may produce a number of smaller bands whose individual sizes combine to 1.5 kb total (Fig. 5.5). Thus, each clone may be characterized based on the restriction pattern, or DNA fingerprint, generated by the EcoRI digest.

Once confirmed that the clones contain DNA inserts of a size consistent with that expected for 16S rDNA, they are sequenced (see Unit 3 for review) and a phylogenetic tree is constructed, using the same gene (16S rDNA) for all members of the bacterial community. Although the details of tree construction are not discussed until Unit 7, the end result of the community sampling approach is a phylogenetic "snapshot" of many members of the community (for example, see the evolutionary tree in Fig. 2.14). One exciting, additional outcome is the potential identification of novel community members—those that have no close relatives in existing DNA sequence databases (e.g., GenBank) or the literature. Moreover, since we now have a more representative picture of the total phylogenetic diversity present in a particular community, we can go on to devise culture methods that will capture a more diverse representation of the microbial community or specifically target lineages that may otherwise be difficult to cultivate; detection of the gene implies that the organism is actually present in the sample, hence justifying subsequent investment of time and resources in development of a proper cultivation strategy.

DGGE fingerprinting: an alternative community sampling strategy that bypasses cloning procedures

New technologies mentioned in Unit 3 facilitate sequencing of DNA directly from environmental samples, bypassing the creation of a library by traditional cloning steps (Hall, 2007; Medini et al., 2008; Schuster, 2008). These technologies, however, are quite expensive and not readily available to most instructional and many research laboratories. As one alternative, the process can be streamlined using a variation of polyacrylamide gel electrophoresis (described in Unit 3) called denaturing gradient gel electrophoresis (DGGE). These gels contain a gradient of a DNA denaturant such as urea or formamide that causes the gradual separation of a DNA duplex into single stranded DNA (Fig. 5.6). This technique permits resolution of a heterogeneous mix of DNA fragments of equivalent length based on differences in melting temperature (T_m) or thermal denaturation profile, producing a DGGE fingerprint of the mixture (Temmerman et al., 2003; Vanhoutte et al., 2005). As discussed in Unit 3, the melting properties are dependent upon the primary sequence, or nucleotide composition (largely by GC content), of the DNA fragment. Regions of a DNA fragment with low GC content melt at lower concentrations

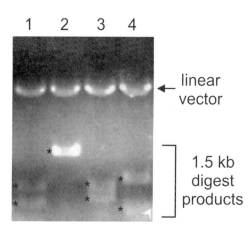

FIGURE 5.5 Representative results following restriction digestion of recombinant pCR2.1-TOPO clones. A 0.8% agarose 1× TAE gel was run for 30 minutes at 95 mA. Lanes 1 to 4 show complete EcoRI digests of four different plasmids containing 16S rDNA inserts. Note how some of the genes produce multiple insert bands (asterisks), suggesting the presence of one or two internal EcoRI sites. Gel picture courtesy of Brian Kirkpatrick and Areerat Hansanugrum (University of California Los Angeles).

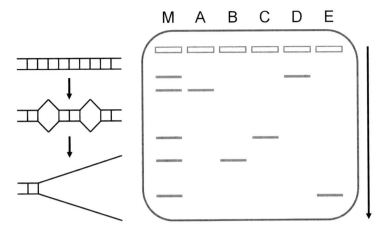

FIGURE 5.6 Principles behind DGGE. Lanes are labeled as follows: M, mix of five organisms (A to E); A, organism A only; B, organism B only; etc. Diagrams to the left of the gel picture depict the changes in DNA structure as molecules pass through the gel matrix. The vertical arrow to the right of the gel picture indicates the direction of electrophoresis. In the gel, the lowest concentration of denaturant is at the top and the highest concentration is at the bottom. Illustration by Erin Sanders.

of denaturant than do regions of high GC content. Melting of the DNA duplex causes a decrease in the mobility of the DNA fragment; therefore, DNA fragments with low GC content display a reduced mobility compared to fragments with higher GC content.

Figure 5.7 outlines how the DGGE procedure can be used to characterize microbial communities; comparing it to Fig. 5.1, one sees that the starting material and end product are the same. The first two steps of the procedure require extraction of metagenomic DNA from an environmental sample followed by PCR-based amplification of a target gene, in this example 16S rDNA. If one subjects the PCR products to standard agarose gel electrophoresis, as shown on the left side of the flow diagram in Fig. 5.7, then for each sample one obtains a single band with a migration reflecting its expected size (e.g., 1.5 kb for 16S rDNA). The band intensity may vary between samples if the samples were collected from different environments in which the total number of cells is not the same, different amounts of sample (e.g., soil) were used during the metagenomic DNA isolation step, there was variability in the efficiency of DNA extraction, or the PCR conditions were not optimal for all samples tested. However, because each band on an agarose gel represents a population of 16S rRNA genes of essentially equivalent length but variable primary-sequence composition, subjecting these same samples to DGGE facilitates resolution of the individual phylotypes according to their melting temperature. Thus, as shown on the right side of the flow diagram in Fig. 5.7, each band visualized after DGGE represents a population of genes of equivalent length and primary-sequence composition. The number of bands in each lane corresponds to the number of phylotypes in the environmental sample from which the metagenomic DNA was obtained, and the band intensity reflects the abundance of a particular phylotype relative to all others in the sample.

The DGGE results parallel those obtained with T-RFLP (Section 1.2), in that microbial communities are depicted as a pattern of bands, or DNA fingerprint, generated by two different types of sample processing. Microbial communities from different environmental sources can be characterized and compared based on the unique distribution (i.e., migration distance) and abundance (i.e., relative intensity) of bands on the gel. The advantage is that one does not need a fluorescently tagged primer and specialized detection equipment to perform DGGE fingerprinting; however, because the technique relies upon postelectrophoresis staining (e.g., ethidium bromide and SYBR Gold [Molecular Probes]) for band visualization, DGGE may be potentially less sensitive than T-RFLP, particularly with regard to phylotypes present at very low abundance. Like T-

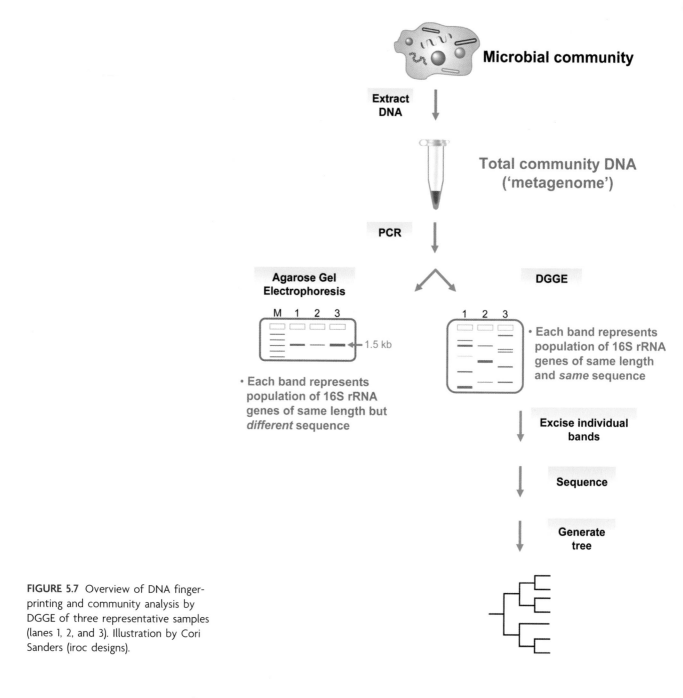

Microbial community

Extract DNA

Total community DNA ('metagenome')

PCR

Agarose Gel Electrophoresis

M 1 2 3

← 1.5 kb

• Each band represents population of 16S rRNA genes of same length but *different* sequence

DGGE

1 2 3

• Each band represents population of 16S rRNA genes of same length and *same* sequence

Excise individual bands

Sequence

Generate tree

FIGURE 5.7 Overview of DNA finger-printing and community analysis by DGGE of three representative samples (lanes 1, 2, and 3). Illustration by Cori Sanders (iroc designs).

RFLP, however, DGGE analysis offers clues to the complexity of a microbial community, providing a tool for screening different environments prior to initiating the sequencing phase of the project. As shown in Figure 5.7, the individual bands obtained following DGGE can be excised and extracted from the gel matrix and purified in preparation for DNA sequencing. The data generated then can be used to construct a phylogenetic tree, allowing identification of the phylotypes based on primary-sequence information.

Phylogenetic staining using fluorescent in situ hybridization

Fluorescent in situ hybridization (FISH) is a technology used widely in both microbial ecology and clinical microbiology (Amann et al., 1990, 1995, 1996). Upon microscopic examination of an environmental sample or a specimen from a patient, FISH allows rapid detection and identification of specific microorganisms, including pathogens in a clinical laboratory. This technique relies on the use of highly specific nucleic acid probes, which are DNA or RNA oligonucleotides complementary to signature sequences within the target gene or gene product (e.g., 16S rRNA). As regions that are unique to either a single organism or a group of organisms, signature sequences can be deduced from examination of full-length or near-full-length gene sequences. Thus, the DNA sequence information generated by metagenomic gene surveys can be used to monitor and further characterize complex microbial communities directly in a natural environment. Oligonucleotide probes designed to hybridize to signature sequences are tagged with fluorescent dyes and then allowed to penetrate cells, where they hybridize directly to the target gene or gene product (e.g., 16S rRNA phylogenetic stains anneal to rRNA directly on the ribosome). Following association of the two complementary strands of DNA or RNA, cells become uniformly fluorescent and can be viewed under a fluorescence microscope (Fig. 5.8). To phylogenetically characterize and quantify the abundance of microorganisms within natural habitats, environmental samples can be treated with multiple probes simultaneously. If each probe is unique to a particular microorganism and each probe is labeled with a different colored dye, an overall picture of community structure may be obtained, providing information about cell morphology and community member interactions and abundance. That the signature sequences to which the probes hybridize can be derived from microorganisms without having to culture them makes FISH a very powerful and useful technique in microbial ecology (DeLong et al., 1989).

16S rDNA libraries provide insights into the human gut microbiome

16S rDNA libraries reveal phylogenetic information about the diversity of microorganisms inhabiting any natural environment, including the human gastrointestinal tract, which implies that gene surveys can serve as tools to address health-related questions. One group of researchers recently tested this hypothesis by exploring how the normal body flora affects human health as it relates to obesity (Ley et al., 2006). Their work was an extension of another study comparing genes derived from bacterial communities in the gastrointestinal tracts of obese versus lean mice (Turnbaugh et al., 2006). These results

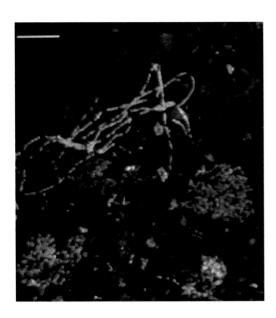

FIGURE 5.8 FISH analysis of a sewage sludge sample by confocal microscopy. The sample was treated with three probes, each tagged with a different dye (green, red, or purple) and each targeting a different bacterial lineage. Those cells containing signature sequences that reacted with only a single probe appear green, red, or purple, while those cells that reacted with two different probes appear yellow or blue. The yellow sizing bar is 10 μm. (Reprinted from Amann et al. [1996] with permission.)

demonstrated that between the two mouse populations, there was a difference in the abundance of two major bacterial divisions, the *Firmicutes* and *Bacteroidetes*. Furthermore, when microbial communities were harvested from both the obese and lean mice and transplanted into lean (germfree) mice, the mice that were given microbes from obese mice gained more fat than did those receiving microbes from lean mice. These data suggest that the significance of caloric intake from the food we eat may vary depending on the composition of the microbiome within the gut of an individual.

The human health study showed a correlation between body weight loss and the relative abundance of these two bacterial divisions (Ley et al., 2006). A 16S rRNA gene survey approach was used to monitor the composition of the human gut microbiomes for 12 obese individuals, who were randomly assigned to either a fat- or carbohydrate-restricted low-calorie diet, over the course of 1 year. With time, the relative abundance of *Firmicutes* decreased while the abundance of *Bacteroidetes* increased, irrespective of diet type (Fig. 5.9). The characterization of obesity as a pathological condition potentially associated with the ecological distribution of bacteria in the human gut is very interesting, suggesting realistic consideration of alternative approaches to the treatment and prevention of obesity.

Targeting alternative genes for library construction

Libraries with genes other than the 16S rRNA gene have been constructed and used for phylogenetic analysis as well as another type of FISH called in situ reverse transcription (ISRT). This technique permits the detection of genes expressed by cells in natural samples (Chen et al., 1997; also known as recognition of individual genes [RING] FISH [Zwirglmaier et al., 2004]). Gene surveys that target specific, highly conserved metabolic genes reveal information about microbial activities in various habitats. For example, detection of *nifH* genes, which encode the enzyme nitrogenase, suggests that the community contains microorganisms capable of nitrogen fixation (Valdés et al., 2005). One could imagine how ISRT FISH could be used to identify and enumerate bacterial cells actually expressing the genes required for this process. This approach has been used successfully to visualize chemoautotrophic nitrifying bacteria in aged Sargasso seawater samples expressing ribulose-bisphosphate carboxylase (RuBisCO), a key enzyme in the carbon fixation pathway (Sinigalliano et al., 2001). Thus, metabolic gene libraries provide

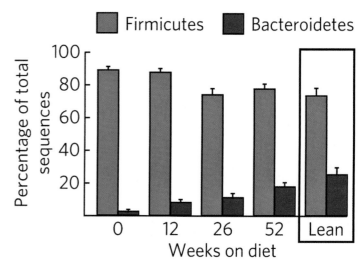

FIGURE 5.9 Correlation between body weight loss and human gut microbial ecology. Note the change in relative abundance of the two bacterial lineages, *Firmicutes* and *Bacteroidetes*, over time. For each time point, values for all available samples were averaged (*n* = 11 or 12 per time point). Lean subject controls include four samples taken 1 year apart from two lean people plus three additional samples representing healthy human subjects from the Eckburg et al. (2005) study. (Reprinted from Ley et al. [2006] with permission.)

some preliminary information about the functional role of microbes within the community.

Recall that there may be multiple (up to 15) copies of RNA operons *(rrn)* per bacterial genome, a trend that complicates T-RFLP and DGGE data by either underestimating or overestimating phylotype diversity and biases the construction of 16S rRNA gene libraries to favor detection of genes derived from rapidly growing bacteria (Klappenbach et al., 2000). Alternatively, housekeeping genes, which are involved in storage and processing of genetic information, are sometimes targeted instead of 16S rDNA because they are present in single copy within the genome. Examples include *rpoB*, which encodes the β subunit of RNA polymerase, and *recA*, which encodes an enzyme that facilitates genetic recombination (Venter et al., 2004; Dahllöf et al., 2000; Santos and Ochman, 2004). In addition to using housekeeping genes for phylogenetic analysis, these genes can serve as phylogenetic anchors, or markers, linking functional genes to phylotypes (Handelsman, 2004; Riesenfeld et al., 2004; Venter et al., 2004; Tyson et al., 2004). This latter application is necessary during analysis of fragmented genome data gathered from shotgun sequencing, when assembly of an entire genome may not feasible. Table 5.1 provides a list of markers that are considered reliable indicators of phylogenetic affiliation.

Environmental shotgun sequencing

Metagenomics allows scientists to access the genome of a community without relying on pure cultures of single species, thereby transcending the limitations of classical genomics and microbiology (Streit and Schmitz, 2004). However, one major disadvantage to the gene survey approach discussed above is that it provides no additional genetic information about bacteria from which 16S rDNA was detected, and typically only little functional information can be derived by using genes other than the 16S rRNA gene. To overcome these limitations, microbial ecologists began cloning random DNA fragments of variable size and gene composition into vectors for subsequent sequencing or functional analysis, a process called shotgun sequencing. The goal was to detect as many open reading frames (ORFs) as possible and then determine the phylogenetic scaffold to which they belong by using phylogenetic anchors such as the 16S rRNA gene as well as the genes listed in Table 5.1. The strategy mimicked that used in genomic studies, where the objective is to generate the complete and finished DNA sequence of the entire genome for a single microorganism. As shown in Fig. 5.10, although the two approaches ask different questions and produce distinct outcomes, metagenomic studies of the collective genome for a microbial community (i.e., the microbiome) employ the same shotgun sequencing techniques as are used in genomic studies of single microorganisms.

Overall, shotgun sequencing involves isolating DNA from a source, whether an environmental sample (in the case of metagenomic analysis) or a pure culture of a bacterial isolate (in the case of genomic analysis), followed by cloning into a suitable vector. Recall, however, that DNA sequencing technology is rapidly advancing, with new methodologies under development that eventually will render cloning into a vector before sequencing obsolete (Hall, 2007; Medini et al., 2008; Schuster, 2008). Currently, analysis of shotgun libraries relies upon capillary-based Sanger sequencing of cloned DNA fragments. Before being cloned, the DNA must be mechanically sheared or digested by restriction enzymes into smaller fragments, typically less than 10 kb, and then ligated into a standard sequencing vector (Streit and Schmitz, 2004). The recombinant plasmids are transformed into a host strain such as *Escherichia coli*, generating a library that must be screened by conventional methods such as those used in the "I, Microbiologist" project.

Table 5.1 Examples of markers used to link gene function to phylotype in *Bacteria*[a]

Gene	Protein	Function	PCR primers (5′ to 3′)	Reference(s)[b]
rpoB	RpoB	β subunit of RNA polymerase	1698F (AACATCGGTTTGATCAAC) 2041R (CGTTGCATGTTGGTACCCAT)	**Dahllöf et al. (2000)**, Santos and Ochman (2004)
recA	RecA	DNA repair protein	BDUP1 (CCCGAGTCCTCCggnaaracnac) BGDN2 (CGTTGCCGCCGgkngtnryytc)	**Santos and Ochman (2004)**, Venter et al. (2004)
atpAB		α and β subunits of ATPase	NA	Brown and Doolittle (1997), Paulsen et al. (2000), Ludwig et al. (1993)
fusA	EF-G	Elongation factor for protein synthesis	F (CATCGGCATCATGgcncayathga) R (CAGCATCGGCTGCaynccyttrtt)	**Santos and Ochman (2004)**, Brown and Doolittle (1997)
tufA	EF-Tu	Elongation factor for protein synthesis	F (CATYGGHCACGTBGACCA) R (TCNCCNGGCATNACCAT)	Brown and Doolittle (1997), **Ludwig et al. (1993)**
hsp60	Chaperonin-60	Heat shock protein	NA	Brown and Doolittle (1997)
rplB	L2	Large-subunit ribosomal protein	BDUP1 (CAAGGTGGAGCGCATCsantaygaycc) BHDN1 (GCCGCCGCCGwdnggrtgrtc)	**Santos and Ochman (2004)**, Brown and Doolittle (1997)
ileRS		tRNA synthetase for isoleucine	BCUP1 (GCCCGGCTGGgaywsncaygg) BKDN1 (TGGAGCCGGAGTCGawccanmmntc)	**Santos and Ochman (2004)**, Brown and Doolittle (1997), Woese et al. (2000)

[a]The list is not comprehensive. Primer sequences are provided if the gene is used in PCR-based surveys; for markers used in reports based on ESS or other cloning methodologies, no primer sequences are known to be available (NA).

[b]References from which the primer sequences are reported are in bold type. Lowercase letters denote degenerate nucleotides, with abbreviations that follow IUPAC ambiguity codes. Primers with UP or F in the name are forward primers, while those with DN or R in the name are reverse primers.

Plasmids are purified from recombinant clones and DNA inserts are sequenced using primers that anneal to sites on the vector that flank the inserted DNA fragment (Fig. 5.11). Computer-based methods are used to align the resulting sequences end-to-end in a process called genome assembly, which involves finding homologous regions on different DNA inserts and then assembling them into larger, contiguous, double-stranded DNA fragments (i.e., contigs). To facilitate the construction of a complete genome, PCR is employed to fill the gaps between contigs by generating primers that are complementary to sequences within DNA fragments that are now known and flank the gap. Computational analysis of the assembled genome results in the identification of ORFs and, when possible, preliminary assignment of gene functions; this process is referred to as annotation. Depending on the depth of coverage and overall complexity of the microbial community under study, it often is not possible to assemble complete genomes with metagenomic DNA sequencing data, so that individual contigs are annotated and screened for the presence of the 16S rRNA gene or other phylogenetic marker such that the putative function of the surrounding genes can be linked to a particular phylotype (Tringe et al., 2005). This method is unique in that it may reveal new genes or novel functions associated with known phylotypes.

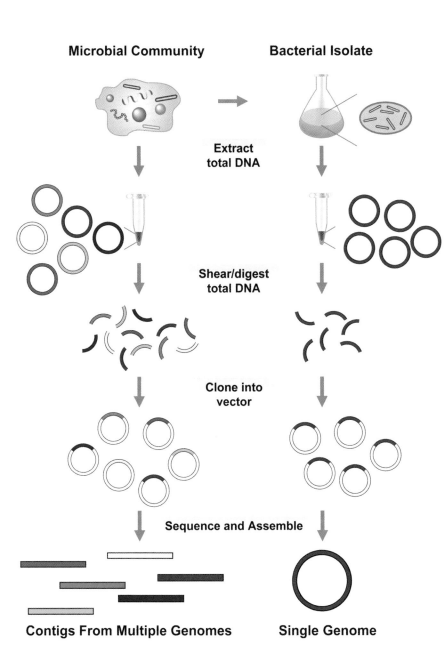

Microbial Community　　　**Bacterial Isolate**

Extract
total DNA

Shear/digest
total DNA

Clone into
vector

Sequence and Assemble

Contigs From Multiple Genomes　　　**Single Genome**

FIGURE 5.10 Overview of the shotgun sequencing approach comparing its use in metagenomic (left side) and genomic (right side) studies. The procedure involves extraction of total DNA from samples derived from either an environmental microbial community (microbiome) or a single, purified isolate. Individual genomes are represented as colored circles. In either case, the genomic DNA is fragmented by shearing techniques or restriction digestion and then cloned into a suitable vector. The plasmids are transformed into competent cells, which are screened; recombinant plasmids are then selected and purified for DNA sequence analysis (not shown). Following sequencing of the DNA inserts, the fragments are assembled into longer contigs and eventually into complete genomes if sequencing coverage is deep enough. Illustration by Cori Sanders (iroc designs).

Using environmental shotgun sequencing to investigate biodiversity in a complex microbial community, the Sargasso Sea

One of the most famous examples of gene discovery came from a large-scale shotgun sequencing project launched by Craig Venter and colleagues in 2003. This group sought to reconstruct the genomes of uncultured microorganisms in the Sargasso Sea, a sprawling but well-characterized ecosystem in the Atlantic Ocean near Bermuda (Venter et al., 2004). The depth of the Sargasso Sea exceeds 500 m. As an open-ocean environment, the Sargasso Sea exhibits seasonal oligotrophic (nutrient-limiting) conditions, with annual convective mixing that introduces nutrient-rich deep water into the upper 150- to 300-m depths, causing blooms of phytoplankton including diatoms, cyanobacteria, and dinoflagellates, in the spring (Fig. 5.12). This photosynthetic biomass is thought to play an integral role in global CO_2 exchange.

Between 170 and 200 liters of surface water was collected and filtered to ensure that the genomic DNA was microbial and not viral or eukaryotic. Metagenomic DNA libraries

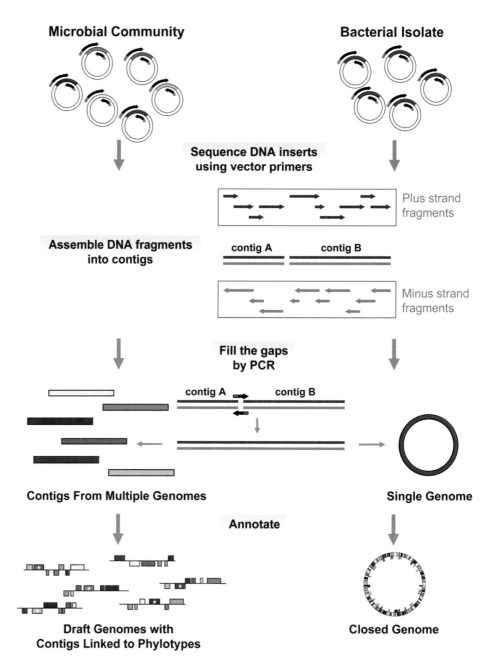

Microbial Community

Bacterial Isolate

Sequence DNA inserts
using vector primers

Plus strand
fragments

Assemble DNA fragments
into contigs

contig A   contig B

Minus strand
fragments

Fill the gaps
by PCR

contig A   contig B

Contigs From Multiple Genomes

Single Genome

Annotate

Draft Genomes with
Contigs Linked to Phylotypes

Closed Genome

FIGURE 5.11 Detailed overview of the sequencing and assembly processes (the last step in Fig. 5.10) followed by annotation of coding sequences. The DNA inserts are sequenced with primers that anneal to flanking vector sites. Computer programs assemble overlapping DNA fragments into contigs. In this example, the gaps between contigs A and B are closed by generating a forward primer that anneals to the 3′ end of the minus strand of contig A and a reverse primer that anneals to the 5′ end of the plus strand of contig B. Since most bacterial chromosomes are circular, this process usually allows completion of the entire, closed genome as shown. For metagenomic samples, it may not be possible to assemble multiple, closed genomes, so the project stops at the draft stage, leaving contigs representative of multiple genomes. Both draft and closed genomes can be annotated. The color of each ORF typically corresponds to a functional category to which the ORF has been assigned. For instance, red may indicate metabolic genes, yellow may indicate protein synthesis, blue may indicate cell division, etc. Asterisks within ORFs denote the phylogenetic anchor for that particular contig, linking the DNA fragment to a particular phylotype as denoted by the color corresponding to the original genome from which it was derived. If contigs are assembled into a circular diagram representing a chromosome, the outer circle represents genes encoded in the forward direction (i.e., plus strand) while the inner circle corresponds to genes encoded in the reverse direction (i.e., minus strand). For draft genomes displayed as linear diagrams, the upper line represents genes encoded by the plus strand while the bottom line corresponds to genes encoded by the minus strand. Illustration by Cori Sanders (iroc designs).

FIGURE 5.12 Image of water in the Sargasso Sea with lines of sargassum, or dense masses of seaweed (brown algae), floating in surface waters. Famous for ensnaring ships in nautical legends about the Bermuda Triangle, sargassum and the marine phytoplankton, which are phototrophic microorganisms in aquatic systems, absorb atmospheric CO_2 via photosynthesis and then convert it to organic carbon. From http://commons.wikimedia.org/wiki/Image:Lines__of__sargassum__Sargasso__Sea.jpg.

were constructed as described above with insert sizes ranging from 2 to 6 kb. ESS produced more than 1.9 million DNA sequence reads averaging about 800 bp in length. This monumental effort resulted in approximately 265 Mbp worth of DNA sequence data, which, when released to GenBank, increased the total number of microbial DNA sequences in the database by several orders of magnitude (Fig. 5.13).

Although far too many bacterial lineages were represented in the metagenomic DNA sample from the Sargasso Sea to produce closed genomes of individual community members, annotation of the contigs assembled after sequence analysis revealed novel linkages between phylogeny and microbial metabolic activities. For example, more than 782 new genes encoding rhodopsin-like photoreceptors were discovered within several bacterial lineages not previously known to contain light-energy-harvesting functions. Rhodopsins are proton pumps that facilitate energy production through nonchlorophyll pathways and at one time were thought to exist only in *Archaea*, not marine bacteria (Gartner and Losi, 2003; Béjà et al., 2000, 2001). Previous work by Béjà and colleagues confirmed the light-harvesting functions of rhodopsin-like photoreceptors found in bacteria (also called bacteriorhodopsin), demonstrating heterologous expression of the genes in *E. coli* (Man et al., 2003)(Fig. 5.14). The work by Venter et al. (2004) revealed that bacteriorhodopsin diversity was much greater than previously imagined, suggesting that bacterial prototrophy by this mechanism may be a globally significant oceanic microbial process.

Although we stand to benefit from the wealth of information provided by the Sargasso Sea study, one major drawback to using this type of large-scale sequencing approach to study complex microbial communities is the cost and time required to process so many data. In addition, the annotation procedure was completely automated, which could have led to an overestimation of the number of ORFs assigned to any particular contig. Manual curation of the data could address the latter concern, but the sheer number of data

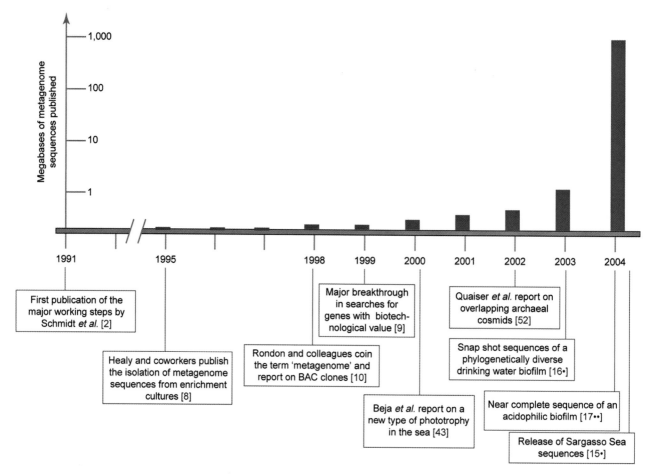

FIGURE 5.13 Timescale of published DNA sequences from 1991 to 2004, noting the impact of depositing metagenomic DNA sequences between 2003 and 2004. Image and references therein from Streit and Schmitz (2004), reprinted with permission.

under consideration presents a tremendous task. To overcome the current ESS roadblocks, development of new sequencing methods and bioinformatics tools geared specifically for analysis of metagenomic data are under way (Mardis, 2008; NIH News Release, 2004).

Using environmental shotgun sequencing to reconstruct the metagenome of a microbial community of low biodiversity: an acidophilic biofilm caused by acid mine drainage

While the Sargasso Sea was an example of a highly diverse, complex microbial community that was recalcitrant to efforts aimed to produce closed genomes, the bacterial community thriving in acid mine drainage (AMD) represents one of the most extreme environments on our planet and is dominated by only a few major bacterial and archaeal lineages (Tyson et al., 2004; Baker and Banfield, 2003; Allen and Banfield, 2005). As shown in Fig. 5.15, AMD microbial communities produce pink biofilms that hover on the water surface, which can have a pH as low as 0 and is high in toxic heavy metals such as iron (Fe), zinc (Zn), copper (Cu), and arsenic (As) as well as sulfide minerals including pyrite (FeS_2). The water beneath and immediately surrounding the biofilm has a temperature of approximately 42°C and an oxygen content of less than 21%, which is reduced from that in the atmosphere. Furthermore, the only source of carbon and ni-

FIGURE 5.14 Red Sea proteorhodopsin variants tuned to different wavelengths (from Man et al. [2003], published on the cover of *EMBO Journal*). Reprinted with permission from Oded Beja, Gazalah Sabehi, and Dikla Man (© 2003).

trogen for hardy survivors comes from the air. These microoxic, mesophilic, oligotrophic conditions reduce the range of microorganisms capable of growth in this environment. Comprising only a few major lineages as determined from a 16S rRNA gene survey, this low-diversity community became the focus of efforts by Jillian Banfield and colleagues to reconstruct entire genomes by using an ESS approach (Tyson et al., 2004).

Before this study, only one biofilm community member, the archaean *Ferroplasma acidarmanus*, had been cultured and its genome sequenced. Tyson and colleagues used ESS to sequence the genomes of the remaining uncultured (or unculturable) members,

FIGURE 5.15 Acid mine drainage in the Cheat River watershed in West Virginia, the result of polluted runoff from abandoned coal mines. The panels show two sites in need of further remediation. (Photographs courtesy of the Friends of the Cheat, a conservation organization formed to address the severe problems caused by AMD. For more information, students are encouraged to visit the website listed in Web Resources.)

which included *Leptospirillum* groups II and III as well as species of *Sulfobacillus,* "A-plasma," and "G-plasma" (Baker and Banfield, 2003; Tyson et al., 2004). Metagenomic DNA was extracted directly from samples of the biofilm and used to construct a DNA library with an average insert size of around 3 kb. ESS produced more than 100,000 DNA sequence reads, resulting in approximately 76 Mbp worth of DNA sequence data. Subsequent assembly and annotation produced nearly complete genomes for two community members and partial genomes for three other community members. As shown in Fig. 5.13, the sequence data from AMD was deposited in GenBank at about the same time as the Sargasso Sea data, making annotation of new DNA sequences laborious and further skewing genomic analysis (Riesenfeld et al., 2004). In other words, without an appropriate filter, database users are less likely to find query matches to sequences due to ecological similarities than to sequences from metagenomic studies due to the sheer abundance of the data.

Armed with the ESS sequence information, the authors next tried to determine which ecological functions were tied to each of the prokaryotic lineages (Tyson et al., 2004). They hypothesized that because the only source of nitrogen for the community was the atmosphere, at least one of the community members must contain genes required for nitrogen fixation (e.g., *nif* genes encoding nitrogenase). Furthermore, because nitrogen fixation would have to be an essential biochemical process for this community, the most abundant member of the community would most likely be responsible for this process. Annotation of the genome sequences revealed the presence of ORFs with similarity to the *nif* genes; however, the genes were not found in the genome of the most numerically dominant member of the community. Instead, the ORFs resided in the genome of *Leptospirillum* group III, which was in low abundance relative to other members in the community, signifying its role as a keystone species in the AMD ecosystem. The genome data for many of the other community members indicates the presence of ORFs encoding putative transporters for nitrogenous compounds such as NH_4^+ and NO_3^-, suggesting only one of many cooperative interactions between community members necessary for survival in this extreme environment.

Functional metagenomic analysis as a strategy for bioprospecting

Annotation and mining of metagenomic sequence data derived from ESS studies has led to the discovery of new genes that later may be heterologously expressed and functionally characterized. However, this approach demands that the genes of interest be identifiable exclusively through analysis of their primary sequence. An alternative, more direct strategy involves the development of functional assays for screening clones that exhibit a particular phenotype due to expression of gene products (Daniel, 2005; Handelsman, 2004). Referred to as functional metagenomic analysis, this strategy is especially suited for detection of secreted gene products including antibiotics and hydrolytic enzymes, to name only a few (Gillespie et al., 2002; MacNeil et al., 2001; Wang et al., 2000; and Healy et al., 1995). Genes that encode novel molecules or biocatalysts may have potential applications in the pharmaceutical or biotechnological industries (Streit and Schmitz, 2004; Allen et al., 2008). For instance, metagenome searches have uncovered enzymes involved in the biosynthesis of vitamin C and biotin, which may be useful in development of new, large-scale production processes (Streit and Entcheva, 2003; Eschenfeldt et al., 2001).

As outlined in Fig. 5.16, construction of DNA libraries proceeds in similar fashion to that described for sequence-based ESS above, with one major difference. Standard se-

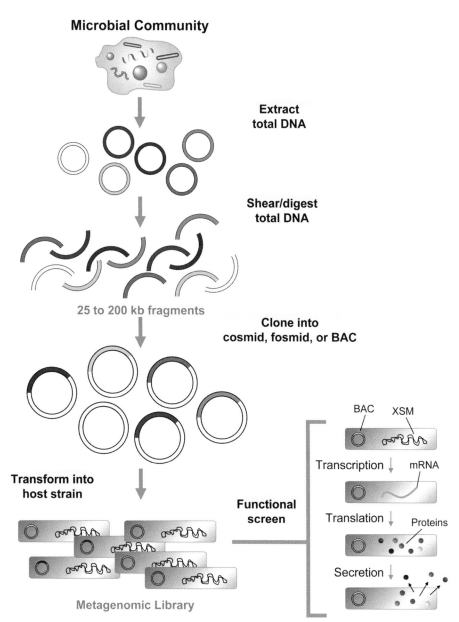

Microbial Community

Extract total DNA

Shear/digest total DNA

25 to 200 kb fragments

Clone into cosmid, fosmid, or BAC

Transform into host strain

Functional screen

Metagenomic Library

BAC   XSM

Transcription ↓ mRNA

Translation ↓ Proteins

Secretion ↓

FIGURE 5.16 Construction and functional screening of metagenomic DNA libraries. The procedure involves extraction of total DNA from environmental samples. Individual genomes are represented as colored circles. Genomic DNA is fragmented by mechanical shearing or restriction digestion and then cloned into vectors that accommodate large DNA fragments (e.g., BACs). The recombinant plasmids are transformed into an appropriate host strain, generating a library that is screened by a particular functional assay. Although the host strain, depicted as a grey rectangle, contains both its own chromosome (XSM) and the recombinant plasmid (BAC), the screen is designed specifically to assay for products encoded by a BAC. After transcription of genes residing on a BAC into mRNA, followed by translation of these mRNAs into proteins, the cell secretes a subset of the proteins, whose activity is monitored via indicator media on which the cells have been plated. Illustration by Cori Sanders (iroc designs).

quencing vectors accommodate only relatively small insert sizes (e.g., less than 10 kb) and therefore do not have the capacity to incorporate operons or large gene clusters, which may be required for heterologous expression of the metagenomic genes of interest (Streit and Schmitz, 2004). Thus, to get around this issue, large insert libraries are constructed using specialized vectors including cosmids for 25- to 35-kb DNA fragments, fosmids for up to 40-kb fragments, and bacterial artificial chromosomes (BACs) for up to 200-kb fragments. After transformation into a suitable host strain and plating on the desired indicator medium, colonies that exhibit specific traits or enzyme activities are selected. For example, if the goal is to find genes involved in the biosynthesis of novel antibiotics, screening transformed cells for their ability to develop into colonies surrounded by a zone of growth inhibition against a sensitized strain may lead to such discoveries.

The plasmids can be isolated from colonies exhibiting the desired phenotype and then sequenced and annotated to uncover the gene or genes responsible for that activity.

Because such large DNA inserts are obtained in each plasmid, the initial use of flanking vector primers provides sequence information for only the ends of the DNA insert. To determine the primary sequence of the entire DNA fragment, a process called primer walking may be employed, in which new primers are systematically generated that are complementary to the newly sequenced regions within the DNA insert (Fig. 5.17). One continues to design new pairwise forward and reverse primers with the latest set of sequence reads until the two produce a complementary region of overlap, indicating that the DNA sequence for the whole DNA fragment has been obtained. The individual sequence reads are aligned end to end in a process similar to what is done for genome assembly, resulting in the compilation of a single, contiguous DNA fragment that is subsequently annotated. If the contig contains a phylogenetic marker as well as metabolic genes of interest, function once again can be linked to phylotype. More commonly, however, a hierarchical clone-based shotgun sequencing approach is used because it is

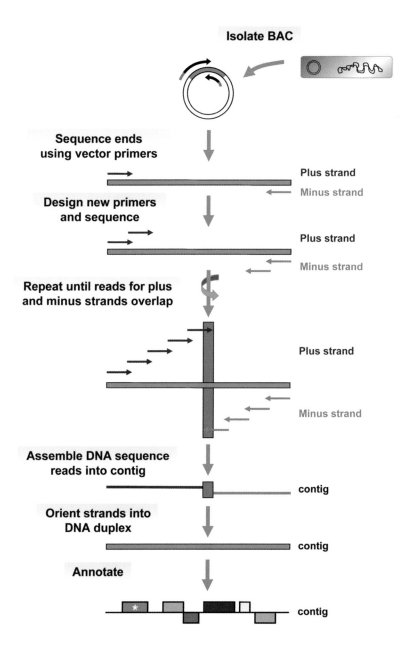

FIGURE 5.17 Overview of primer walking. After the vectors are purified from the host strain, the DNA insert initially is sequenced with primers that are complementary to the flanking vector sequences. New primers are designed that are complementary to the newly sequenced 5′ plus strand and 3′ minus strand. This process is repeated multiple times, such that new forward and reverse primers are generated based on the latest set of plus- and minus-strand sequence reads until the two produce a complementary region of overlap (highlighted by the grey bar). The individual sequence reads are assembled into a contig, which is a hybrid of plus- and minus-strand data at this point. One may continue to independently sequence both the plus and minus strands to the ends of the DNA insert. Alternatively, prior to annotation, the reverse complement of each strand may be examined for ORFs. Either way, both the sequenced DNA strands and their complements are represented in the contig. Note that in the above schematic of the annotated contig, the colors and orientation of ORFs are as described for Fig. 5.11. Illustration by Cori Sanders (iroc designs).

more cost-effective and less time-consuming (Fig. 5.18). With this methodology, individual BAC clones with inserts as large as 200 kb are isolated and then sheared into smaller fragments (<10 kb), which are cloned into a suitable sequencing vector. DNA fragments are sequenced and assembled into longer contigs, representing the original, full-length BAC insert.

There are a couple of practical obstacles preventing the efficient use of metagenomic functional searches. The frequency of detecting a metagenomic clone that displays the desired activity is low, such that several hundred thousand clones must be analyzed in a single screen to detect fewer than 10 active clones (Daniel, 2005; Handelsman, 2004; Streit and Schmitz, 2004). High-throughput screening technology, which relies on sophisticated and expensive "picking and pipetting" robots, is often required. The low frequency of active clone discovery may be attributed to use of inappropriate host expression strains. Consequently, there may be a lack of transcription of the metagenomic

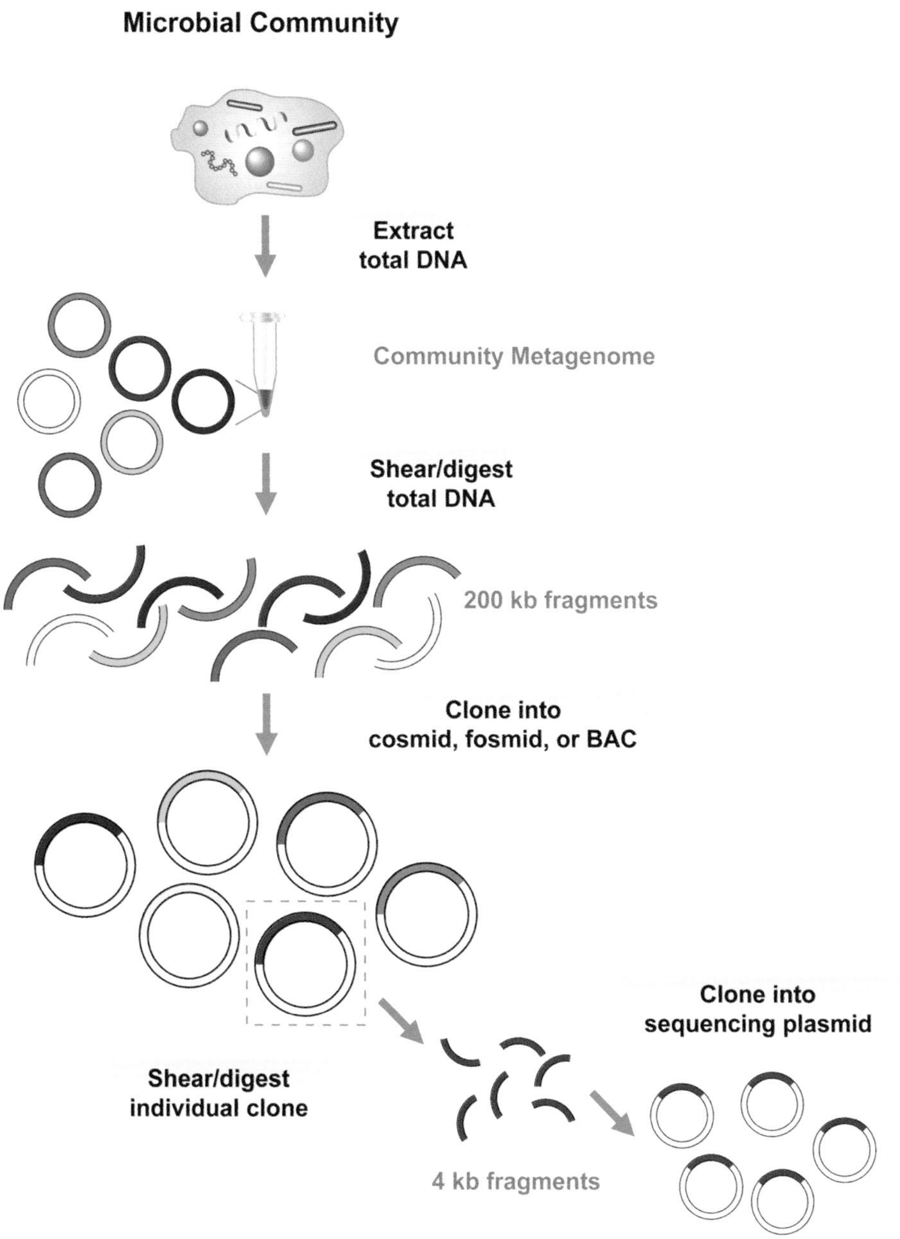

Microbial Community

Extract total DNA

Community Metagenome

Shear/digest total DNA

200 kb fragments

Clone into cosmid, fosmid, or BAC

Shear/digest individual clone

4 kb fragments

Clone into sequencing plasmid

FIGURE 5.18 Overview of hierarchical clone-based shotgun sequencing. Metagenomic DNA is extracted from the microbial community and then sheared into ~200-kb DNA fragments, which are subsequently cloned into a suitable vector such as a BAC. Individual clones are isolated and purified; they are then sheared into ~4-kb DNA fragments, which are ligated into a small-scale vector suitable for DNA sequencing. Illustration by Cori Sanders (iroc designs).

genes due to missing promoter or regulatory sequences and absence of the appropriate sigma factor. This effect could be exacerbated by minimal translation due to use of rare codons or misfolding of polypeptides due to absence of necessary chaperones or cofactors. Even if the protein is expressed, the host cell may recognize it as foreign and target the products for degradation. In addition, the host cell may lack the transport machinery necessary for secretion of the foreign protein, rendering its activity undetectable in functional assays. Efforts to find solutions to these expression problems are under way. For instance, since *E. coli* is the preferred host strain for ESS, strains are being engineered to express alternative sigma factors or rare codons. Microorganisms other than *E. coli* have been investigated for use as host strains. For example, *Streptomyces lividans* has served as a host strain in the search for genes involved in the biosynthesis of novel antibiotics and *Pseudomonas* strains have worked as host strains in the hunt for genes implicated in the production of other natural products (Wang et al., 2000; Courtois et al., 2003; Martinez et al., 2004).

Another approach to performing functional analyses of metagenomic DNA libraries involves the use of microarrays to monitor gene expression (Dennis et al., 2003; Sebat et al., 2003). The major impediment to the use of microarrays to profile environmental gene expression is the low sensitivity of this method compared to PCR, hindering the ability to detect genes expressed by microorganisms in low relative abundance within a community (Zhou and Thompson, 2002). In addition, unlike the highly conserved 16S rRNA gene targeted in gene surveys, there is a lack of sequence conservation among non-rRNA genes, making hybridization specificity a challenge in microarray analyses. Nonetheless, with the advent of high-throughput sequencing technology, the expressed genetic information (mRNA) of a microbial ecosystem (i.e., metatranscriptome) has become accessible, making it an appealing approach for microbial ecology studies (Gilbert et al., 2008).

KEY TERMS

Annotation Conversion of raw DNA sequence data into list of ORFs.

Bacterial artificial chromosome (BAC) A circular DNA construct derived from the *E. coli* fertility plasmid that accommodates DNA fragments of up to 200 kb.

Biofilm An assemblage of bacterial cells that adhere to a surface and to each other by secreting polysaccharides that enclose and protect the community, trapping nutrients required for growth.

Chimera A PCR artifact that generates a sequence product comprised of two or more phylogenetically distinct parent sequences; thought to occur when a prematurely terminated amplicon reanneals to a DNA strand belonging to a different organism during a subsequent PCR cycle and is copied to completion in the second cycle, albeit using a foreign DNA template for the remaining sequence; also called a chimeric sequence.

Clone A population of plasmids that arise from replication in bacterial cells propagating by binary cell division such that all descendents are genetically identical to original plasmid.

Confocal microscopy A type of fluorescent microscopy that generates a three-dimensional image of a sample.

Cosmids Phage lambda particles that package linear DNA fragments of 25 to 35 kb.

Depth of coverage The amount of sequencing that must be done to ensure that all unique regions of a given genome or microbiome are represented in the sequence reads. It depends upon the genome size and relative abundance of individual community members. The depth of coverage in unique regions should approximate a Poisson distribution, $P = e^{-m}$, where P is the probability that a base is not sequenced and m is the sequence coverage, which is equal to the number of bases sequenced divided by the genome size.

Fidelity In DNA replication, the accuracy of a copy with respect to the DNA template strand.

Fosmids Episomal, circular F-plasmids, present in single copy within a cell, that hold pieces of DNA up to 40 kb in length.

Genome assembly Arrangement of DNA fragments produced by shotgun sequencing into an order reflecting either a draft or complete genome suitable for subsequent annotation.

Heteroduplex molecules A PCR artifact in which two homologous DNA segments that differ by only a point mutation or a small insertion or deletion are amplified and then reanneal to each other rather than to their native partner strand.

Heterologous expression Transcription and translation of genes originating from a microorganism that is completely different from the host strain in which the gene now resides exogenously on a plasmid; expression of foreign (e.g., nonself) DNA.

Hierarchical clone-based shotgun sequencing A stepwise approach to sequencing large-insert libraries, in which large DNA fragments ($\geq$200 kb) are cloned into BAC vectors; subsequently, independent BAC clones are sheared into smaller DNA fragments ($\leq$4 kb), which are cloned into sequencing vectors; the small-insert libraries are sequenced and assembled to recreate the large DNA insert originating from a single BAC vector; the sequence information derived from multiple BAC vectors can be assembled into contigs and scaffolds, which later can be annotated and assigned to a phylotype.

Keystone species An organism which has low numerical representation within a community but plays a specialized role in the environment that is essential to community function and survival.

Mesophilic In the middle range of temperature conditions; conducive to growth of mesophiles, which thrive in temperatures ranging from about 20 to 45°C.

Microbiome The genomes of all microbial community members living within a particular environment (e.g., soil microbiome, human gut microbiome); also called metagenomic DNA.

Microoxic Having oxygen present at levels reduced from that in the atmosphere, where full oxygen tension is 21% O_2. A microoxic environment is conducive to growth of microaerophiles, which are aerobes that require oxygen for respiration but can use it only when its proportion is lower than that found in air.

Oligotrophic environment An environment in which nutrient levels are low, limiting growth and diversification of organisms.

Open reading frames (ORFs) Sequences that potentially encode proteins, in that they have start and stop codons separated by at least 800 to 900 bp; further annotation and functional analysis is required to verify that the ORF actually represents a gene.

Phylotype A group of microorganisms with 16S rRNA genes that are very similar to one another (either >97% or >99% identity as typical cutoff, corresponding to definitions used for genus or species level designations).

Primer walking Progressive sequencing along the length of a DNA fragment, using primers designed from the latest sequence read; necessary for sequencing DNA fragments longer than a single sequence read, which typically is only 800 to 1,000 bp.

Satellite colony A bacterial colony that grows in the immediate vicinity of a second bacterial colony owing to the expression of an antibiotic resistance gene by the latter group of cells; such colonies arise because the antibiotic in the medium has been broken down or deactivated in the area adjacent to the colony comprised of cells expressing antibiotic-resistance genes whose products are secreted into the medium.

Shotgun sequencing Sequencing of cloned DNA derived from genome that was randomly sheared into fragments of variable size; computational methods are subsequently used to assemble the DNA fragments in order, resulting in the reconstruction of an entire genome.

Signature sequence A short nucleotide sequence that is unique to a certain group of microorganisms.

REFERENCES

Acinas, S. G., R. Sarma-Rupavtarm, V. Klepac-Caraj, and M. F. Polz. 2005. PCR-induced sequence artifacts and bias: insights from comparison of two 16S rRNA clone libraries constructed from the same sample. *Appl. Environ. Microbiol.* **71:**8966–8969.

Allen, E. E., and J. F. Banfield. 2005. Community genomics in microbial ecology and evolution. *Nat. Rev.* **3:**489–498.

Allen, H. K., L. A. Moe, J. Rodbumrer, A. Gaarder, and J. Handelsman. 2008. Functional metagenomics reveals diverse β-lactamases in a remote Alaskan soil. *ISME J.* **3:**243–251.

Amann, R., J. Snaidr, M. Wagner, W. Ludwig, and K.-H. Schleifer. 1996. In situ visualization of high genetic diversity in a natural microbial community. *J. Bacteriol.* **178:**3496–3500.

Amann, R. I., L. Krumholz, and D. A. Stahl. 1990. Fluorescent-oligonucleotide probing of whole cells for determinative, phylogenetic, and environmental studies in microbiology. *J. Bacteriol.* **172:**762–770.

Amann, R. I., W. Ludwig, and K.-H. Schleifer. 1995. Phylogenetic identification and in situ detection of individual microbial cells without cultivation. *Microbiol. Rev.* **59:**143–169.

Baker, B. J., and J. F. Banfield. 2003. Microbial communities in acid mine drainage. *FEMS Microbiol. Ecol.* **44:**139–152.

Béjà, O., L. Aravind, E. V. Koonin, M. T. Suzuki, A. Hadd, L. P. Nguyen, S. B. Jovanovich, C. M. Gates, R. A. Feldman, J. L. Spudich, E. N. Spudich, and E. F. DeLong. 2000. Bacterial rhodopsin: evidence for a new type of phototrophy in the sea. *Science* **289:**1902–1906.

Béjà, O., E. N. Spudich, J. L. Spudich, M. Leclerc, and E. F. DeLong. 2001. Proteorhodopsin phototrophy in the ocean. *Nature* **411:**786–789.

Brown, J. R., and W. F. Doolittle. 1997. *Archaea* and the prokaryote-to-eukaryote transition. *Microbiol. Mol. Biol. Rev.* **61:**456–502.

Chen, F., J. M. González, W. A. Dustman, M. A. Moran, and R. E. Hodson. 1997. In situ reverse transcription, an approach to characterize diversity and activities of prokaryotes. *Appl. Environ. Microbiol.* **63:**4907–4913.

Courtois, S., C. M. Cappellano, M. Ball, F. X. Francou, P. Normand, G. Helynck, A. Martinez, S. J. Kolvek, J. Hopke, M. S. Osburne, P. R. August, R. Nalin, M. Guérineau, P. Jeannin, P. Simonet, and J. L. Pernodet. 2003. Recombinant environmental libraries provide access to microbial diversity for drug discovery from natural products. *Appl. Environ. Microbiol.* **69:**49–55.

Dahllöf, I., H. Baillie, and S. Kjelleberg. 2000. *rpoB*-based microbial community analysis avoids limitations inherent in 16S rRNA gene intraspecies heterogeneity. *Appl. Environ. Microbiol.* **66:**3376–3380.

Daniel, R. 2005. The metagenomics of soil. *Nat. Rev.* **3:**470–478.

DeLong, E. F., G. S. Wickham, and N. R. Pace. 1989. Phylogenetic stains: ribosomal RNA-based probes for identification of single cells. *Science* **243:**1360–1363.

Dennis, P., E. A. Edwards, S. N. Liss, and R. Fulthorpe. 2003. Monitoring gene expression in mixed microbial communities by using DNA microarrays. *Appl. Environ. Microbiol.* **69:**769–778.

Eisen, J. A. 2007. Environmental shotgun sequencing: its potential and challenges for studying the hidden world of microbes. *PLoS Biol.* **5:**e82. doi:10.1371/journal.pbio.0050082.

Eschenfeldt, W. H., L. Stols, H. Rosenbaum, Z. S. Khambatta, E. Quaite-Randall, S. Wu, D. C. Kilgore, J. D. Trent, and M. I. Donnelly. 2001. DNA from uncultured organisms as a source of 2,5-diketo-D-gluconic acid reductases. *Appl. Environ. Microbiol.* **67:**4206–4214.

Fortin, N., D. Beaumier, K. Lee, and C. W. Greer. 2004. Soil washing improves the recovery of total community DNA from polluted and high organic content sediments. *J. Microbiol. Methods* **56:**181–191.

Gartner, W., and A. Losi. 2003. Crossing the borders: archaeal rhodopsins go bacterial. *Trends Microbiol.* **11:**405–407.

Gilbert, J. A., D. Field, Y. Huang, R. Edwards, W. Li, P. Gilna, and I. Joint. 2008. Detection of large numbers of novel sequences in the metatranscriptomes of complex marine microbial communities. *PLoS ONE* **3:**e3042. doi:10.1371/journal.pone.0003042.

Gillespie, D. E., S. F. Brady, A. D. Bettermann, N. P. Cianciotto, M. R. Liles, M. R. Rondon, J. Clardy, R. M. Goodman, and J. Handelsman. 2002. Isolation of antibiotics turbomycin A and B from a metagenomic library of soil microbial DNA. *Appl. Environ. Microbiol.* **68:**4301–4306.

Gyllensten, U. B., and M. Allen. 1993. Sequencing of *in vitro* amplified DNA. *Methods Enzymol.* **218:**3–16.

Hall, N. 2007. Advanced sequencing technologies and their wider impact in microbiology. *J. Exp. Biol.* **290:**1518–1525.

Handelsman, J. 2004. Metagenomics: application of genomics to uncultured microorganisms. *Microbiol. Mol. Biol. Rev.* **68:**669–685.

Healy, F. G., R. M. Ray, H. C. Aldrich, A. C. Wilkie, L. O. Ingram, and K. T. Shanmugam. 1995. Direct isolation of functional genes encoding cellulases from microbial consortia in a thermophilic, anaerobic digester maintained on lignocellulose. *Appl. Microbiol. Biotechnol.* **43:**667–674.

Invitrogen Corporation. 2006. *TOPO TA Cloning® User Manual, Version U, 25-0184.* Invitrogen Corp., Carlsbad, CA.

Klappenbach, J. A., J. M. Dunbar, and T. M. Schmidt. 2000. rRNA operon copy number reflects ecological strategies of bacteria. *Appl. Environ. Microbiol.* **66:**1328–1333.

Lakay, F. M., A. Botha, and B. A. Prior. 2007. Comparative analysis of environmental DNA extraction and purification methods from different humic acid-rich soils. *J. Appl. Microbiol.* **102:**265–273.

Lawrence, J. G., D. L. Hartl, and H. Ochman. 1993. Sequencing products of polymerase chain reaction. *Methods Enzymol.* **218:**26–35.

Lawyer, F. C., S. Stoffel, R. K. Saiki, S.-Y. Chang, P. A. Landre, R. D. Abrarnson, and D. H. Gelfand. 1993. High-level expression, purification, and enzymatic characterization of full-length *Thermus aquaticus* DNA polymerase and a truncated form deficient in 5′ to 3′ exonuclease activity. *Genome Res.* **2:**275–287.

Ley, R. E., P. J. Turnbaugh, S. Klein, and J. I. Gordon. 2006. Human gut microbes associated with obesity. *Nature* **444:**1022–1023.

Ludwig, W., J. Neumaier, N. Klugbauer, E. Brockmann, C. Roller, S. Jilg, K. Reetz, I. Schachtner, A. Ludvigsen, M. Bachleitner, U. Fischer, and K. H. Schleifer. 1993. Phylogenetic relationships of *Bacteria* based on comparative sequence analysis of elongation factor Tu and ATP-synthase β-subunit genes. *Antonie Leeuwenhoek* **64:**285–305.

MacNeil, I. A., C. L. Tiong, C. Minor, P. R. August, T. H. Grossman, K. A. Loiacono, B. A. Lynch, T. Phillips, S. Narula, R. Sundaramoorthi, A. Tyler, T. Aldredge, H. Long, M. Gilman, D. Holt, and S. Osburne. 2001. Expression and isolation of antimicrobial small molecules from soil DNA libraries. *J. Mol. Microbiol. Biotechnol.* **3:**301–308.

Madigan, M. T., and J. M. Martinko. 2006. *Brock Biology of Microorganisms,* 11th ed. Pearson Prentice Hall, Pearson Education, Inc., Upper Saddle River, NJ.

Man, D., W. Wang, G. Sabehi, L. Aravind, A. F. Post, R. Massana, E. N. Spudich, J. L. Spudich, and O. Béjà. 2003. Diversification and spectral tuning in marine proteorhodopsins. *EMBO J.* **22:**1725–1731.

Mardis, E. R. 2008. The impact of next-generation sequencing technology on genetics. *Trends Genet.* **24:**133–141.

Martinez, A., S. J. Kolvek, C. L. T. Yip, J. Hopke, K. A. Brown, I. A. MacNeil, and M. S. Osburne. 2004. Genetically modified bacterial strains and novel bacterial artificial chromosome shuttle vectors for constructing environmental libraries and detecting heterologous natural products in multiple expression hosts. *Appl. Environ. Microbiol.* **70:**2452–2463.

Martin-Laurent, F., L. Philippot, S. Hallet, R. Chaussod, J. C. Germon, G. Soulas, and G. Catroux. 2001. DNA extraction from soils: old bias for new microbial diversity analysis methods. *Appl. Environ. Microbiol.* **67:**2354–2359.

Medini, D., D. Serruto, J. Parkhill, D. A. Relman, C. Donati, R. Moxon, S. Falkow, and R. Rappuoli. 2008. Microbiology in the post-genomic era. *Nat. Rev. Microbiol.* **6:**419–430.

National Institutes of Health. 2004. NIH News Release. National Human Genomic Research Initiative (NHGRI) seeks next generation of sequencing technologies. http://www.genome.gov/12513210.

National Research Council. 2007. *The New Science of Metageno-mics: Revealing the Secrets of Our Microbial Planet.* Committee on Metagenomics: Challenges and Functional Applications, National Academies Press, Washington, DC.

Pace, N. R. 1997. A molecular view of microbial diversity and the biosphere. *Science* **276**:734–740.

Paulsen, I. T., L. Nguyen, M. K. Sliwinski, R. Rabus, and M. H. Saier, Jr. 2000. Microbial genome analyses: comparative transport capabilities in eighteen prokaryotes. *J. Mol. Biol.* **301**:75–100.

Riesenfeld, C. S., P. D. Schloss, and J. Handelsman. 2004. Meta-genomics: genomic analysis of microbial communities. *Annu. Rev. Genet.* **38**:525–552.

Santos, S. C., and H. Ochman. 2004. Identification and phyloge-netic sorting of bacterial lineages with universally conserved genes and proteins. *Environ. Microbiol.* **6**:754–759.

Schuster, S. C. 2008. Next-generation sequencing transforms to-day's biology. *Nat. Methods* **5**:16–18.

Sebat, J. L., F. S. Colwell, and R. L. Crawford. 2003. Metagenomic profiling: microarray analysis of an environmental genomic li-brary. *Appl. Environ. Microbiol.* **69**:4927–4934.

Shaw, K. J., P. H. Rather, R. S. Hare, and G. H. Miller. 1993. Molecular genetics of aminoglycoside resistance genes and familial relationships of the aminoglycoside-modifying enzymes. *Microbiol. Rev.* **57**:138–163.

Shuman, S. 1991. Recombination mediated by vaccinia virus DNA topoisomerase I in *Escherichia coli* is sequence specific. *Proc. Natl. Acad. Sci. USA* **88**:10104–10108.

Shuman, S. 1994. Novel approach to molecular cloning and poly-nucleotide synthesis using vaccinia DNA topoisomerase. *J. Biol. Chem.* **269**:32678–32684.

Sinigalliano, C. D., D. N. Kuhn, R. D. Jones, and M. A. Guerrero. 2001. In situ reverse transcription to detect the *cbbL* gene and visualize RuBisCO in chemoautotrophic nitrifying bacteria. *Lett. Appl. Microbiol.* **32**:388–393.

Smith, C. A., and E. N. Baker. 2002. Aminoglycoside antibiotic resistance by enzymatic deactivation. *Curr. Drug Targets Infect. Dis-orders* **2**:143–160.

Stratagene. 2007. *Pfu DNA polymerase,* revision B. *Stratagene In-struction Manual.* www.stratagene.com.

Streit, W. R., and P. Entcheva. 2003. Biotin in microbes, the genes involved in its biosynthesis, its biochemical role and perspectives for biotechnological production. *Appl. Microbiol. Biotechnol.* **61**: 21–31.

Streit, W. R., and R. A. Schmitz. 2004. Metagenomics—the key to the uncultured microbes. *Curr. Opin. Microbiol.* **7**:492–498.

Temmerman, R., I. Scheirlinck, G. Huys, and J. Swings. 2003. Culture-independent analysis of probiotic products by denaturing gradient gel electrophoresis. *Appl. Environ. Microbiol.* **69**:220–226.

Tindall, K. R., and T. A. Kunkel. 1988. Fidelity of DNA synthesis by the *Thermus aquaticus* DNA polymerase. *Biochemistry* **27**:6008–6013.

Tringe, S. G., C. Von Mering, A. Kobayashi, A. A. Salamov, K. Chen, H. W. Chang, M. Podar, J. M. Short, E. J. Mathur, J. C. Detter, P. Bork, P. Hugenholtz, and E. M. Rubin. 2005. Com-parative metagenomics of microbial communities. *Science* **308**: 554–557.

Turnbaugh, P. J., R. E. Ley, M. A. Mahowald, V. Magrini, E. R. Mardis, and J. I. Gordon. 2006. An obesity-associated gut micro-biome with increased capacity for energy harvest. *Nature* **444**: 1027–1031.

Tyson, G. W., J. Chapman, P. Hugenholtz, E. E. Allen, R. J. Ram, P. M. Richardson, V. V. Solovyev, E. E. Rubin, D. S. Rokhsar, and J. F. Banfield. 2004. Community structure and metabolism through reconstruction of microbial genomes from the environ-ment. *Nature* **428**:37–43.

Valdés, M., N. O. Pérez, P. Estrada-de los Santos, J. Caballero-Mellado, J. J. Peña-Cabriales, P. Normand, and A. M. Hirsch. 2005. Non-*Frankia* actinomycetes isolated from surface-sterilized roots of *Casuarina equisetifolia* fix nitrogen. *Appl. Environ. Micro-biol.* **71**:460–466.

Vanhoutte, L. T., G. Huys, and I. S. Cranenbrouck. 2005. Explor-ing microbial ecosystems with denaturing gradient gel electro-phoresis (DGGE). *BCCM Newsl.* **17**(2). http://bccm.belspo.be/newsletter/17-05/bccm02.htm.

Venter, J. C., K. Remington, J. F. Heidelberg, A. L. Halpern, D. Rusch, J. A. Eisen, D. Wu, I. Paulsen, K. E. Nelson, W. Nelson, D. E. Fouts, S. Levy, A. H. Knap, M. W. Lomas, K. Nealson, O. White, J. Peterson, J. Hoffman, R. Parsons, H. Baden-Tillson, C. Pfannkoch, Y. Rogers, and H. O. Smith. 2004. Environmental ge-nome shotgun sequencing of the Sargasso Sea. *Science* **304**:66–74.

Wang, G. Y., E. Graziani, B. Waters, W. Pan, X. Li, J. McDermott, G. Meurer, G. Saxena, R. J. Andersen, and J. Davies. 2000. Novel natural products from soil DNA libraries in a streptomycete host. *Org. Lett.* **2**:2401–2404.

Woese, C. R., O. Kandler, and M. L. Wheelis. 1990. Towards a natural system of organisms: proposal for the domains *Archaea, Bacteria,* and *Eucarya. Proc. Natl. Acad. Sci. USA* **87**:4576–4579.

Woese, C. R., G. J. Olsen, J. Ibba, and D. Soll. 2000. Aminoacyl-tRNA synthetases, the genetic code, and the evolutionary process. *Microbiol. Mol. Biol. Rev.* **64**:202–236.

Zhou, J., and D. K. Thompson. 2002. Challenges in applying mi-croarrays to environmental studies. *Curr. Opin. Biotechnol.* **13**: 204–207.

Zhou, M.-Y., and C. E. Gomez-Sanchez. 2000. Universal TA clon-ing. *Curr. Issues Mol. Biol.* **2**:1–7.

Zwirglmaier, K., W. Ludwig, and K. H. Schleifer. 2004. Recogni-tion of individual genes in a single bacterial cell by fluorescence in situ hybridization—RING FISH. *Mol. Microbiol.* **51**:89–96.

Web Resource

Friends of the Cheat http://www.cheat.org (see gallery for more photographs of AMD)

ANALYSIS OF THE PRIMARY LITERATURE: THE HUMAN GUT MICROBIOME

READING ASSIGNMENT
Gill, S. R., M. Pop, R. T. DeBoy, P. B. Eckburg, P. J. Turnbaugh, B. S. Samuel, J. I. Gordon, D. A. Relman, C. M. Fraser-Liggett, and K. E. Nelson. 2006. Metagenomic analysis of the human distal gut microbiome. *Science* **312:**1355–1359.

Human body surfaces contain at least 10-fold more microorganisms than do somatic or germ cells, with the majority inhabiting the gastrointestinal tract (Fig. 5.19) (Gill et al., 2006; Bäckhed et al., 2005; Madigan and Martinko, 2006). Between 10 trillion and 100 trillion microorganisms (10^{13} to 10^{14} cells) are estimated to inhabit this strictly anaerobic niche, with the majority being found in the distal gut, which is made up of the ileum and the colon. The study by Gill and colleagues provides a paradigm for how the data generated from DNA sequences encoding 16S rRNA genes (16S rDNA) and ESS libraries can be applied to address ecological questions related to human health (Gill et al., 2006). Specifically, the authors performed a metagenomic analysis of the human distal gut microbiome in an effort to evaluate phylotype diversity, assess gene content, and understand the assortment of microbial metabolic functions present within the community.

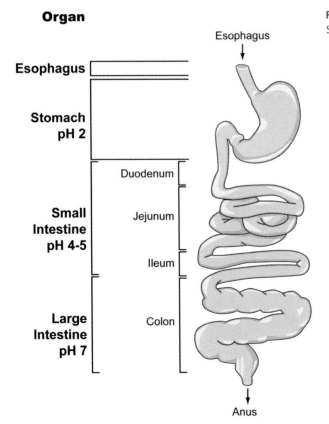

Organ

Esophagus

Esophagus

Stomach pH 2

Duodenum

Small Intestine pH 4-5

Jejunum

Ileum

Large Intestine pH 7

Colon

Anus

FIGURE 5.19 Human gastrointestinal tract. Illustration by Cori Sanders (iroc designs).

Role of the composite metagenome of the distal gut in human health

Before the study by Gill et al. (2006), authors from the same laboratory used a 16S rDNA community sampling approach to establish that the human distal gut is dominated by two bacterial divisions, the *Bacteroidetes* and the *Firmicutes,* as well as by one prominent methanogenic archaean, *Methanobrevibacter smithii* (Eckburg et al., 2005) (Fig. 5.20). Many of the sequences represented bacterial phylotypes that either had not yet been cultivated or were completely novel. The analysis was based on data from three healthy adult individuals, with samples derived from mucosal tissue, obtained from several sites within the colon during a colonoscopy, and feces, collected 1 month after the colonoscopy. As depicted in Fig. 5.20, there was obvious variation in microbial-community composition among the three subjects, in terms of phylotype distribution and evenness. For example, upon examining the population of *Bacteroidetes* for subject A, which is circled in Fig. 5.20, one notices that only three major phylotypes are represented in six of the seven sampling sites. Furthermore, the diversity of the *Bacteroidetes* population as found in subject A was much lower than that found in subjects B and C, although there are clear differences in both distribution and abundance between subjects B and C as well. It is exciting to imagine how a simplified DNA fingerprint of the complex microbial communities inhabiting human body surfaces could enable us to discern patterns and irregularities that ultimately may be correlated to human health or disease progression. For example, one may wonder if the differences in gender, genotype, age, diet, or lifestyle for the three individuals in the study could explain the dissimilarity in community structure. A recent study has demonstrated that bacterial diversity in the mammalian gut is influenced by host diet as well as the phylogeny of the community itself (Ley et al., 2008). Specifically, it would appear that bacterial communities actually coevolve with their host, suggesting vertical transmission of gut microbiota from parent to offspring in a number of different mammalian species. Preceding the work of Eckburg and colleagues, studies of mouse models and humans suggest a mutualistic relationship between mammals and their associated gut microbiota, where it appears that community members influence physiological processes such as maturation of the immune system (Mazmanian et al., 2005), control of intestinal epithelial cell homeostasis and response to epithelial cell injury (Rakoff-Nahoum et al., 2004), regulation of dietary energy harvest and fat storage (Bäckhed et al., 2004, 2005), and metabolizing xenobiotics (Nicholson et al., 2005). One might hypothesize that disruptions in microbial-community composition as a result of changes in diet or following administration of antibiotics could contribute to variations in human physiology.

Recall, as discussed in Section 5.1, that it was not long after the publication of the work by Eckburg and colleagues that two studies from the laboratory of Jeffery Gordon (Washington University) found correlations between obesity and the ecological distribution of microbes in the mammalian distal gut (Ley et al., 2006; Turnbaugh et al., 2006). One also might hypothesize that changes in the relative distribution of community members may predispose individuals to other gastrointestinal diseases such as ulcerative colitis and Crohn's disease. It stands to reason that a comprehensive understanding of the human microbiome is in order before being able to explain such complex and seemingly intractable diseases (Eckburg and Relman, 2007).

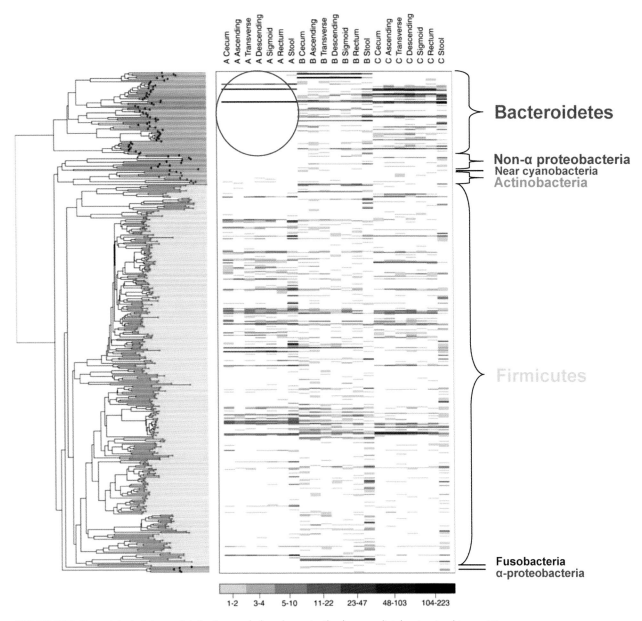

FIGURE 5.20 Bacterial phylotype distribution and abundance in the human distal gut microbiome. Displayed in the left panel is a neighbor-joining phylogenetic tree containing one representative sequence for each of the 395 bacterial phylotypes revealed in the Eckburg et al. (2005) study. Major phyla are color coded top to bottom *(Bacteroidetes, non-Alphaproteobacteria, unclassified near-Cyanobacteria, Actinobacteria, Firmicutes, Fusobacteria, and Alphaproteobacteria)*. The right panel describes the relative community composition for each subject (A, B, and C), including the anatomical site from which the sample was obtained (six major subdivisions of the human colon [cecum, ascending colon, transverse colon, descending colon, sigmoid colon, and rectum] and one fecal stool). The band in each row below the subject ID represents the presence of the phylotype corresponding to the same position on the phylogenetic tree, with the band intensity corresponding to the abundance of that particular phylotype. The number of individuals per phylotype was determined, and values are reported using the grey scale at the bottom. (Reprinted from Eckburg et al. [2005] with permission.)

Summary of the experimental strategy used to study the human distal gut microbiome

Although one may examine community composition with 16S rRNA gene surveys, the functional contributions of each phylotype to the community remain unknown since the information obtained from the studies is limited. Biotransformations that humans are not equipped to perform on their own will probably be the most informative, since either their loss or excess production stands to be most detrimental to the mammalian recipient of the purported mutualism. Thus, the authors of the Gill et al. (2006) study set out to characterize the human distal gut microbiome by using parallel community sampling and ESS approaches, not only hypothesizing that the analysis should reveal genetic diversity representing the three dominant microbial lineages (*Bacteroidetes, Firmicutes,* and *M. smithii*) consistently found in previous community sampling studies, but also predicting that genes encoding metabolic activities not present in the human genome will be identified. This combinatorial approach was used to circumvent the biases potentially introduced by the sole use of PCR-based gene surveys, which tend to overrepresent lineages with greater numbers of rRNA operons or underrepresent lineages with relatively higher GC content. However, because both approaches rely on a cloning step, the most abundant lineages still may be overrepresented.

Figure 5.21 shows an outline of the overall methodology used by Gill et al. (2006). Total DNA was extracted from fecal samples derived from two healthy individuals, one 28-year-old male and one 37-year-old female. Neither subject had used antibiotics during the year prior to sample collection. One individual had an unrestricted diet, while the other followed a vegetarian diet. The sampling protocol was simplified compared to that of the previous study by Eckburg et al. (2005), in which they examined both fecal and mucosal samples from three individuals. Using the fecal DNA samples, 16S rDNA libraries were constructed by standard PCR-based cloning and sequencing procedures, and phylogenetic analysis was performed to determine which phylotypes were represented in

FIGURE 5.21 Overview of the project conducted by Gill and colleagues.

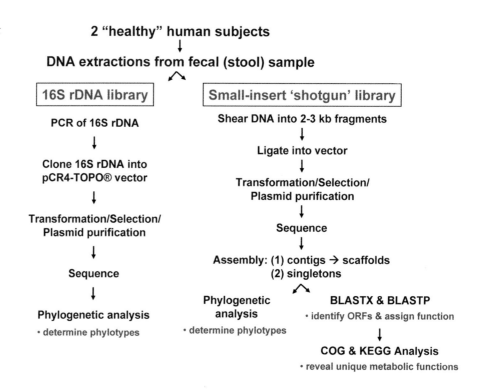

each of the two samples. The fecal DNA extractions also were used to construct small-insert ESS libraries, in which the DNA was sheared into fragments of 2 to 3 kb. Bidirectional sequencing with flanking vector primers was performed on each randomly selected recombinant plasmid. Sixty percent of the sequence reads were assembled into contigs, representing 92- to 44,747-bp stretches of contiguous DNA sequence assembled from plus and minus strands. Multiple contigs were aligned into longer stretches of continuous DNA sequence called scaffolds, which were between 1,000 and 57,894 bp. Scaffolds, which essentially are big contigs, are assembled based on gene order in a reference genome. In total, there was approximately 33.7 Mbp of unique DNA sequence derived from the combined distal gut microbiome. The assemblies were validated by comparison to reference genomes including the complete genomes of *Bifidobacterium longum* and *Bacteroides thetaiotaomicron*, as well as the draft genome of *M. smithii*. Due to the low depth of coverage, the remaining 40% of the sequence reads could not be assembled into contigs; these singletons represented another 45 Mbp of unique DNA sequence.

Gene survey and ESS approaches both produce phylogenetic profiles with discrepancies attributed to methodological bias

To assess the diversity of phylotypes represented in the ESS assemblies, the sequences were scanned for the presence of a phylogenetic anchor, in this case partial 16S rRNA gene segments. The shotgun reads produced 237 candidates, although only 132 were selected for further analysis; these represented good-quality sequences that were equal to or greater than 500 bp. Eight partial-length archaeal sequences between 291 and 714 bp also were analyzed. The 16S rDNA sequences were aligned to the Ribosomal RNA Database Project II (RDP-II) and NCBI GenBank to find nearest neighbors, or those sequences in the database most closely related to the queried sequences (see Unit 6 and Table 5.3 for further descriptions of RDP-II and BLASTN). Sequences then were grouped into phylotypes from which a phylogenetic tree was created using a neighbor-joining algorithm (see Unit 7 for further discussion of tree construction). The analysis revealed 72 bacterial phylotypes as well as *M. smithii* as the only archaeal phylotype. Although several bacterial phylotypes were considered novel, and the majority of phylotypes represented uncultivated species, all 72 were assigned to one of two major phyla, the *Firmicutes* (62 phylotypes total) and the *Actinobacteria* (10 phylotypes total). Surprisingly, no 16S rDNA sequences from *Bacteroidetes* were identified, although *Bacteroides fragilis* and *Bacteroides uniformis* were detected using species-specific primers. The authors suggest that the lysis and DNA extraction methods used to generate the ESS library may have led to the apparent inconsistency between the findings in this study and in the Eckburg et al. (2005) study (see McOrist et al. [2002]).

The authors also generated a PCR-based 16S rDNA clone library using the broad-range bacterial primers Bact-8F and Bact-1510R listed in Table 5.2. These sequences, of which there were approximately 1,000 nearly full-length genes representing each stool sample, were subjected to phylogenetic analysis in a manner identical to that described for the ESS library above. The same fecal lysis and extraction methods were used to generate the genomic DNA used for the clone library as were used for the ESS library; therefore, the same discrepancy observed with the ESS data in terms of overall community structure was expected. However, the authors utilized a different combination of PCR primers in the 2006 study from that used in the 2005 study. As described in Table

Table 5.2 Primers used to amplify the 16S rRNA gene

Target gene	Primer name	PCR primer (5′ to 3′)	Reference(s)
16S rRNA	Bact-8F[a]	AGAGTTTGATCCTGGCTCAG	Eckburg et al. (2005), Gill et al. (2006)
16S rRNA	Bact-1391R[b]	GACGGGCGGTGTGTRC	Eckburg et al. (2005)
16S rRNA	Bact-1510R[a]	CGGTTACCTTGTTACGACTT	Gill et al. (2006)

[a]Baker et al. (2003).

[b]Lane et al. (1985).

5.2, the studies shared the forward primer Bact-8F but used distinct reverse primers (Bact-1391R versus Bact-1510R). Thus, the outcomes of the PCR-based analyses from the two studies could exhibit differences that could be attributed to primer bias.

Of the 151 phylotypes identified in the Gill et al. (2006) study, 150 were assigned to the *Firmicutes* phylum while only 1 represented the *Actinobacteria,* a distribution that is consistent with the results of the ESS library analysis. Further inspection of the results raises a critical question about this and previous studies, the first of which the authors also acknowledge—do biases in the genomic DNA preparatory methods or PCR primer usage reflect the apparent underrepresentation of *Actinobacteria* and *Proteobacteria* in the human distal gut? In other words, is the apparent overrepresentation of *Firmicutes* and *Bacteroidetes* seen in the Eckburg et al. (2005) study a reflection of the bias inherent to the methods used to construct the library? The discrepancy observed between the two studies highlights the importance of issues pertaining to experimental bias.

Functional analysis reveals the unique metabolic transformations of the microbial community

Functional analysis of the human distal gut microbiome in the Gill et al. (2006) study began with annotation of the sequencing reads, in which two programs, BLASTX and BLASTP, were used to identify putative ORFs and assign them hypothetical functions. As explained in Table 5.3, three general types of BLAST searches were utilized in this study. Recall, BLASTN was used to assign contigs and singletons generated from the DNA sequence reads to phylotypes. In this case, the queries, which were comprised of the nucleotide sequences in the ESS assembly, were compared to the nucleotide sequences

Table 5.3 BLAST programs used to detect similarities

Program[a]	Query[b]	Database[c]
BLASTN	Nucleotide	Nucleotide
BLASTP	Protein	Protein
BLASTX	Nucleotide (translated into 6 frames)	Protein

[a]Search engines that make inquiries, or queries, specific to DNA and RNA databases such as GenBank/EMBL/DDBJ (Benson et al., 2006) or protein databases such as Swiss-Prot, UniProt and PDB (Apweiler, 2005).

[b]Nucleotide or amino acid sequence for which one is performing an inquiry-based search of available sequences in a particular database. The goal is to determine the identity or function of the query sequence based on observable similarities between it and the sequences in the database for which the identity or function is known.

[c]Major collection, or repository, of nucleotide or amino acid sequences that are available to the public and can be accessed using BLAST programs.

available in the RDP-II and NCBI GenBank databases containing known 16S rDNA sequences for *Bacteria* and *Archaea*. A BLASTP search uses an amino acid sequence as the query, searching protein databases for a match between the query and a protein of known function. BLASTX is a variation of the BLASTP search, except that a nucleotide sequence is submitted as the query, which is then translated into each of the six potential reading frames, which (as diagrammed in Fig. 5.22) includes three plus-strand and three minus-strand possibilities. The amino acid sequences deduced from the six-frame translation are then subjected to a search of the protein databases by the BLASTX algorithm, where a match can be made in any one of the six reading frames. For the BLASTP and BLASTX searches performed in the Gill et al. (2006) study, only matches that were a minimum of 50 amino acids long and that exhibited at least 35% identity (and E-values less than 10^{-15}) to proteins of known function in the databases were considered for further functional analysis. Based on these criteria, more than 25,000 ORFs were predicted for each subject.

Recall that the authors hypothesized that the ESS analysis would reveal unique functions encoded by the distal gut microbiome, and that these genes would support metabolic transformations that humans are not able to carry out because such genes are not represented in the human genome. Thus, if one compares the ORFs found in the distal

FIGURE 5.22 Translation of six possible reading frames deduced from the plus and minus strands of a nucleotide sequence. Note that the DNA strands are first transcribed into mRNA and that translation occurs in the 5'-to-3' direction along the mRNA strands. Thus, frames +1, +2, and +3 reflect translation of the mRNA sequence in each of the three reading frames derived from the plus strand. The orientation of the amino and carboxy termini of the polypeptide chain is written from right to left. Frames −1, −2, and −3 reflect translation of the mRNA sequences in each of the three reading frames derived from the minus strand, where the orientation of the amino and carboxy termini of the polypeptide chain is written from left to right. The only translation that gives in-frame start (Met) and stop (asterisk) codons is highlighted in blue. Note that this example is not to scale, in that the start and stop codons would be separated by at least 50 amino acids in the Gill et al. [2006] study. Illustration by Erin Sanders.

```
Frame +1   aa uac uag cgu cuu gcg acc uga cgu cgu uca auu ggu
              His Asp Cys Phe Ala Pro Ser Cys Cys Thr Leu Trp

Frame +2   a aua cua gcg ucu ugc gac cug acg ucg uuc aau ugg u
              Ile Ile Ala Ser Arg Gln Val Ala Ala Leu  *  Thr

Frame +3   aau acu agc guc uug cga ccu gac guc guu caa uug gu
            *  Ser Arg Leu Val Ser Ser Asp Leu Leu Asn Val
```

```
3`- aauacuagcgucuugcgaccugacgucguucaauuggu - 5` mRNA
                            ↑
5`- ttatgatcgcagaacgctggactgcagcaagttaacca - 3` Plus strand
3`- aatactagcgtcttgcgacctgacgtcgttcaattggt - 5` Minus strand
                            ↓
5`- uuaugaucgcagaacgcuggacugcagcaaguuaacca - 3` mRNA
```

```
Frame -1   uua uga ucg cag aac gcu gga cug cag caa guu aac ca
           Leu  *  Ser Gln Asn Ala Gly Leu Gln Gln Val Asn

Frame -2   u uau gau cgc aga acg cug gac ugc agc aag uua acc a
             Tyr Asp Arg Arg Thr Leu Asp Cys Ser Lys Leu Thr

Frame -3   uu aug auc gca gaa cgc ugg acu gca gca agu uaa cca
             Met Ile Ala Glu Arg Trp Thr Ala Ala Ser  *
```

gut microbiome to all those found in the human genome, one might expect to find an overrepresentation of ORFs that are found exclusively in the distal gut microbiome. To do such a genome-wide comparison, one needs to refine the annotation scheme in a way that specifically targets the metabolic potential in the genomes under consideration. Fortunately, such bioinformatics tools exist and were used to perform a functional analysis in the Gill et al. (2006) study. The authors investigated the metabolic capabilities of the distal gut microbiome by categorizing the ORFs into COGs (clusters of orthologous groups) and KEGG pathways (Kyoto Encyclopedia of Genes and Genomes) (Tatusov et al., 2003; Kanehisa et al., 2004).

COGs are generated from complete, curated genomes of *Bacteria*, *Archaea*, and even eukaryotes in which the amino acid sequences for each gene from several different genomes are compared. Genes that have the greatest sequence identity are assumed to have the same function and consequently are recognized as orthologs. Thus, the functional categories in the COG database account for the evolutionary origins, or phylogeny, of the genes. As shown in Fig. 5.23, at a minimum there must be at least three genomes that contain the gene. In similar fashion to performing a BLAST search, one submits a protein query to the COG database (via NCBI-CDD, listed below in Web Resources), where it is compared to individual sequences within each COG (group of three or more orthologous genes) from all functional categories (lipid metabolism, energy production, and transport). If best hits for the query belong to the same COG, one may conclude that the query is a member of that COG.

KEGG maps are constructed by superimposing ORFs encoding putative metabolic enzymes onto known biochemical pathways derived from finished genomes. Like COG analysis, reconstructing KEGG pathways results in the organization of ORFs into gene clusters, although the relationship between genes is based strictly on functional similarity within the context of a reference genome. Thus, one should only treat the KEGG map assignments as hypotheses until at the very least one can verify that the genes are concomitantly expressed within the organism of interest.

To generate a KEGG map of a biochemical pathway, each ORF first must be assigned an enzyme commission (EC) number, which describes its activity as well as the substrate upon which it functions. ORFs with the same EC number as those in a particular ref-

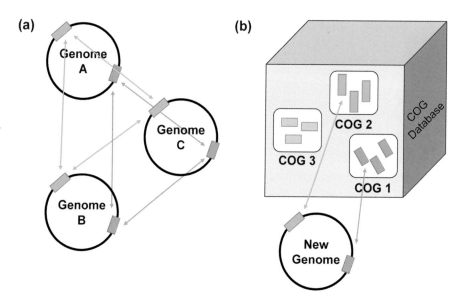

FIGURE 5.23 Overview of COG analysis. Orthologous genes are denoted as rectangles of the same color. Panel a shows three genomes (A, B, and C) containing genes that are of sufficient identity to be grouped into COGs 1 and 2 within the NCBI Conserved Domain Database (CDD) shown in panel b. The ORF queries from the new genome match those genes in COGs 1 and 2. Illustration by Erin Sanders.

erence pathway are assumed to have the same function and consequently are included as an enzymatic component in the reconstruction of a particular KEGG pathway. As shown in the example in Fig. 5.24, the result of this analysis can be superimposed onto a KEGG map, detailing the metabolic functions represented in the genome of the query.

For the Gill et al. (2006) study, the two types of functional analyses allowed the authors to assign 31% and 17% of the ORFs to COGs and KEGG pathways, respectively, corresponding to a total of approximately 24,000 genes for the two subjects in the study. Although this is a very large number of genes, it is nowhere near saturation in terms of the total number of genes present in the distal gut microbiome, which is expected to contain up to 1,000 different phylotype genomes (Eckburg et al., 2005). Assuming that the average bacterial genome contains around 6,000 genes, the distal gut microbiome would be expected to contain at least 6 million genes, which is 250 times more than was

FIGURE 5.24 KEGG map of the reductive carboxylate cycle used for CO_2 fixation in photosynthetic bacteria such as the phytoplankton in the Sargasso Sea. Enzymes are identified in blue boxes by their EC number. CoA, coenzyme A. (From the KEGG Pathway Database at http://www.genome.jp/kegg/.)

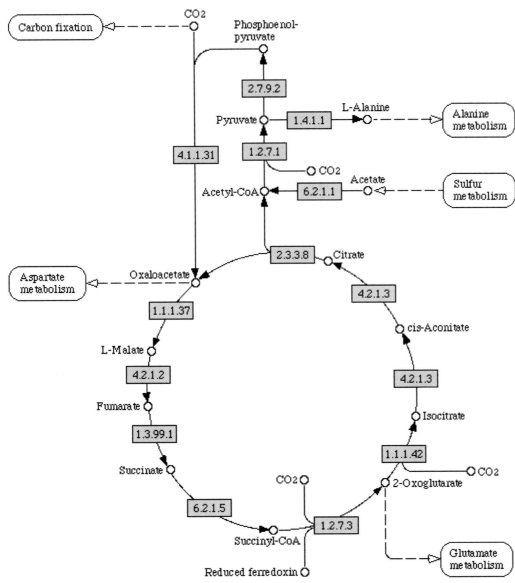

categorized in the Gill et al. (2006) study. Thus, the authors could not be confident that the mere absence of a component in a KEGG map or lack of a hit in the COG database meant that the gene was not present in the distal gut microbiome. To get around this problem, odds ratios were used to determine whether the genes assigned to either a COG category or KEGG pathway were overrepresented or underrepresented in the distal gut microbiome relative to previously sequenced reference genomes. An odds ratio equal to 1 implies that the distal gut community has the same number of hits as the reference data set. An odds ratio greater than 1 indicates that there were more hits for the distal gut community than in the reference genome (enriched groups). An odds ratio less than 1 suggests that there were fewer hits in a particular category for the distal gut community than in the reference genome (underrepresented groups).

Figure 5.25 shows the results of the odds ratio analysis for eight COG categories, listed across the *x* axis. For this analysis, the genes from the human distal gut microbiome that had been categorized into COGs were compared to genes in a database comprised of 163 different microbial genomes. The overall pattern displayed by the two subjects is similar. For instance, genes implicated in nucleotide transport and metabolism are enriched in samples from both subjects (odds ratio > 2), while genes involved in ion transport and metabolism are underrepresented in samples from both subjects (odds ratio < 1). Thus, it appears that there are metabolic capabilities for which the distal gut microbiome may be more specialized as compared to other microorganisms, at least those for which the genome sequence is known. There also were several COG categories, denoted by asterisks on the graph, for which there were statistically significant differences in the odds ratio between the two subjects, who differed in age, gender, and diet. The results lead us to ask whether the differences in COG category representation can be

FIGURE 5.25 COG analysis reveals metabolic functions that are enriched for or underrepresented in the human distal gut microbiome. Bars above both dashed lines denote enrichment, while bars below both dashed lines signify underrepresentation ($P < 0.05$). Asterisks indicate COG categories that are significantly different between the two subjects ($P < 0.05$). (Reprinted from Gill et al. [2006] with permission.)

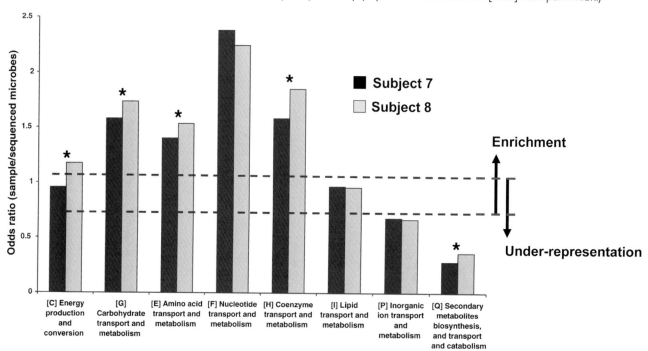

attributed to these differences. This is an unresolved question but an excellent hypothesis for future experiments.

Although the COG analysis is informative, the results do not specifically address the hypothesis under consideration. The authors wanted to find evidence to support the idea that the distal gut microbiome would encode unique functions that were not encoded by the human genome. This question required a direct comparison of the metabolic potential of the human distal gut microbiome with that of the human genome. The KEGG analysis permitted the authors to make such an assessment, comparing the distal gut microbiome to three different data sets, which included the genome of *Homo sapiens*, 202 bacterial genomes in KEGG, and 21 archaeal genomes in KEGG. The results of this analysis for 11 different KEGG pathways are shown in Fig. 5.26. It is immediately obvious that 9 of the 11 pathways show enzymes in the distal gut microbiome that are not present in the human genome (e.g., red bars with an odds ratio of >1). The unique metabolic capabilities attributed to the distal gut microbiome span a variety of KEGG categories, including the biosynthesis or catabolism of carbohydrates, energy, lipids, nucleotides, amino acids, peptides, vitamins, and secondary metabolites. Perhaps surprisingly, the enzymes required to degrade xenobiotics were underrepresented in the distal gut microbiome, suggesting a less predominant role for the microbial community in this metabolic process than previously postulated by Nicholson et al. (2005). However, the fact that there were clearly biases in the genomes represented in the Gill et al. (2006)

FIGURE 5.26 KEGG pathway reconstructions uncover metabolic functions that are unique to the human distal gut microbiome. The analysis combined the genes from both subjects to create an aggregate distal gut microbiome, which was then compared to the three genome data sets noted. Asterisks indicate enrichment (odds ratio > 1, $P < 0.05$) or underrepresentation (odds ratio < 1, $P < 0.05$). (Reprinted from Gill et al. [2006] with permission.)

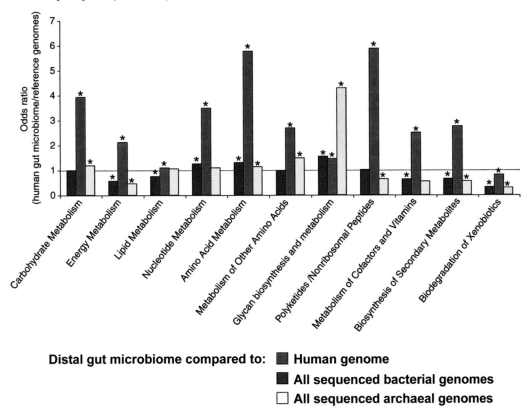

study, due to the fecal lysis and DNA extraction methods employed, leaves open the possibility that genes from microorganisms responsible for xenobiotic metabolism were present in low abundance in the KEGG pathway analysis.

There also were groups of enzymes present in the distal gut microbiome that are not present in archaeal genomes (shown in Fig. 5.26 by yellow bars with odds ratio > 1), of which those involved in glycan biosynthesis and catabolism are most obvious. An abundant repository for glycans, which serve as a source of energy for some members of the distal gut microbiome, is mucus produced by the human host of the microbes. Interestingly, the authors note that *Bacteroidetes,* a phylotype that was not represented in this study, previously was implicated as a prime consumer of host glycans as an energy source (Bäckhed et al. [2004, 2005]). Perhaps the results of the Gill et al. (2006) study suggest that the metabolic potential for this process is shared among numerous microbial lineages.

This study certainly succeeds at highlighting both the strengths and limitations of metagenomic community analysis, whether the libraries are constructed from targeted genes such as 16S rRNA genes or from ESS. This field clearly is in its infancy but stands to unveil an entirely new body of knowledge, which, with the right sequencing technology and bioinformatics tools, can be harnessed and applied to solve previously intractable problems spanning a variety of disciplines.

KEY TERMS

Bidirectional sequencing Simultaneous sequencing of the plus and minus strands of a DNA template, using two differentially labeled primers in the same reaction mixture.

Contig A continuous sequence of DNA produced from the assembly of overlapping sequence reads.

Curated genome A genome in which automated annotation of ORFs has been manually validated in silico such that gene calls are all evidence based (e.g., start and stop codons mapped, functional assignments verified in multiple databases); note that function has not necessarily been proven experimentally.

E-values Expectation values that provide an indication of the statistical significance of a pairwise sequence alignment produced in BLAST search, signifying whether the alignment portrays a biological relationship or similarity that is due solely to chance.

Identity Quantitative assessment of the degree of relatedness of two sequences; measured as the total number of exact matches between two sequences when aligned along their length.

Microbiota The community of microorganisms present in an environmental habitat.

Odds ratio The possibility of observing a given term in the sample relative to the data set to which the sample is being compared; used to portray the relative enrichment or under-representation of COG categories or KEGG map components.

Orthologs Homologous genes derived from a common ancestor; genes found in one organism that are functionally similar to those found in another organism but differ because of speciation.

Phylogeny The evolutionary history of a group of organisms.

Reference genome A completely sequenced and annotated genome used for comparative analysis of gene order or content with respect to those of unfinished genomes or metagenomes, to aid in the reconstruction of the latter into contigs and scaffolds.

Scaffold A continuous sequence of DNA produced from the assembly of overlapping contigs; sometimes generated by aligning the contigs along a related reference genome that has been finished.

Singletons Sequence reads that cannot be assembled into contigs because of a lack of overlapping regions with other DNA sequences produced during an ESS analysis.

Xenobiotics Completely synthetic chemical compounds.

REFERENCES

Apweiler, R. 2005. Sequence databases, p. 3–24. *In* A. D. Baxevanis and B. F. F. Ouellette (ed.), *Bioinformatics: A Practical Guide to the Analysis of Genes and Proteins,* 3rd ed. John Wiley & Sons, Inc., Hoboken, NJ.

Bäckhed, F., H. Ding, T. Wang, L. V. Hooper, G. Y. Koh, A. Nagy, C. F. Semenkovich, and J. I. Gordon. 2004. The gut microbiota as an environmental factor that regulates fat storage. *Proc. Natl. Acad. Sci. USA* **101**:15718–15723.

Bäckhed, F., R. E. Ley, J. L. Sonnenburg, D. A. Peterson, and J. I. Gordon. 2005. Host-bacterial mutualism in the human intestine. *Science* **307**:1915–1920.

Baker, G. C., J. J. Smith, and D. A. Cowan. 2003. Review and re-analysis of domain-specific 16S primers. *J. Microbiol. Methods* **55**: 541–555.

Benson, D. A., I. Karsch-Mizrachi, D. J. Lipman, J. Ostell, and E. W. Sayers. 2009. GenBank. *Nucleic Acids Res.* **37**(Database Issue):D26–D31. doi:10.1093/nar/gkn723.

Eckburg, P. B., E. M. Bik, C. N. Bernstein, E. Purdom, L. Dethlefsen, M. Sargent, S. R. Gill, K. E. Nelson, and D. A. Relman. 2005. Diversity of the human intestinal microbial flora. *Science* **308**:1635–1638.

Eckburg, P. B., and D. A. Relman. 2007. The role of microbes in Crohn's disease. *Clin. Infect. Dis.* **44**:256–262.

Gill, S. R., M. Pop, R. T. DeBoy, P. B. Eckburg, P. J. Turnbaugh, B. S. Samuel, J. I. Gordon, D. A. Relman, C. M. Fraser-Liggett, and K. E. Nelson. 2006. Metagenomic analysis of the human distal gut microbiome. *Science* **312**:1355–1359.

Kanehisa, M., S. Goto, S. Kawashima, Y. Okuno, and M. Hattori. 2004. The KEGG resource for deciphering the genome. *Nucleic Acids Res.* **32**:D277–D280.

Lane, D. J., B. Pace, G. J. Olsen, D. A. Stahl, M. L. Sogin, and N. R. Pace. 1985. Rapid determination of 16S ribosomal RNA sequences for phylogenetic analyses. *Proc. Natl. Acad. Sci. USA* **82**: 6955–6959.

Ley, R. E., P. J. Turnbaugh, S. Klein, and J. Gordon. 2006. Human gut microbes associated with obesity. *Nature* **444**:1022–1023.

Ley, R. E., M. Hamady, C. Lozupone, P. J. Turnbaugh, R. R. Ramey, J. S. Bircher, M. L. Schlegel, T. A. Tucker, M. D. Schren-zel, R. Knight, and J. I. Gordon. 2008. Evolution of mammals and their gut microbes. *Science* **320**:1647–1651.

Madigan, M. T., and J. M. Martinko. 2006. *Brock Biology of Microorganisms,* 11th ed. Pearson Prentice Hall, Pearson Education, Inc., Upper Saddle River, NJ.

Mazmanian, S. K., C. H. Liu, A. O. Tzianabos, and D. L. Kasper. 2005. An immunomodulatory molecule of symbiotic bacteria directs maturation of the host immune system. *Cell* **122**:107–118.

McOrist, A. L., M. Jackson, and A. R. Bird. 2002. A comparison of five methods for extraction of DNA from human faecal samples. *J. Microbiol. Methods* **50**:131–139.

Nicholson, J. K., E. Holmes, and I. D. Wilson. 2005. Gut microorganisms, mammalian metabolism, and personalized health care. *Nat. Rev. Microbiol.* **3**:431–438.

Rakoff-Nahoum, S., J. Paglino, F. Eslami-Varzaneh, S. Edberg, and R. Medzhitov. 2004. Recognition of commensal microflora by Toll-like receptors is required for intestinal homeostasis. *Cell* **118**: 229–241.

Tatusov, R. L., N. D. Fedorova, J. D. Jackson, A. R. Jacobs, B. Kiryutin, E. V. Koonin, D. M. Krylov, R. Mazumder, S. L. Mekhedov, A. N. Nikolskaya, B. S. Rao, S. Smirnov, A. V. Sverdlov, S. Vasudevan, Y. I. Wolf, J. J. Yin, and D. A. Natale. 2003. The COG database: an updated version includes eukaryotes. *BMC Bioinformatics* **4**:41.

Turnbaugh, P. J., R. E. Ley, M. A. Mahowald, V. Magrini, E. R. Mardis, and J. I. Gordon. 2006. An obesity-associated gut microbiome with increased capacity for energy harvest. *Nature* **444**: 1027–1031.

Web Resources

CDD (Conserved Domain Database) used to query COGs http://www.ncbi.nlm.nih.gov/Structure/cdd/wrspb.cgi

KEGG pathway database http://www.genome.jp/kegg/pathway.html. (1995–2008 Kanehisa Laboratories)

NCBI-COGs http://www.ncbi.nlm.nih.gov/COG/new/

READING ASSESSMENT

1. DNA extraction using a direct-lysis approach results in the copurification of prokaryotic and eukaryotic nucleic acids from lysed cells as well as the extracellular environment. Whereas this type of contamination can interfere with the construction of bacterial shotgun libraries, why is this method suitable for the libraries generated for the "I, Microbiologist" project?

2. Does transformation of recombinant pCR2.1-TOPO vectors into competent *E. coli* cells followed by plating onto agar media containing X-Gal and kanamycin result in a selection and/or screening of recombinant colonies? Briefly explain your answer.

3. On the DGGE gel diagram below, draw bands representing 16S rDNA as you would expect to find them after performing PCR on communities derived from the Great

Salt Lake in Utah (lane 1) versus Lake Michigan in the midwestern United States (lane 2). Briefly explain your reasoning for the depicted patterns.

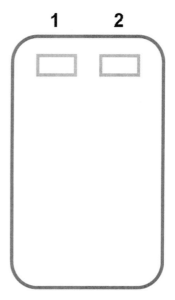

4. Microbes can be detected in the environment by microscopic, cultivation, and molecular approaches. Discuss one advantage and one disadvantage of each technique.

5. From a soil sample you isolate a facultative anaerobe that is capable of reducing nitrate (NO_3^-) to nitrite (NO_2^-), ascertained using a standard biochemical test in which a medium containing a large amount of nitrate (KNO_3) is inoculated with the organism. After incubation, α-naphthylamine and sulfanilic acid are added, and the medium turns red upon interaction with nitrite produced by the organism. Two distinct enzyme complexes are responsible for anaerobic respiration of nitrate, NAP and NAR. The genes encoding the catalytic subunits for each complex are *napA* and *narG*, respectively. Devise a microscopy-based community sampling approach to visualize the cells expressing these genes in the soil sample.

6. List some of the challenges encountered when trying to characterize the gene content of an ESS library.

7. What distinguishes a keystone species from other inhabitants of an ecosystem?

8. Why are strategies such as primer walking and hierarchical clone-based shotgun sequencing employed in the analysis of BAC libraries?

9. If the average size of a bacterial gene is 1,000 bp, why is it necessary to use BAC vectors when constructing metagenomic DNA libraries used for functional analysis?

The remaining questions pertain to the Gill et al. (2006) study:

10. What is the purpose or goal of this study?

11. Find the hypothesis (or hypotheses) tested in this study, and rephrase it as a conditional proposition (i.e., an "if...then..." statement).

12. Do the results support or refute the hypothesis or hypotheses?

13. Identify and briefly explain the key result that enabled you to draw the conclusion stated in question 12. If there was more than one hypothesis, there likely will be more than one result to discuss; specify which result addressed which hypothesis.

14. On the basis of your own evaluation of the data, do you agree with the conclusions reached by the authors? Why or why not? Identify problems or ambiguities in their results that could lead you to question their analysis.

15. Thinking about future directions, suggest one experiment the authors should do next as a follow-up to this study.

UNIT 5

EXPERIMENTAL OVERVIEW

In Experiments 5.1 through 5.9, students will work directly with a soil sample to extract genomic DNA from the microorganisms in this environmental niche. Now referred to as the soil microbiome, or metagenomic DNA sample, it will serve as the template for PCR with primers specific for the 16S rRNA gene. If agarose gel electrophoresis verifies that a product consistent with the size expected for 16S rDNA has been generated, the pool of amplicons will be purified using a gel extraction technique. If no product is made, students should use the Decision Guide at the end of Experiment 5.3 to work through this part of the procedure by using some alternative approaches.

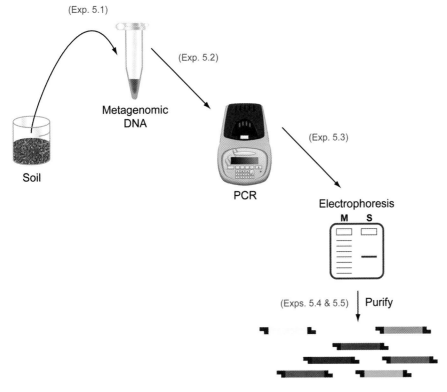

To create a template for DNA sequencing, however, the purified PCR products first must be cloned into a vector, which will separate the pool of 16S rRNA genes by producing plasmids that incorporate only a single gene. Individual recombinant plasmids can be purified from E. coli cells following transformation. A colorimetric screen enables cells containing recombinant plasmids to be distinguished from those containing nonrecombinant plasmids, and purification of individual colonies from the transformation plates by a streak plate procedure ensures that the cells contain the correct selection markers. Once single colonies of candidate 16S rDNA clones are obtained, they can be used to inoculate liquid cultures in which the plasmid may be amplified to a concentration necessary for subsequent biochemical analysis. A standard cell lysis procedure coupled to column chromatography is used to recover and purify plasmids from the cells. Note that two tubes (A and B) are produced per clone, each for a specific purpose as described below.

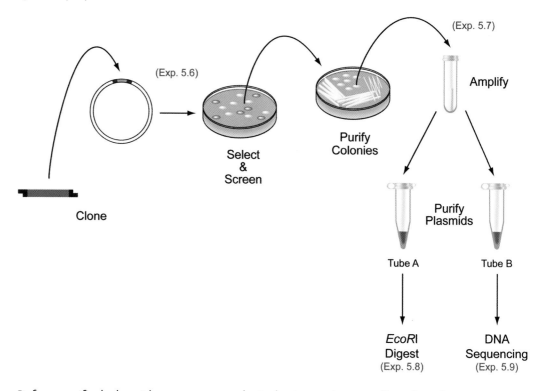

Before purified plasmids are sequenced, students need to confirm that they contain a DNA fragment of a size consistent with that expected for the 16S rRNA gene. The purified plasmids in tube A are used in a restriction digest, whereby the endonuclease EcoRI excises the DNA fragments from the vectors, and gel electrophoresis is once again used to resolve the resulting DNA bands based on size. As long as the assay produces the expected restriction pattern, the sample may be submitted for DNA sequencing. However, the purified plasmids in tube B are used for this purpose, as the remaining volume in tube A should be stored in freezer boxes. The sequences are used for bioinformatics and phylogenetic analyses as described in Units 6 and 7.

EXPERIMENT 5.1 Isolation of Metagenomic DNA Directly from Soil

In this experiment, students extract the genetic material from soil microorganisms in the sample collected in Experiment 1.1. The procedure described is for the MoBio UltraClean soil DNA isolation kit.

MATERIALS

MoBio UltraClean soil DNA isolation kit
"Boil-proof" microcentrifuge tubes
Vortex Genie II
Horizontal multitube vortex adaptor
1.8-ml microcentrifuge tubes

60°C heat block or water bath
Phenol-chloroform-isoamyl alcohol
Chloroform
Analytical scale
P200 pipette with beveled sequencing tips

METHODS

Wear gloves throughout the DNA isolation procedure. The procedure does not work optimally with extremely moist soil (e.g., after a heavy rain), sediment, or soils with high humic acid content. The PowerSoil kit is recommended for these types of soils (see Box 5.1).

1. Working with your partner, prepare a total of four metagenomic DNA samples. Add 0.25 to 1.0 g of soil sample to the 2.0-ml bead solution tubes provided. The bead solution is a buffer that disperses the soil particles and begins to dissolve humic acids as the first part of the lysis procedure. Since you have a total of four samples, try a range of soil amounts (e.g., 0.25, 0.5, 0.75, and 1.0 g). The amounts may be approximate (e.g., 0.24 or 0.26 g is fine if targeting 0.25 g); just be sure to record the exact amount in your laboratory notebook.

2. Gently vortex to mix the soil sample and bead solution.

3. Check solution S1, which contains sodium dodecyl sulfate (SDS), a detergent that aids in cell lysis by breaking down fatty acids and lipids associated with the cell membrane. If the solution gets cold, the SDS will precipitate. If solution S1 is precipitated, heat it to 60°C until the SDS precipitate dissolves.

4. Add 60 μl of solution S1, and invert several times or vortex briefly. The solution S1 may be used while it is still hot.

5. Add 200 μl of IRS (inhibitor removal solution), which is required because the DNA is to be used for PCR. IRS is a proprietary reagent designed to precipitate humic acids and other PCR inhibitors. Humic acids are generally (but not always) brown. They belong to a large group of organic compounds associated with most soils that are high in organic content.

6. Secure the bead tubes horizontally, using the MoBio multitube vortex adaptor. The tube holder secures 2.0-ml tubes horizontally on a flat-bed vortex pad. Vortex at maximum speed for 10 minutes, introducing mechanical lysis at this step.
    *   The protocol uses a combination of mechanical and chemical lysis. By randomly shaking the beads, they collide with one another and with microbial cells, causing them to break open.

Experiment continues

- The method you use to secure tubes to the vortex is critical. MoBio has designed the vortex adaptor as a simple tool that keeps tubes tightly attached to the vortex. It should be noted that although you can attach tubes with tape as an inexpensive alternative to using the adaptor, often the tape becomes loose and not all tubes will shake evenly or efficiently. This may lead to inconsistent results or lower yields.

7. Make sure the 2.0-ml tubes rotate freely in your centrifuge without rubbing. Centrifuge the tubes at 13,000 rpm for 30 seconds. *Caution:* Be sure not to exceed 13,000 rpm, or the tubes may break. Particulates including cell debris, soil, beads, and humic acids form a pellet at this point. The DNA is in the liquid supernatant at this stage.

8. Transfer the supernatant containing DNA to a clean microcentrifuge tube. The supernatant may still contain some soil particles and humic acids. With 0.25 g of soil (depending upon soil type), between 400 and 450 μl of supernatant is expected.

9. Add 250 μl of solution S2, which contains a protein precipitation reagent, and vortex for 5 seconds. Incubate at 4°C for 5 minutes. It is important to remove contaminating proteins that may reduce DNA purity and inhibit downstream applications for the DNA.

10. Centrifuge the tubes for 1 minute at 13,000 rpm.

11. Avoiding the pellet, transfer the entire volume of supernatant to a clean microcentrifuge tube. The pellet at this point contains residues of humic acid, cell debris, and proteins. For the best DNA yields and quality, avoid transferring any of the pellet.

12. Perform a phenol-chloroform extraction twice as follows. *Caution:* Wear gloves because phenol-chloroform is highly corrosive to skin. Perform extractions under a fume hood because phenol-chloroform is highly volatile with noxious fumes.

    a. Add an equal volume of phenol-chloroform to your sample (e.g., if you have 700 μl of supernatant, add 700 μl of phenol-chloroform). *Note:* **There are two layers in the bottle; the top layer contains isoamyl alcohol whereas the bottom layer is a solubilized mixture of phenol and chloroform; make sure you withdraw liquid for extraction from the bottom layer.**

    b. Gently mix by inverting the tube for 2 to 3 minutes by hand (the layers should be thoroughly mixed). *Note:* **Exercise care and invert the tubes gently to avoid shearing the genomic DNA, as that will cause problems for subsequent experiments.**

    c. Centrifuge the tubes for 15 minutes at high speed (13,000 rpm).

    d. Set the P200 pipette at maximum volume (200 μl), and, using a beveled tip (sequencing gel-loading tips work well), collect the top (aqueous) layer and transfer the extract into a clean microcentrifuge tube. You will have to aspirate two or three times to collect the entire aqueous layer. Use a fresh tip each time. Avoid the interface by tipping the tube at a 45° angle when aspirating aqueous layer into the pipette tip.

    e. Repeat steps a to d.

13. With the extract from step 12, perform a chloroform extraction twice as follows. *Caution:* Wear gloves because chloroform is a probable carcinogen. Perform extractions under a fume hood because chloroform is harmful and highly volatile although it has a sweet odor.

    a. Add an equal volume of chloroform to your sample (e.g., if you have 700 μl of supernatant, add 700 μl of chloroform).

    b. Gently mix by inverting the tube for 2 to 3 minutes by hand (the layers should be thoroughly mixed). *Note:* **Exercise care and invert the tubes gently to avoid shearing the genomic DNA, as that will cause problems for subsequent experiments.**

    c. Centrifuge the tubes for 15 minutes at high speed (13,000 rpm).

    d. Using beveled tips, collect the top, aqueous layer and transfer the extract into a clean microcentrifuge tube as described above.

    e. Repeat steps a to d.

14. Add 1.3 ml of solution S3, a DNA binding salt solution, to the supernatant. DNA binds to silica in the presence of high salt concentrations. Be careful not to spill the solution as the volume will reach the rim of the microcentrifuge tube. Close the lid, and vortex for 5 seconds.

15. Load approximately 700 μl onto a spin filter, and centrifuge at 13,000 rpm for 1 minute. Discard the flowthrough, then add the remaining supernatant to the spin filter, and centrifuge at 13,000 rpm for 1 minute. Repeat until all supernatant has passed through the spin filter. A total of three loads is required for each sample processed. The DNA is selectively bound to the silica membrane in the spin filter device. Almost all contaminants pass through the filter membrane, leaving only the desired DNA behind.

16. Add 300 μl of solution S4, an ethanol-based wash solution, and centrifuge for 30 seconds at 13,000 rpm. Solution S4 is used to further clean the DNA that is bound to the silica filter membrane in the spin filter. This wash solution removes residues of salt, humic acid, and other contaminants while allowing the DNA to stay bound to the silica membrane.

17. Discard the flowthrough from the collection tube. This flowthrough is just waste containing ethanol wash solution and contaminants that did not bind to the silica spin filter membrane.

18. Repeat the wash steps (steps 16 and 17) two more times (a total of three solution S4 washes should be done).

19. Centrifuge again for 1 minute after the final wash, to remove residual solution S4. It is critical to remove all traces of wash solution because it can interfere with downstream applications for the DNA.

20. Carefully place the spin filter in a new clean microcentrifuge tube. Avoid splashing any solution S4 onto the spin filter. Once again, it is important to avoid any traces of the ethanol-based wash solution.

Experiment continues

21. Add 50 μl of solution S5, a sterile elution buffer, to the center of the small white filter membrane, ensuring that the entire membrane is wet, which will result in more efficient release of the desired DNA. *Caution:* Do not touch the pipette tip to the membrane.

22. Centrifuge for 30 seconds. As solution S5 passes through the silica membrane, DNA is released, flowing through the membrane and into the collection tube. The DNA is released because it can bind to the silica spin filter membrane only in the presence of salt. Solution S5 is 10 mM Tris buffer (pH. 8) and does not contain salt.

23. Discard the spin filter. The DNA in the tube (metagenomic DNA) is now application ready. No further steps are required (unless troubleshooting becomes necessary [see Box 5.1]).

24. Store DNA at −20°C.

ACKNOWLEDGMENT

The protocol for the UltraClean soil DNA isolation kit has been modified and reproduced with permission from MoBio Laboratories, Inc.

BOX 5.1 Troubleshooting Metagenomic DNA Purification

Isopropanol precipitation

ADDITIONAL MATERIALS

Store the following reagents at ambient temperature:
Sterile distilled water
Chloroform–2-pentanol
Tris-EDTA (TE) buffer (pH 8)

Store the following reagents at 4°C:
7.5 M ammonium acetate
Isopropanol (2-propanol)
70% (vol/vol) ethanol (dilute 200-proof ethanol with sterile distilled H_2O)

METHODS

Repeat Experiment 5.1, except include an isopropanol precipitation as the final purification step.

1. To 50 μl of metagenomic DNA eluted in step 23, first add 300 μl of sterile water and then add 150 μl of 7.5 M ammonium acetate. Vortex gently to mix.

2. Keep on ice for 10 minutes.

3. Add 500 μl of chloroform–2-pentanol, and mix by gentle vortexing. Centrifuge for 10 minutes at maximum speed (13,000 rpm).

4. Transfer the aqueous phase (top layer) to a new microcentrifuge tube. Be careful not to transfer any of the lower, organic layer.

5. Add 1.1× volume of cold isopropanol (e.g., if you have 500 μl of aqueous phase, add 540 μl of cold isopropanol). To mix, invert the tube gently but completely for 1 minute.

6. Centrifuge for 5 minutes at high speed (13,000 rpm).

7. Pour off the supernatant into a waste container.

BOX 5.1 Troubleshooting Metagenomic DNA Purification *(continued)*

8. Add 200 μl of cold 70% ethanol to wash the pellet, and centrifuge for 5 minutes at 13,000 rpm.

9. Pour off the supernatant, and then repeat steps 8 and 9 two more times.

10. Dry the DNA pellet completely by storing the tube with cap open in a fume hood at ambient temperature until the ethanol wash has evaporated completely and no fluid is visible (2 to 5 minutes).

11. Resuspend the DNA pellet in 50 μl of TE buffer.

12. Be sure the tubes are properly labeled. The DNA samples must be stored in a freezer at −20°C.

PowerSoil DNA isolation kit

MoBio Laboratories has developed a kit intended for use with difficult soil samples with a high humic acid content, including compost, sediment, and even manure. This kit is distinguished from the UltraClean soil DNA isolation kit by inclusion of a humic-substance removal procedure, which removes the brown coloration of samples and PCR inhibitors. Instructions for this kit are provided by the manufacturer.

MATERIALS

16S rRNA PCR primer 1492R

16S rRNA PCR primer 27F

10× PCR buffer

0.5 M KCl

Taq DNA polymerase

PCR tubes and caps

PCR tube rack

1.8-ml microcentrifuge tubes (sterile)

Ice bucket

DNA template (from Experiment 5.1)

10 mM deoxynucleoside triphosphates (dNTPs)

Dimethyl sulfoxide (DMSO) sterilized with a nylon filter

Sterile distilled water

Thermal cycler

Ethanol-resistant markers

METHODS

Wear gloves when performing all PCR procedures. Always run negative controls (and positive controls when possible).

1. Use the following PCR primers for 16S rRNA gene amplification of all four metagenomic DNA samples:

$5' \rightarrow 3'$ sequence[a]	Primer name	Position	T_m (°C) (100 mM KCl)
GGT TAC CTT GTT ACG ACT T	16S_1492R	1492–1510	55
AGA GTT TGA TCM TGG CTC AG	16S_27F	8–27	61

[a] M = A or C.

2. Prepare the PCR master mix by adding the following reagents to a 1.8-ml microcentrifuge tube.

Note 1 Keep all PCR reagents and reaction mixtures on ice throughout the setup process. Do not discard PCR reagent stocks that are given to you—these will be your working stocks throughout the project. Keep these reagent stocks in a −20°C freezer box when not in use.

Note 2 To minimize cross-contamination, always use a fresh aliquot of sterile water in the PCR master mix. Each time you perform PCR, use a sterile 5.0-ml pipette (not a pipettor) to aseptically transfer approximately 1.0 ml of water from the stock bottle to a sterile microcentrifuge tube.

Note 3 Label the sides rather than the caps of PCR tubes, using an ethanol-resistant marker. Markings on the caps may rub off during the PCR run.

Experiment continues

Volumes of reactants		Final concentration in 50-μl reaction mixture
5.0 μl × (no. of reactions + 2) =	μl of 10× PCR buffer[a]	1×
2.5 μl × (no. of reactions + 2) =	μl of DMSO[b]	5% (vol/vol)
5.0 μl × (no. of reactions + 2) =	μl of 0.5 M KCl	50 mM KCl
2.0 μl × (no. of reactions + 2) =	μl of 25-pmol/μl 1492R primer[c]	1 pmol/μl (1 μM)
2.0 μl × (no. of reactions + 2) =	μl of 25-pmol/μl 27F primer[c]	1 pmol/μl (1 μM)
28.0 μl × (no. of reactions + 2) =	μl of sterile distilled H_2O	NA[d]
1.0 μl × (no. of reactions + 2) =	μl of 10 mM dNTPs	0.2 mM each dNTP
45.5 μl × (no. of reactions + 2) =	μl of PCR master mix	

[a]Contains 15 mM $MgCl_2$ and 500 mM KCl, so a final concentration of 1.5 mM $MgCl_2$ and 50 mM KCl is obtained in a 50-μl reaction volume. Since a KCl concentration of 100 mM is desired, an additional 50 mM KCl must be added to the reaction mix.

[b]Wear gloves when handling DMSO; use it in the fume cabinet only. DMSO facilitates denaturation of the DNA template, thereby decreasing the apparent T_m of the primer-template DNA duplex.

[c]Prevent the formation of primer dimers by first incubating the primer stocks at 80°C for 2 minutes and then immediately plunging the tubes into ice. Keep the primer stocks cold, and add aliquots to the PCR master mix as needed. The heat treatment should resolve existing secondary structure (e.g., hairpins and nonspecific heteroduplexes), while the cold incubation temporarily prevents nonspecific priming events and stops secondary structures from forming again before the primers are added to the PCR mix.

[d]NA, not applicable.

3. Gently vortex the tube containing the PCR master mix several times to mix. Return the tube to the ice bucket. Note that the *Taq* DNA polymerase will be added to the reaction mixture after the initial hot-start denaturation step of the thermal cycling procedure.

4. Add 4 μl of metagenomic DNA template (should be ≤1 μg per reaction), one for each soil amount used for extractions in Experiment 5.1, to four separate PCR tubes. For a negative control, add 4 μl of sterile water (same as that used in the PCR master mix) to one PCR tube. Keep a PCR tube rack on ice such that the DNA in the five tubes remains cold at 4°C. (*Remember:* nonspecific priming at ambient temperature.)

5. Add 45.5 μl of PCR master mix to each PCR tube. Pipette up and down to mix. Discard leftover master mix.

   Note: Make sure the DNA-PCR mix solution is at the bottom of each tube before placing the tubes in the thermal cycler. If drops cling to the sides of the tubes, use a pipette tip to guide the drops to the bottom.

6. Put the caps on the PCR tubes. Place the tubes into a 96-well thermal cycler (PCR machine), and start the Metagenomic PCR program, which includes a hot start in the thermal cycling protocol. The PCR machine may be paused to allow time for addition of *Taq* and then restarted to commence the cycling procedure:

   a. Hot-start denaturation: 94.0°C for 5:00 minutes (without *Taq*)
      Add *Taq* polymerase (step 7 below).

   b. Initial denaturation: 94.0°C for 3:00 minutes (with *Taq*)

   c. Three-step cycling procedure repeated 35 times:
      Denaturation: 94.0°C for 1 minute
      Annealing: 48.0°C for 30 seconds (~5°C below the T_m of primers)
      Extension: 72.0°C for 1 minute

   d. Final extension: 72.0°C for 7 minutes

   e. Storage: 4.0°C for infinite time (∞)

7. At the conclusion of the hot-start denaturation step, add 0.5 μl of 5-U/μl *Taq* polymerase to each tube (you should have a final concentration of 2.5 U per 50-μl reaction mixture). You will notice that the *Taq* enzyme, which typically is stored in a 50% glycerol buffer, initially descends to the bottom of the tube upon expulsion from the pipette tip. Gently but quickly vortex the tube once *Taq* has been added to ensure that it is dispersed evenly within the master mix. Immediately return the tube to the heat block of the PCR machine.

Note 1 Due to the high viscosity of enzyme stock solutions, students often struggle to pipette small volumes accurately without extensive practice. Thus, it is advised that the instructor or teaching assistant (TA) demonstrate how to add the *Taq* to the tubes and ensure that it is properly mixed before continuing with the PCR cycling steps. It is also worthwhile to have students practice pipetting 50% glycerol solutions before trying to work with *Taq* on their own.

Note 2 Because DMSO inhibits *Taq* activity by ~50%, the amount of *Taq* used for these reactions is twice the amount typically used for standard PCR.

8. When the program has finished, the PCR tubes must be removed from the thermal cycler and stored at −20°C until the next lab period.

BOX 5.2 Troubleshooting PCR with Metagenomic DNA

1. Try different PCR conditions such as touchdown PCR described by Don et al. (1991) (see Unit 3 references).

2. Try different primer combinations (16S rDNA 8F instead of 27F in combination with 1492R, or 1510R instead of 1492R in combination with 27F, etc.).

5′ → 3′ sequence	Primer name
CGG TTA CCT TGT TAC GAC TT	16S_1510R
AGA GTT TGA TCC TGG CTC AG	16S_8F

MATERIALS

Agarose
125-ml Erlenmeyer flask
Hot mitts
1× Tris-acetate-EDTA (TAE) buffer
Loading dye
PCR products from Experiment 5.2

Spatula
Analytical scale
Power supply
Minigel apparatus
1-kb DNA ladder
Glass petri dish lid

METHODS

Confirm that PCR was successful before performing large-scale purification by gel extraction.

1. Prepare a 0.8% agarose gel as described in Experiment 3.3 (steps 1a to j).

2. Prepare a 1-kb DNA ladder as follows for gel electrophoresis: 2 μl of 6× dye + 9 μl of TE buffer + 1 μl of 1-kb ladder stock (12 μl, total volume)

3. Prepare each metagenomic PCR sample as follows for gel electrophoresis: 2 μl of 6× dye + 6 μl of TE buffer + 4 μl of PCR sample (12 μl, total volume)

4. Load 12 μl of the 1-kb ladder into one of the wells, and then load your PCR samples (12 μl each) into the remaining wells. Subject the samples to electrophoresis for approximately 30 minutes at 95 mA.

5. When electrophoresis is complete, turn off the power supply and then remove the gel from the electrophoresis apparatus. Be sure to wear gloves when manipulating the gel. View the DNA bands using a UV gel-imaging system (a transilluminator); consult the TA or instructor for instructions on how to use the system. Take a picture of the illuminated gel for your laboratory notebook. Save the gel image as a JPEG (.jpg) file for upload to the CURL Online Lab Notebook (http://ugri.lsic.ucla.edu/cgi-bin/loginmimg.cgi).

6. Dispose of your gel, your gloves, and any other products contaminated with ethidium bromide (EtBr) into the appropriate waste container.

7. Examine your gel picture. Your DNA bands corresponding to the 16S rRNA amplicon should align with the 1.5-kb marker on the 1-kb ladder. Confirm that your metagenomic PCR product forms a tight band of approximately 1.5 kb. If the PCR product is smeared, the final yield will be lower and potentially refractory to subsequent cloning steps. Make sure you affix the picture to a page in your laboratory notebook with tape or glue and immediately label each lane with the identity of the sample loaded as well as the size of each DNA fragment.

Experiment continues

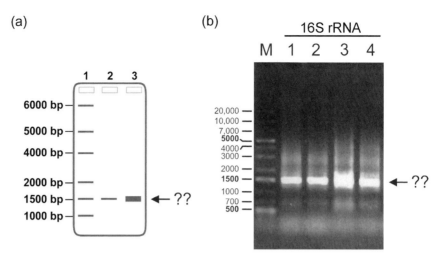

Results of gel electrophoresis. (a) Diagrammatic representation of an agarose gel following electrophoresis. Lanes: 1, 1-kb ladder; 2, PCR product from metagenomic template DNA extracted from 0.25 g of soil; 3, PCR product from metagenomic template DNA extracted from 0.75 g of soil. (b) A 0.8% agarose 1× TAE gel run for 30 min at 95 mA. Lanes: M, marker (Fermentas GeneRuler 1-kb DNA Ladder Plus); 1 to 4, PCR products from metagenomic template DNA extracted from two different locations at the Mildred E. Mathias Botanical Gardens at the University of California, Los Angeles: 1 and 2 from 1.3 g of soil at the base of an *Aloe reitzeii* (aloe) plant and 3 and 4 from 0.45 g of soil at the base of a *Chorisia insignis* tree (drunken tree). Gel picture courtesy of Brian Kirkpatrick and Areerat Hansanugrum.

> *Note:* When you read a gel, it should be oriented such that the wells in which samples were loaded are placed at the top and the order can be read from left to right. The larger fragments should be at the top, and the smaller fragments should be near the bottom. You should label the size of each band in the DNA ladder, since this can vary depending on the manufacturer. Your TA or instructor will provide this information. You should also label each lane with the sample that was loaded and note the size of each fragment in each lane, which can be deduced by comparing relative migration of bands to that in the DNA ladder. In the examples in panels a and b above, what is the approximate size of each of the bands noted by the arrow and question marks?

8. If your electrophoresis results indicate that you have successfully amplified the 16S rRNA gene, then purify the PCR products with the QIAquick gel extraction kit as described in Experiment 5.4. If no PCR product is obtained, you should follow the Decision Guide at the end of this experimental section. The guide will steer you through troubleshooting strategies for this part of the project.

BOX 5.3 Alternative Project Strategies

"In the event of failure": what to do if the cultivation-independent experiments go wrong

By the very nature of science, experiments do not always go as planned. The cultivation-independent part of the "I, Microbiologist" research experience is occasionally problematic for inexperienced students, as it presents technical challenges that are less predictable than typically encountered during the cultivation-dependent part of the project. Although perseverance is a valuable asset when troubleshooting, sometimes it becomes necessary to adapt and take on new challenges, moving the project in a different direction altogether,

BOX 5.3 Alternative Project Strategies *(continued)*

which can be very exciting. Below are two experimental approaches designed to serve as backup plans should students encounter seemingly insurmountable obstacles with the metagenomic DNA isolation and PCR procedures (Experiments 5.1 to 5.3). Note that the materials presented also could be used to expand both the cultivation-dependent and cultivation-independent surveys of soil bacterial communities to include genes other than the 16S rRNA gene.

1. Use the same isolates obtained by students performing the cultivation-dependent part of the "I, Microbiologist" to construct a phylogenetic tree based on a different gene (e.g., *rpoB*, encoding the β subunit of RNA polymerase). Compare the RpoB tree (based on protein sequences) to the 16S rRNA tree (based on DNA sequences). Are the two trees congruent, or do they have the same topology?

    Primers
    5′-AAC ATC GGT TTG ATC AAC-3′   rpoB_1698F
    5′-CGT TGC ATG TTG GTA CCC AT-3′   rpoB_2041R

2. Expand the culture-dependent analysis by using an N_2-BAP medium exclusively, amplifying the nitrogenase gene *(nifH)* with nested primer sets. Generate a phylogenetic tree based on the protein sequences for NifH.

    Primers (first step, 1.2-kb fragment comprising the entire *nifH* gene, the intergenic spacer region, and the 5′ end of the *nifD* gene)

    5′-TAC GGY AAR GCB GGY ATC GG-3′   IGK
    5′-TTG GAG CCG GCR TAN GCR CA-3′   NDR-1

    Primers (second step, 360-bp fragment of *nifH*)
    5′-TGC GAY CCS ARR GCB GGY ATC GG-3′   PolF
    5′-ATS GCC ATC ATY TCR CCG GA-3′   PolR

    Y = C or T; R = A or G; B = T, C, or G; N = A, C, G, or T

To confirm that metabolic activity is consistent with that expected for a true nitrogen fixer, consider performing acetylene reduction assays. Like N_2, acetylene (C_2H_2) contains a triple bond, the reduction of which is catalyzed by nitrogenase, producing ethylene (C_2H_4):

$$HC \equiv CH + 2H^+ + 2\ e^- \rightarrow H_2C = CH_2$$

Both the substrate and the product can be detected by gas chromatography, providing a readout for nitrogenase activity (Murry et al., 1984).

Alternative *nifH* gene primers have also been described (Bürgmann et al., 2004; Minerdi et al., 2001; Zehr et al., 1998).

REFERENCES

Bürgmann, H., F. Widmer, W. Von Sigler, and J. Zeyer. 2004. New molecular screening tools for analysis of free-living diazotrophs in soil. *Appl. Environ. Microbiol.* **70**:240–247.

Dahllöf, I., H. Baillie, and S. Kjelleberg. 2000. *rpoB*-based microbial community analysis avoids limitations inherent in 16S rRNA gene intraspecies heterogeneity. *Appl. Environ. Microbiol.* **66**:3376–3380.

Minerdi, D., R. Fani, R. Gallo, A. Roarino, and P. Bonfante. 2001. Nitrogen fixation genes in an endosymbiotic *Burkholderia* strain. *Appl. Environ. Microbiol.* **67**:725–732.

Murry, M. A., M. S. Fontaine, and J. G. Torrey. 1984. Growth kinetics and nitrogenase induction in *Frankia* sp. HFPArI 3 grown in batch culture. *Plant Soil* **78**:61–78.

Experiment continues

Poly, F., L. Jocteur-Monrozier, and R. Bally. 2001. Improvement in the RFLP procedure for studying the diversity of *nifH* genes in communities of nitrogen fixers in soils. *Res. Microbiol.* **152:**95–103.

Santos, S. C., and H. Ochman. 2004. Identification and phylogenetic sorting of bacterial lineages with universally conserved genes and proteins. *Environ. Microbiol.* **6:**754–759.

Valdés, M., N. O. Pérez, P. Estrada-de los Santos, J. Caballero-Mellado, J. J. Peña-Cabriales, P. Normand, and A. M. Hirsch. 2005. Non-*Frankia* actinomycetes isolated from surface-sterilized roots of *Casuarina equisetifolia* fix nitrogen. *Appl. Environ. Microbiol.* **71:**460–466.

Zehr, J. P., M. T. Mellon, and S. Zani. 1998. New nitrogen-fixing microorganisms detected in oligotrophic oceans by amplification of nitrogenase (*nifH*) genes. *Appl. Environ. Microbiol.* **64:**3444–3450.

Decision Guide

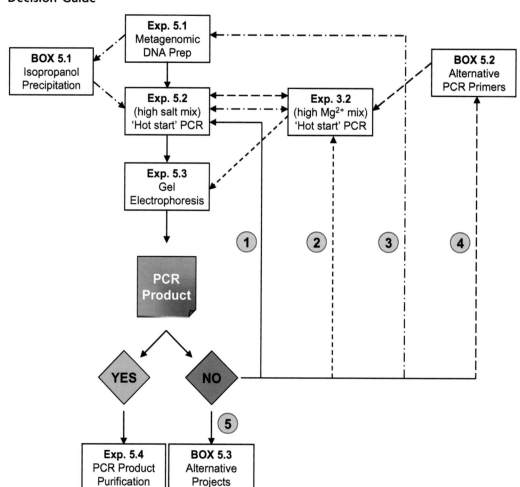

Decision Guide Summary. (1) If no PCR product is obtained after Experiment 5.3, repeat Experiments 5.2 and 5.3. (2) If still no PCR product is obtained, try the PCR master mix from Experiment 3.2 (high Mg^{2+}), which provides lower-stringency conditions compared to Experiment 5.3, but use hot-start PCR conditions. (3) No PCR product at this point suggests that the metagenomic DNA isolation (Experiment 5.1) was not successful. Repeat the entire isolation procedure by performing Experiment 5.1 again, but include the isopropanol precipitation step (Box 5.1). Note that it is recommended at this point to perform the PCR step on the new metagenomic DNA sample under both master mix conditions (Experiments 5.2 and 3.2). Do not discard any of the metagenomic DNA preparations yet. (4) If the first three troubleshooting strategies still yield no PCR products, you might repeat the PCR step by using alternative primer combinations described in Box 5.2. Note that these are the same primers used in the Gill et al. (2006) study discussed in Section 5.2. Again, it would be wise to try both Experiment 5.2 and 3.2 master mix conditions with the substitute primers. (5) If you exhaust all recommended troubleshooting strategies, you may either start over with a new soil sample, or proceed to Box 5.3 for alternative project options. The latter choice facilitates a continued collaboration with the rest of your project team.

EXPERIMENT 5.4 Purification of Metagenomic 16S rRNA Genes

Once amplification of the 16S rRNA gene has been confirmed via electrophoresis, the PCR products should be cleaned using a protocol designed to purify double-stranded DNA fragments produced by PCR. Smearing, multiple banding, primer-dimer artifacts, and PCR products larger than 1 kb necessitate gel purification before cloning. The procedure described is for the QIAquick gel extraction kit from Qiagen. Other kits are commercially available, including the DNA gel extraction kit (Fermentas), the GenElute gel extraction kit (Sigma-Aldrich), the QuickClean 5M gel extraction kit (GenScript Corporation), the PowerPrep Express gel extraction system (Marligen Biosciences), the AxyPrep DNA gel extraction kit (Axygen Biosciences), and the PureLink gel extraction kit (Invitrogen). The manufacturer of each kit supplies instructions.

MATERIALS

QIAquick gel extraction kit   P1000 and P200 pipettes
Microcentrifuge   Pipette tips
50°C heat block   "Wide-well" combs
Clean, sharp scalpel (razor blade)   Safety glasses
Isopropanol (2-propanol)   Analytical scale
3 M sodium acetate (pH 5.0)   Gel documentation system

METHODS

1. Prepare a 0.8% agarose gel as described in Experiment 3.3, except that when you assemble the minigel casting tray, place a wide-well comb at the top of the tray. The small lane is for loading the 1-kb size marker, while the four wide wells are for loading the PCR samples.

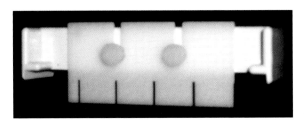

2. Prepare a 1-kb DNA ladder as follows for gel electrophoresis: 2 μl of 6× dye + 9 μl of TE buffer + 1 μl of 1-kb ladder stock = 12 μl (total volume).

3. Prepare each metagenomic PCR sample as follows for gel electrophoresis: 9 μl of 6× dye + 46 μl of PCR sample = 55 μl (total volume).

4. Load 12 μl of the 1-kb ladder into the small well, and then load one PCR sample (55 μl) into a single wide well. Repeat the loading procedure for the other three PCR samples. Electrophorese the samples for approximately 75 minutes at 95 mA.

5. When electrophoresis is complete, turn off the power supply and remove the gel from the electrophoresis apparatus. Be sure to wear gloves when manipulating the gel. View the DNA bands using a UV gel-imaging system (a transilluminator). Take a picture of the illuminated gel for your laboratory notebook. Make sure to print the actual size of the gel image.

Experiment continues

6. Weigh a colorless microcentrifuge tube on analytical scale.

7. Next, align the gel itself with the printed picture of the gel image, and excise the DNA fragments from the agarose gel with a clean, sharp scalpel or razor blade. Minimize the size of the gel slice by removing extra agarose. Return the gel to imager and view to confirm that all DNA has been excised.

   Alternative procedure (not recommended for students): Wear a safety shield and gloves to protect your eyes and skin. Viewing the DNA bands in the gel with UV light from the transilluminator, excise the DNA fragment with a scalpel or razor blade. You must work quickly if using this method, since UV light damages DNA, causing intrastrand linkages of adjacent pyrimidines (e.g., thymine dimers), which ultimately distorts the DNA helix and affects subsequent replication steps during PCR.

8. Place each excised gel slice containing the desired DNA fragment into a previously weighed microcentrifuge tube. Next, weigh each tube and gel slice on an analytical scale. Determine the mass of each gel slice by difference:

   mass of gel slice = (mass of tube + gel slice) − mass of tube alone

9. Choose only one DNA sample for further purification; temporarily store the remaining gel slices at 4°C until you confirm that you have successfully recovered a purified DNA product (Experiment 5.5). If the product is lost during the purification procedure, you may return to the other samples as a backup plan.

10. Add 3 volumes of buffer QG to 1 volume of gel slice (100 mg = 100 μl). For example, add 300 μl of buffer QG to each 100 mg of gel. The maximum amount of gel slice per single QIAquick column is 400 mg; for gel slices of >400 mg, you must use more than one QIAquick column.

11. Incubate at 50°C for 10 minutes (or until the gel slice has completely dissolved). To help dissolve the gel slice, mix by vortexing the tube every 2 to 3 minutes during the incubation.

12. Once the gel slice has dissolved completely, check that the mixture is yellow (similar to buffer QG itself without dissolved agarose). The adsorption of DNA to the QIAquick column membrane is efficient only at pH ≤ 7.5. Buffer QG contains a pH indicator that is yellow at pH ≤ 7.5 and orange or violet at higher pH. Thus, if the color of the mixture is orange or violet, add 10 μl of 3 M sodium acetate (pH 5.0) and mix. The mixture should turn yellow.

13. Add 1 gel volume of isopropanol (2-propanol) to the sample, and mix by vortexing. For example, if the agarose gel slice is 100 mg, add 100 μl of isopropanol.

14. Do not centrifuge the sample yet. Place a QIAquick spin column in a 2.0-ml collection tube (provided with the kit).

15. To bind DNA to the column membrane, apply the sample to the QIAquick column. Do not touch the pipette tip to the column membrane. Centrifuge for 1 minute at 13,000 rpm. The maximum volume of the column reservoir is 800 μl. For sample volumes greater than 800 μl, simply load the column multiple times and centrifuge between loads.

16. Discard the flowthrough, and place the QIAquick column back in the same collection tube. Collection tubes are reused to reduce plastic waste.

17. Add 0.5 ml of buffer QG to the QIAquick column, and centrifuge for 1 minute at 13,000 rpm. This step removes all traces of agarose. It is required if DNA is to be used subsequently for sequencing.

18. Wash the DNA bound to the membrane by adding 0.75 ml of buffer PE to the QIAquick column. Let stand for 3 minutes, and then centrifuge for 1 minute at 13,000 rpm.

    Important: Be sure to use buffer PE to which ethanol has been added (check the lid). You must use 200-proof ethanol in preparation of the buffer, or the DNA will not stay bound to the column during the wash steps.

19. Discard the flowthrough, and centrifuge the QIAquick column for an additional 1 minute at 13,000 rpm. This step is essential for removing residual ethanol from buffer PE, which impedes the subsequent elution step.

20. Place the QIAquick column into a clean 1.5-ml microcentrifuge tube.

21. To elute the DNA, add 30 μl of buffer EB (10 mM Tris-Cl [pH 8.5]) to the center of the QIAquick membrane. Ensure that the elution buffer is dispensed directly onto the QIAquick membrane for complete elution of bound DNA. Let the column stand for 1 minute, and then centrifuge the column for 1 minute at 13,000 rpm. The average eluate volume is 28 μl (from the 30 μl that was applied). Elution efficiency is dependent upon pH; the maximum elution efficiency is achieved between pH 7.0 and 8.5.

22. Store the purified metagenomic PCR product at −20°C.

ACKNOWLEDGMENT

This protocol has been adapted from that provided with the QIAquick gel extraction kit as described in the *QIAquick® Spin Handbook for PCR Purification Kit, Nucleotide Removal Kit, and Gel Extraction Kit*, November 2006.

EXPERIMENT 5.5 Quantification of Purified Metagenomic PCR Products

After gel purification, an aliquot of the DNA should be run on an agarose gel alongside a DNA ladder to confirm that a purified metagenomic PCR product was recovered successfully (method A below). At the same time, this step provides a qualitative assessment of the total yield after purification. If a DNA band of the expected size is observed, proceed to method B for a more quantitative determination of the purified product.

Wear gloves throughout DNA analysis.

Method A Qualitative comparative analysis

1. Run a gel as described in Experiment 3.3, except prepare the DNA sample as follows. Add 2.4 μl of DNA eluate from Experiment 5.4 to 2 μl of 6× TAE loading dye plus 7.6 μl of distilled water (dH$_2$O) for a total volume of 12 μl. Prepare a 1-kb DNA ladder as described above, and load the sample and ladder onto a 0.8% agarose gel.

2. Use the picture you obtain from this gel to make a qualitative comparison of how much DNA you have recovered relative to that visualized in the first picture taken in Experiment 5.3. The intensity of the band in the second picture should be as high as the first if you recovered the equivalent amount.

Method B Quantitative spectrophotometric analysis

MATERIALS

UV/visible spectrophotometer such as a NanoDrop (Thermo Scientific), a SpectraMax Plus384 Microplate Reader (Molecular Devices), or equivalent

METHODS

1. Obtain optical density (OD) measurements at wavelengths of 260 and 280 nm for an appropriate dilution of your DNA sample. The dilution and/or volume to be tested will be determined by your TA and instructor in accordance with the specifications and sensitivity of the UV/visible spectrophotometer used for this project.

2. Calculate the molar DNA concentration by using the OD at 260 nm (OD$_{260}$) and check for purity by using the OD$_{260}$/OD$_{280}$ ratio: one absorbance unit at a wavelength (λ) of 260 nm is equivalent to a double-stranded DNA molar concentration of 0.15 mM (0.15 mmol/liter). In other words:

$$\frac{0.15 \text{ mM}}{1 \text{ Å}} = \frac{x \text{ mM}}{\text{Measured OD}_{260}}$$

Solve for x, and then multiply by your dilution factor. This calculation will give you the molar concentration of your "purified metagenomic PCR product" in millimolar (mM) units.

Hint: You will find it wise to convert from mM (millimoles per liter; milli = 10^{-3}) to nM (nanomoles per liter; nano = 10^{-9}) to make later calculations more straightforward.

3. To check for purity, divide the OD$_{260}$ by the OD$_{280}$. For pure DNA, the OD$_{260}$/OD$_{280}$ should be between 1.7 and 1.8. Ratios lower than 1.7 indicate protein contamination in your sample. Ratios higher than 1.8 indicate RNA contamination in your sample. A ratio of 2 indicates pure RNA.

MATERIALS

TOPO TA cloning kit (Invitrogen no. K4500-01)

Fresh purified metagenomic PCR product from Experiment 5.5

One Shot TOP10 chemically competent *E. coli*

LB–Kan–X-Gal plates	Sterile glass beads (~12 per 13-mm tube)
LB-Kan plates (50 μg/ml)	42°C water bath
LB-Kan broth (50 μg/ml)	37°C shaker
SOC medium	42°C shaker
2.0-ml microcentrifuge tubes	P10 pipette and tips
1.8-ml microcentrifuge tubes	

METHODS

Keep all reagents, including PCR products, on ice throughout the setup process.

PERIOD 1
Part I Ligation Reaction and TOPO Cloning Reaction

Note: Because TOPO kits are expensive and the time in which to complete the project may be very limited, each student should attempt the TOPO cloning procedure one time per the protocol described below. If poor results are obtained (e.g., few or no recombinant colonies), it is recommended that the TA or instructor use an aliquot of the undiluted purified metagenomic PCR product to repeat the cloning reaction on behalf of the student.

1. Use the purified metagenomic PCR product to set up a maximum of two cloning reactions. Additional reactions require discussion with and approval by the TA or instructor before proceeding. Before diluting your purified metagenomic PCR product in the next step, remove 5 μl and transfer to a fresh 1.8-ml microcentrifuge tube, clearly labeled with your initials, date, and sample identification number (ID). Submit this undiluted sample to your TA or instructor.

2. For a cloning reaction to work efficiently and reproducibly, using a 1:1 molar ratio of purified metagenomic PCR product to TOPO vector is recommended by the manufacturer. A dilution of the PCR product in the range of 1:25 to 1:100 is typically appropriate.

    Prepare 1:25 and 1:50 dilutions of your purified metagenomic PCR product as described below for use in subsequent TOPO cloning reactions.

    5 μl of PCR product + 120 μl of buffer EB = 125 μl of a 1:25 dilution

    5 μl of PCR product + 245 μl of buffer EB = 250 μl of a 1:50 dilution

    Dilution tubes should be labeled and stored at −20°C. Do not discard undiluted stock; store it at −20°C together with dilution tubes.

    If troubleshooting becomes necessary, the actual molar ratio must be calculated and utilized for subsequent TOPO cloning reactions. The concentration of the pCR2.1-TOPO vector is 10 ng/μl, or 3.9 nM (length, 3.9 kb). You may use the molar concentration data from Experiment 5.5 and the molar concentration of the TOPO vector to calculate the dilution required to bring the purified metagenomic PCR product to the appropriate concentration for the TOPO cloning reaction.

Experiment continues

3. Using a P10 pipette, mix the components in a microcentrifuge tube, one for each diluted metagenomic PCR product (be sure to add them in the order given): PCR product, salt, and TOPO vector. Because it is difficult to pipette a small volume of a highly viscous solution accurately, the TA or instructor may aliquot the TOPO vector for the student.

   Reminder: Make sure that you keep all components on ice throughout the setup procedures. The TOPO vector should not be removed from the −20°C freezer until all other components of the TOPO cloning reaction have been mixed. The topoisomerase enzyme that is covalently attached to vector backbone is temperature labile.

Reaction component	Volume (μl)
Fresh PCR product (diluted 1:25 or 1:50)	4
Salt solution	1
pCR2.1-TOPO vector	1
Final volume	6

4. Mix the TOPO cloning reaction components by gently swirling the pipette tip. Incubate the mixture for 30 minutes at ambient temperature (22 to 23°C), then proceed immediately to Part II.

Part II Transformation of *E. coli* with TOPO Cloning Reaction Mixture

One Shot TOP10 chemically competent *E. coli* genotype: F⁻ *mcrA* Δ(*mrr-hsdRMS-mcrBC*) φ80*lacZ*ΔM15 Δ*lacX74 recA1 araD139* Δ(*ara-leu*)7697 *galU galK rpsL* (Strʳ) *endA1 nupG.*

For each transformation reaction, you will need one vial of competent cells and three LB–Kan–X-Gal plates. Prewarm the SOC medium to room temperature and the LB–Kan–X-Gal plates to 37°C for at least 30 minutes. Thaw on ice one vial of competent cells for each transformation reaction.

Check the temperature of the water bath. Look at both the setting and the thermometer. Make sure that both read 42°C before proceeding with transformation steps

1. Add all of the TOPO cloning reaction mixture from Part I to a vial with 50 μl of One Shot TOP10 chemically competent *E. coli* cells, and mix gently. Do not mix by pipetting up and down; mix by gently flicking the tube.

2. Incubate on ice for 5 minutes.

3. Heat shock the cells for 30 seconds at 42°C without shaking. (*Reminder:* Check the water bath temperature again.)

4. Immediately transfer the tubes back on ice for 1 minute.

5. Place the tubes in your microcentrifuge rack, and add 250 μl of room temperature SOC medium.

6. Cap the tubes tightly, and shake them horizontally (at 200 rpm) at 37°C for 1 hour.

7. Spread 50, 100, and 150 μl of cells from each transformation onto each of three prewarmed LB–Kan–X-Gal plates by using the following glass bead shaking technique:

   • Carefully pour 10 to 12 sterile glass beads onto an agar plate.

   • Add 50 μl of cell suspension to the center of one agar plate, 100 μl to the second plate, and 150 μl to the third plate. To ensure even spreading of small volumes, add another 100 μl of SOC medium to the plate with only 50 μl of cell suspension and add another 50 μl of SOC medium to the plate with 100 μl of cell suspension.

- Close the lid of the plates.

- Gently shake the beads across the surface of the agar six or seven times. To ensure that the cells spread evenly, use a horizontal shaking motion; do not swirl the beads, or else all the cells will end up at the edge of the plate. (*Hint:* The procedure sounds like "shaking maracas" if done properly.) If unsure, have your TA or instructor demonstrate the technique.

- Rotate the plate by 60°, and then shake it horizontally another six or seven times. Rotate it by 60° another time, and shake it horizontally again. By now, you should have achieved even spreading of the cells across the agar surface.

- When the spread plating is done, pour off contaminated beads into a marked collection container. *Do not discard the beads.* The used beads will be autoclaved, washed, and resterilized for repeat usage.

  Note: If the agar surface is still wet after three rounds of shaking, allow the plate to sit for several minutes to allow some drying to occur, and then repeat the shaking procedure until all the medium is absorbed by the agar.

8. Incubate the plates overnight at 37°C. An efficient TOPO cloning reaction should produce several hundred colonies.

9. Discard any remaining transformation mixture into the waste containers.

PERIOD 2
Streak-Purify Transformants

1. Choose 32 colonies for further analysis. Label the candidates with a sample ID as shown in this example:

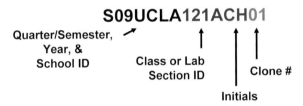

2. Streak purify each of the 32 colonies onto LB-Kan plates (you do not need to use plates containing X-Gal for purification steps). Incubate at 37°C for 24 hours, and then go to Experiment 5.7.

3. Do not discard the transformation plates. Store them in a cold box (4°C). You may need to return to these plates later in the project to obtain additional candidate colonies to screen for 16S rDNA inserts.

Experiment continues

BOX 5.4 Troubleshooting the Cloning Procedure

ADDITIONAL MATERIAL
10 mM dATP (dilute 100 mM stock 1:10 in sterile H_2O)

If the DNA was amplified by a proofreading polymerase (not *Taq*) such that the 3′-A overhangs necessary for TA cloning have been removed, use the following procedures to facilitate cloning of blunt-ended DNA fragments.

This series of steps must be done prior to purification of PCR products.

1. After PCR cycling with the proofreading polymerase is complete, place PCR tubes on ice and add 0.7 to 1 unit of *Taq* polymerase to each tube (you do not need to change the buffer). Mix well by flicking the tubes.

2. Incubate the tubes at 72°C for 10 minutes.

3. Transfer the tubes to ice for immediate use in the TOPO cloning reaction.

This series of steps may be done after PCR products have been amplified with proofreading polymerase and gel purified.

1. Mix the PCR product with *Taq* polymerase buffer (1×), dATP (0.2 mM), and 0.5 unit of *Taq* polymerase.

2. Incubate the tubes at 72°C for 15 minutes.

3. Transfer the tubes to ice for immediate use in the TOPO cloning reaction.

Invitrogen also sells a TOPO TA cloning kit for sequencing, which utilizes the vector pCR4-TOPO containing a DNA insertion site located within the *lacZ-ccdB* gene fusion. The *ccdB* gene encodes a protein that poisons bacterial DNA gyrase, causing degradation of the chromosome and cell death. Ligation of PCR products into the vector disrupts *ccdB* gene expression, enabling only recombinant colonies to grow on either LB-Amp or LB-Kan plates, avoiding the use of X-Gal. This assay results in a positive selection of recombinant cells.

ACKNOWLEDGMENT
This protocol has been adapted from that provided with the TOPO cloning kit (Invitrogen Corp., 2006).

Once unpigmented, kanamycin-resistant, single colonies are obtained using the streak plate procedure, the plasmids must be isolated from each candidate cell population by a method designed to separate plasmids from chromosomal DNA and other cellular materials. The procedure described is for the QIAprep miniprep kit from Qiagen, which involves spin columns in a microcentrifuge. Numerous alternative kits are commercially available, including the PureLink quick plasmid miniprep kit (Invitrogen), the UltraClean standard miniplasmid prep kit (MoBio Laboratories, Inc.), the SpinPrep plasmid kit (Novagen), the Wizard *Plus* miniprep DNA purification system (Promega), the GenElute plasmid miniprep kit (Sigma-Aldrich), and the StrataPrep plasmid miniprep kit (Stratagene). The manufacturer of each kit supplies instructions.

MATERIALS

QIAprep miniprep kit

LB-Kan broth (50 μg/ml)

2.0-ml microcentrifuge tubes (sterile)

Streak plates from Experiment 5.6

18-mm test tubes

37°C rotator or shaking incubator

METHODS

PERIOD 1
Inoculate Cultures

1. Inoculate tubes containing 4.0 ml of LB broth plus 50 μg of kanamycin per ml (LB-Kan) each with a single colony from streak plates (you must have well isolated colonies; if not, you must repeat the streak plate procedure until you obtain them). Label the tubes with the appropriate sample ID number (recall the labeling scheme described in Experiment 5.6).

   Note: Inoculate one tube with a blue colony from the original transformation plates. The plasmid purified from this culture will serve as the negative control in Experiment 5.8.

2. Incubate the cultures in the 37°C rotator or shaking incubator overnight.

PERIOD 2
Isolate Plasmid DNA

1. Decant overnight cultures into two separate 2.0-ml centrifuge tubes, each labeled with an appropriate sample ID number (e.g., F08UCLA121ACH01A and F08UCLA121ACH01B). Treat samples A and B as separate minipreps of the same culture.

2. Centrifuge the cell suspensions at 10,000 rpm for 2 minutes.

3. Pour off supernatant into waste containers.

4. Repeat steps 1 to 3 with the remaining volumes of overnight cultures. Be sure to decant the corresponding culture into each tube. Do not cross-contaminate the samples.

Experiment continues

5. Store all cell pellets in B tubes at −20°C. For now, continue with the miniprep procedure for only cell pellets in A tubes.

   Note: In Experiment 5.8, you will perform a restriction digest to screen your DNA plasmids for the presence of 16S rDNA inserts. If you confirm that a given sample contains an insert of the appropriate size, you should finish the miniprep procedure with the cell pellets in the B tubes. The plasmid DNA from the B tubes is to be used for DNA sequencing in Experiment 5.9.

6. Add 250 μl of resuspension buffer (P1 buffer) (50 mM Tris-HCl [pH 8.0], 10 mM EDTA, 10 μg of RNase A/ml) to the 2.0-ml tubes. P1 buffer is used to solubilize the cell pellet. Tris-HCl is a buffer, EDTA binds divalent cations preventing DNase activity, and RNase removes RNA.

7. Resuspend the cell pellets by pipetting up and down several times or vortexing the tubes until homogeneous cell suspensions are obtained (no cell clumps should be visible).

8. Add 250 μl of lysis buffer (P2 buffer) (100 mM NaOH [pH approximately 12.3], 1% SDS). At the high pH, the sodium hydroxide (NaOH) denatures the DNA and cleaves some of the large RNA fragments. Meanwhile, SDS detergent solubilizes the cell membrane and denatures proteins, resulting in complete lysis of cells.

9. Mix thoroughly but gently, by inverting the tubes four to six times. *Do not vortex,* as this results in shearing of genomic DNA. If necessary, continue inverting the tubes until the solutions become viscous and slightly clear. *Do not allow the lysis reaction to proceed for more than 5 minutes.*

10. Add 350 μl of neutralization buffer (N3 buffer) (3.0 M potassium acetate [pH 5.5]). Note that N3 buffer contains potassium acetate at a very high salt concentration. On addition of the buffer to the cell lysate, the pH of the solution returns to around neutral (pH 7), causing the proteins to precipitate and DNA to rehybridize.

11. Mix immediately and thoroughly by inverting the tube four to six times.

12. Centrifuge for 10 minutes at high speed (13,000 rpm). A compact white pellet will form. Meanwhile, label both the spin columns and collection tubes for subsequent steps.

13. Decant the supernatants into the QIAprep spin column. Avoid the pellet; it will clog your column if it is added.

14. Centrifuge at high speed (13,000 rpm) for 1 minute. Discard the flowthrough into waste containers.

15. Add 750 μl of PE wash buffer to the QIAprep spin column. PE contains ethanol and is used to wash residual salt from the column while allowing DNA to remain bound. Do not touch the pipette tip to the column membrane.

    Important: Be sure to use PE buffer to which ethanol has been added (check the lid). You must use 200-proof ethanol when preparing the buffer, or else DNA will not stay bound to the column during the wash steps.

16. Centrifuge at high speed (13,000 rpm) for 1 minute. Discard the flowthrough into waste containers.

17. Centrifuge at high speed (13,000 rpm) for an additional 1 minute to remove residual wash buffer, which can inhibit subsequent enzymatic reactions (e.g., DNA sequencing) if not entirely eliminated from the column.

    Important: Residual ethanol from the PE wash buffer is not completely removed unless the flowthrough is discarded before this additional centrifugation step.

18. Place the QIAprep spin column into a clean microcentrifuge tube. Be sure to label both the cap and side of the tube (in case the cap should break off in the subsequent centrifugation step).

19. To elute the plasmid DNA, add 40 μl of EB buffer (10 mM Tris-Cl [pH 8.5]) to the center of each membrane in the QIAprep spin columns. Let the columns stand for 1 minute, then centrifuge at high speed (13,000 rpm) for 1 minute. You must use a lid with the centrifuge to prevent the caps of the microcentrifuge tubes from breaking off during the centrifugation.

20. Use the DNA eluate from the A tubes for restriction analysis of recovered plasmids in Experiment 5.8. Later, you will use eluate from the B tubes for DNA sequencing in Experiment 5.9.

21. The DNA samples may be stored in the freezer at −20°C. Be sure to note which tube corresponds to the negative control (plasmid without an insert purified from a blue colony).

ACKNOWLEDGMENT

This protocol has been adapted from that provided with the QIAprep miniprep kit as described in the *QIAprep® Miniprep Handbook for Purification of Molecular Biology Grade DNA*, December 2006.

Once plasmid DNA has been isolated from each clone, students must confirm that the meta-genomic 16S rDNA PCR insert was successfully inserted into the pCR2.1-TOPO vector. To screen the plasmid candidates, restriction analysis is performed with the endonuclease EcoRI, which excises the insert DNA from the vector. Using agarose gel electrophoresis, the digested product is screened for the presence of a DNA band of the expected size (1.5 kb). Note that there may be internal EcoRI restriction sites, giving rise to a number of smaller bands whose sizes add up to 1.5 kb. The restriction pattern (or DNA fingerprint) should be recorded and compared to results of the phylogenetic analysis performed in Unit 7.

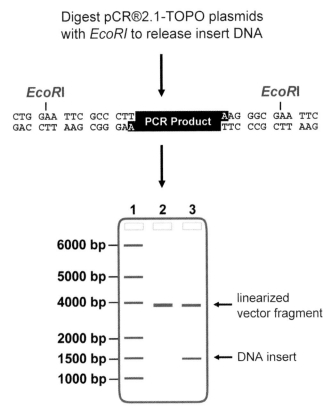

Overview of the secondary screening procedure, giving a dia-grammatic representation of an agarose gel following electro-phoresis; possible products produced by a restriction digest with EcoRI are shown. Lanes: 1, 1-kb ladder; 2, complete digest with EcoRI of a plasmid purified from a blue colony (no DNA insert); 3, complete digest with EcoRI of a plasmid containing a 16S rDNA insert with no internal EcoRI sites.

Note: When you read a gel, it should be oriented such that the wells in which samples were loaded are placed at the top and the order can be read from left to right. The larger fragments should be at the top, and the smaller fragments should be near the bottom. You should label the size of each band in the DNA ladder, since this can vary depending on the manufacturer. You should also label what sample was loaded in each lane, and note the size of each fragment in each lane, which can be deduced by comparing the migration of bands to the migration of those in the DNA ladder. In the example above, what are the approximate sizes of each of the bands indicated by arrows?

Experiment continues

EXPERIMENT 5.8

MATERIALS

Plasmid DNA (tube A) from Experiment 5.7
EcoRI restriction endonuclease
37°C water bath   Sterile distilled H$_2$O
EcoRI 10× buffer   6× TAE load dye with 6% SDS

METHODS

1. Make up a restriction digest master mix, producing enough for each of your clones plus one. Do not vortex to mix. Instead, pipette up and down once all reaction components have been added.

2.0 µl × (no. of reactions + 1) =	µl of 10× EcoRI buffer
12.5 µl × (no. of reactions + 1) =	µl of sterile distilled H$_2$O
0.5 µl × (no. of reactions + 1) =	µl of EcoRI
15.0 µl × (no. of reactions + 1) =	µl of **restriction digest master mix**

2. Add 15 µl of master mix to clean microcentrifuge tubes. Then transfer 5 µl of each of your purified plasmids to the tubes to which master mix has been added. Be sure to label the tubes with the appropriate sample ID so that you know which clone is which. Also, be sure that one tube is set up as the negative control.

3. Incubate the digests for 1 hour at 37°C.

4. At the end of the incubation, add 4 µl of 6× loading dye containing 6% SDS to each tube. This gives you a final concentration of 1% SDS in each tube, which stops the restriction digestion by denaturing the EcoRI enzyme, inhibiting further activity.

5. Prepare a 0.8% agarose gel and a 1-kb DNA ladder as described in Experiment 3.3. Be sure to load a DNA ladder control in one well for each comb.

6. Load only 12 µl of each of your digest reaction mixtures, including the negative control sample. Underlay your samples, including controls, into wells, and then run the gel at 95 mA for 30 to 45 minutes. If two combs are used for one gel, do not allow the fast dye (bromophenol blue) from the top comb samples to enter the bottom comb wells.

7. When finished, take a picture of the illuminated gel for your lab notebook. Save the gel image as a JPEG or TIF file, as this picture will be uploaded to the "I, Microbiologist" database (CURL Online Lab Notebook).

8. Examine your gel picture. Determine the plasmids into which you have cloned the 16S rRNA gene. The band corresponding to the DNA insert should align with the 1.5-kb marker on the 1-kb ladder. Make sure you adhere the picture to a page in your lab notebook with tape or glue, and immediately label each lane with the identity of the sample loaded as well as the size of each DNA fragment.

9. If you have successfully cloned an insert of the expected size (1.5 kb), place the A tubes containing the rest of your plasmid DNA in a box designated by the TA or instructor for long-term storage at −20°C. Then proceed to Experiment 5.9.

   Note: You will sequence only 16 clones for your phylogenetic tree. Thus, if after screening your clones by EcoRI digestion you do not have a total of 16 candidates yet, continue to Experiment 5.9 with those that you have, go back to the transformation plates, and select more colonies to streak purify and screen for 16S rRNA insert DNA (Experiments 5.6, 5.7, and 5.8).

To determine the composition and order of bases in the purified PCR products, an in-house or commercial DNA sequencing service may be used to perform the DNA sequencing reactions and subsequent gel analysis. Alternatively, if the student laboratory is equipped with a DNA analyzer, students may process the samples themselves as part of the project. It is at the discretion of the instructor to determine the most suitable means by which to accomplish this aspect of the project and provide instructions to students accordingly.

The following primers will be used to sequence the 16S rRNA gene within the pCR2.1-TOPO vector:

$5' \rightarrow 3'$ sequence	Primer name	Position[a]	Template
TAA TAC GAC TCA CTA TAG GG	T7_seq	+68 to +87 insert flank	pCR2.1-TOPO
CAG GAA ACA GCT ATG AC	M13R_seq	−89 to −73 insert flank	pCR2.1-TOPO

[a]Refer to vector diagram in Fig. 5.4 (Section 5.1).

These primers are complementary to conserved sequences within regions of the pCR2.1-TOPO vector that flank the 16S rRNA gene inserts. Using bidirectional sequencing technology, the entire DNA insert is sequenced.

MATERIALS

Tube B of purified plasmid DNA from confirmed clones (Experiment 5.7)
Sequencing primers: T7_seq and M13R_seq

METHODS

PERIOD 1

Prepare and submit tube B samples for DNA sequencing or perform the DNA sequencing reactions and gel analysis as directed by the TA or instructor.

Although your team may have successfully cloned as many as 32 candidates, you will sequence only 16 clones—you choose which ones. Store the remaining clones at −20°C. Some of these may be sequenced later if any of the first 16 do not yield usable data.

PERIOD 2

Inspect and manually edit DNA sequences to be used for bioinformatics and phylogenetic analysis.

The quality of the DNA sequence obtained for each of your clones must be assessed as follows before any further analysis takes place.

1. For each cloned gene there should be a pair of T7 and M13R DNA sequences. First examine the chromatograms (they should be .pdf files) for the corresponding DNA sequence pair. A good-quality DNA sequence will have reasonably sharp peaks with height that is uniformly above background for all four bases as shown:

Experiment continues

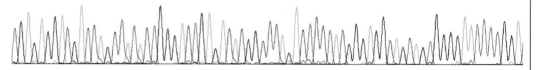

A sequence of substandard quality also is presented below for comparison. Note how difficult it is to resolve any of the base assignments because of overlapping peaks. This sample could have been contaminated with multiple plasmids from different clones, the concentration of the DNA subjected to sequencing may have been too low, or the quality of the DNA sample may have been poor.

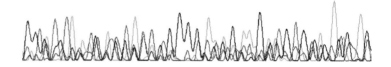

Both the T7 and M13R DNA sequences must be of good quality to continue to the next stage of the data analysis. If only one of the two sequences is of good quality, you should resubmit the plasmid DNA sample for sequencing with the primer that produced the poor-quality sequence. You may have to go back to your streak plates and repeat the plasmid DNA preparation (Experiment 5.6). However, if both the T7 and M13R sequences are of poor quality, you should move on and submit an alternative clone candidate for sequencing instead.

Note: If time and resources permit, one could instead retransform the plasmid DNA into fresh competent cells and then reisolate fresh plasmid DNA for resequencing.

Print the chromatograms, regardless of quality, in color (not black and white or grayscale), and paste/tape the printouts in your laboratory notebook.

2. If visual inspection shows that your T7 and M13R DNA sequences are of a sufficiently reliable quality for further use, open the corresponding FASTA files (they should be plain text files with .txt extension). The nucleotide bases as reported above the peaks in each chromatogram have been transcribed into an accompanying plain text file, retaining the base order as called by the DNA sequencing software. The format of a FASTA nucleotide sequence record is as follows, with a greater-than symbol ($>$) preceding a short description line that is followed by the DNA sequence in uppercase or lowercase letters:

```
>F08UCLA121ACH01_T7
CGGATCGGCTATCTGTGGTACGTCAAACAGCAAGGTATTAACTTACTGCCCTTCCTCCCA
ACTTAAAGTGCTTTACAATCCGAAGACCTTCTTCACACACGCGGCATGGCTGGATCAGGC
TTTCGCCCATTGTCCAATATTCCCCACTGCTGCCTCCCGTAGGAGTCTGGACCGTGTCTC
AGTTCCAGTGTGACTGATCATCCTCTCAGACCAGTTACGGATCGTCGCCTTGGTAGGCCT
TTACCCCACCAACTAGCTAATCCGACCTAGGCTCATCTGATAGCGTGAGGTCCGAAGATC
CCCCACTTTCTCCCTCAGGACGTATGCGGTATTAGCGCCCGTTTCCGGACGTTATCCCCC
ACTACCAGGCAGATTCCTAGGCATTACTCACCCGTCCGCCGCTGAATCCAGGAGCAAGCT
CCCTTCATCCGCTCGACTTGCATGTGTTAGGCCTGCCGCCAGCGTT
```

At both ends of every DNA sequence there likely are N's, each of which represents an ambiguous base call made at a particular position. There also may be a few scattered N's within the internal stretches of the sequences.

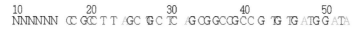

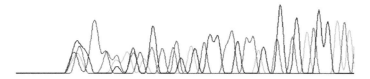

Find the positions of the N's in the appropriate chromatogram to see if you can manually resolve the base calls by inspecting the peaks themselves. Pencil the appropriate base call directly onto the chromatogram printout, and modify the DNA sequence FASTA file accordingly. It is unlikely that you will be able to resolve the ambiguous or incorrect base calls comprising the first 20 bases or so in each sequence, as this is immediately downstream of the position within the gene where either the T7 or M13R primer annealed. These nucleotides may be deleted from the text file (be sure to save your modified text file with a new name; you always want to retain the original FASTA file unchanged).

The length of the DNA sequence expected for each primer should be approximately 700 to 800 bases. The DNA sequence data become less and less reliable beyond this length, so these eventually may be deleted from your FASTA text file. Use the DNA sequence files to establish where the T7 and M13R sequences overlap, assembling them into a single consensus sequence in Experiment 6.1. To facilitate construction of an optimized multiple sequence alignment in Experiment 7.2, it is probable that you will end up deleting additional bases from the beginning and end of the consensus sequence. Be sure to save your modified sequences once more in FASTA format for use in the experiments comprising Units 6 and 7.

UNIT 6

Bioinformatics Analysis of 16S rRNA Genes*

SECTION 6.1
MATHEMATICAL ALGORITHMS AS THE BUILDING BLOCKS OF DNA SEQUENCE ALIGNMENTS

This unit focuses on the methodology used to compare nucleotide sequences and deduce whether the sequences have an evolutionary relationship based on the degree to which any two or more sequences are similar. This type of comparative analysis also can be applied to protein sequences to investigate functional or structural conservation. A presentation of the concepts and tools needed for comparative analysis of nucleotide sequences (DNA and RNA) is provided. Students interested in learning more about inferring evolutionary relationships from protein sequences are encouraged to consult the literature (Baxevanis, 2005; Dunbrack, 2006).

Alignment basics

Obtaining an accurate alignment is the first and most important step in constructing a phylogenetic tree, which is used to depict evolutionary relationships between and among sequences. It is too easy to overlook the importance of troubleshooting alignments. However, if the alignment used to construct the phylogeny is not optimal, the tree is equally unreliable, even when bootstrapping values or other phylogenetic statistics indicate robustness. Although details about phylogenetic tree construction are presented in Unit 7, a conceptual understanding of alignments in an evolutionary context is critical for the responsible use of bioinformatics tools such as the Basic Local Alignment Search Tool (BLAST) and Ribosomal Database Project (RDP-II).

But what is an alignment? This concept actually is not as straightforward as it might seem. An alignment is used to infer evolutionary models, but it also represents an evolutionary model in its own right. Just as our understanding of DNA sequence evolution involves nucleotide substitutions, insertions, and deletions, an accurate alignment con-

---

*This unit was coauthored with Craig Herbold.

tains information about these changes. One can use alignments in analyzing the evolution of a sequence by assigning an alignment position value to each nucleotide. Nucleotides must be identical by descent in order for them to occupy the same alignment position. In most cases, the alignment position is quite different from its actual sequence position.

As shown in Fig. 6.1, the effects of substitutions and a deletion on an alignment of a series of successive sequences are compared. An original, or ancestral, sequence is compared to the sequences that would be produced from the indicated nucleotide changes. In the original sequence (sequence 1), there is a G at nucleotide position 2 and a C at position 6. After two point mutations occur, resulting in a transition at position 2 (G to A) followed by a transversion at position 6 (C to G), the resulting sequence (sequence 3) differs in composition from the original sequence by two nucleotides but the overall sequence length stays the same. In the original sequence and its two descendants (sequences 1 to 3), there is an A at nucleotide position 9. After the deletion event, in which the G at position 8 is lost, the A that was initially in position 9 ends up in position 8 of the sequence and the overall sequence length is shortened by one nucleotide. In the alignment, however, the deleted character is represented by a dash (−), which functions as a placeholder enabling comparison of this sequence to those that do not contain the deletion. Thus, in the alignment, the A in sequence 4 remains in position 9 of the aligned sequence.

Figure 6.2 illustrates the effect of insertions on an alignment. This series of sequences starts with the same original sequence as the series depicted in Fig. 6.1. It shares the first nucleotide transition at position 2 of the second sequence. However, this shared change has been followed by two single-nucleotide insertions as sequentially presented in sequences 5 and 6. Comparison with the original sequence (sequence 1) shows that one insertion precedes position 2 while the other comes before position 9, changes that have the net effect of increasing the overall length of the sequence by two nucleotides (compare sequence 6 to sequence 1). In the alignment, the dash is once again utilized, but this time in the ancestral sequences that do not contain the additional nucleotides. These placeholders facilitate alignment of sequences that vary in length due to the presence of sequence gaps, which are representative of the insertion and deletion events taking place over time.

In Fig. 6.1 and 6.2, the series of sequences are directly related in the sense that one sequence evolved into the next sequence, which evolved into the subsequent sequence and so on through the gradual accumulation of changes in the primary sequence. In a real comparative analysis, these ancestral and intermediate sequences are almost never

FIGURE 6.1 Evolution of sequence 4. Sequences are numbered according to an arbitrary order of descent. Nucleotides affected by the indicated change, either a point mutation or a deletion, are highlighted in color. Numbering of nucleotide positions starts with 1 to 9 in black font, then switches to blue at position 10, which is given as position 0. Gaps are depicted as dashed lines in the sequence alignment (last column). Illustration by Erin Sanders and Craig Herbold.

Sequence Number	Types of changes	Sequences	Alignment
	Nucleotide Position:	1234567890	1234567890
1	Original	AGCTTCGGAT	AGCTTCGGAT
2	Transition (G/A)	AACTTCGGAT	AACTTCGGAT
3	Transversion (C/G)	AACTTGGGAT	AACTTGGGAT
4	Deletion (ΔG)	AACTTGGAT	AACTTGG-AT

Sequence Number	Types of changes	Sequences	Alignment
	Nucleotide Position:	123456789012	123456789012
1	Original	AGCTTCGGAT	A-GCTTCGG-AT
2	Transition (G/A)	AACTTCGGAT	A-ACTTCGG-AT
5	Insertion (+C)	ACACTTCGGAT	ACACTTCGG-AT
6	Insertion (+T)	ACACTTCGGTAT	ACACTTCGGTAT

FIGURE 6.2 Evolution of sequence 6. Sequences are numbered according to an arbitrary order of descent, deliberately avoiding overlap with Fig. 6.1 in numbering of the last two sequences. Nucleotides affected by the indicated change, either a point mutation or an insertion, are highlighted in color. Numbering of nucleotide positions starts with 1 to 9 in black font, then switches to blue at position 10, which is given as position 0, followed by positions 11 and 12, given as positions 1 and 2, respectively. Gaps are depicted as dashed lines in the sequence alignment (last column). Illustration by Erin Sanders and Craig Herbold.

available. Only the two final sequences (e.g., sequences 4 and 6) are available, either because the organisms from which the ancestral or intermediate sequences came are extinct or because they have not yet been discovered. In Fig. 6.3, we follow the evolution of our sequences again. This time, however, the final sequences, representing extant species, are aligned pairwise with respect to one another instead of to their ancestors.

The evolution of both sequences 4 and 6 begins identically, with a single point mutation at position 2 in the shared ancestral sequence. After the common G-to-A transition, several additional nucleotide changes occur that affect one sequence or the other but not both. These are the same changes that were depicted in Fig. 6.1 and 6.2, and here the effects of each sequential change on the pairwise alignments are shown. The order of changes is arbitrarily chosen to represent the two evolutionary pathways from

FIGURE 6.3 Independent evolutionary pathways give rise to sequences 4 and 6. The font colors used to depict sequence changes and the numbering scheme for nucleotide positions are the same as in Fig. 6.1 and 6.2. The ancestral and intermediate sequences for those that gave rise to sequences 4 and 6 are aligned pairwise, with a graphical depiction of each alignment step presented on the right. The grey circles depict the sequences themselves, with the numbers corresponding to those listed in the first column. A dashed line links the two sequences being compared once the ancestral sequences start to diverge or accumulate changes independent of one another. Each vertical step represents a single change in the DNA sequence that takes place along the independent pathways leading to sequences 4 and 6. Illustration by Erin Sanders and Craig Herbold.

Sequence Number	Types of changes	Pathway	Sequences	Alignment	Aligned Sequences (Pathways)
	Nucleotide Position:		123456789012	123456789012	
1	Original	Both 4 & 6	AGCTTCGGAT	AGCTTCGGAT	
1			AGCTTCGGAT	AGCTTCGGAT	
2	Transition (G/A)	Both 4 & 6	AACTTCGGAT	AACTTCGGAT	
2			AACTTCGGAT	AACTTCGGAT	
3	Transversion (C/G)	4 only	AACTTGGGAT	AACTTGGGAT	
2			AACTTCGGAT	AACTTCGGAT	
4	Deletion (ΔG)	4 only	AACTTGGAT	AACTTGG-AT	
2			AACTTCGGAT	AACTTCGGAT	
4	Insertion (+C)	6 only	AACTTGGAT	A-ACTTGG-AT	
5			ACACTTCGGAT	ACACTTCGGAT	
4	Insertion (+T)	6 only	AACTTGGAT	A-ACTTGG--AT	
6			ACACTTCGGTAT	ACACTTCGGTAT	

which sequences 4 and 6 could have descended. While shifting the nucleotide makeup of the sequences, it can be seen that transitions and transversions do not affect the overall length of the alignment. Instead, insertions and deletions exert the greatest influence on alignment structure.

The final alignment between sequences 4 and 6 is depicted in Fig. 6.4. The alignment contains very important information about the level of relatedness of the two descendents of the original ancestral sequence (sequence 1). It is known, for example, that three changes occurred in the evolution of each sequence. One of these changes is shared, and two are independent of one another. Therefore, the final descendents of the ancestral sequence should differ from one another by a total of four changes, two along each independent evolutionary path. An examination of the alignment of sequences 4 and 6 reveals that there are indeed four changes distinguishing the sequences. Three of these changes are represented by gap characters in sequence 4. It is interesting that these gap characters arose from different events: two resulted from insertions into what became sequence 6 while one resulted from a deletion even in what became sequence 4. Therefore, the two gaps represent insertions while one gap represents a deletion. The terms "insertion" and "deletion" indicate that the ancestral state is known. In most cases, the ancestral state is not known with any degree of confidence, and it is impossible to determine which gaps represent insertions and which represent deletions. Thus, the term indel is used to collectively refer to insertions and/or deletions in sequence alignments.

Pairwise alignment methods

The number of changes that separate two sequences can be understood by using a distance metric termed the edit distance, which is simply the number of changes required to transform one sequence into the other sequence (Jones and Pevzner, 2004). Continuing with the example described above, the edit distance between sequences 4 and 6 is four, and the edit distance between either sequence and its ancestor is three. In theory, the most optimal alignment of two sequences produces a lower edit distance because a larger number of identical nucleotides are aligned to one another.

A related but inverse metric is the nominal score (Altschul et al., 1997). For the nominal score, points are awarded for the number of positions at which the nucleotide in one sequence matches the nucleotide in the other sequence. Then, points are subtracted for each mismatch and each gap (−) in the alignment. Therefore, the nominal score (N) for an alignment can be calculated using the following equation:

$$N = aM + bX + cG$$

where M is the match award (a positive number), X is the mismatch penalty (a negative number), and G is the gap penalty (a negative number). The values of M, X, and G may be adjusted, depending on the sequences being aligned or the goal of the analysis. The values for the coefficients a, b, and c reflect the number of matches, mismatches, and gaps in an alignment, respectively. In the alignment of sequences 4 and 6 shown in Fig. 6.4 and again at the bottom of Fig. 6.5a, there are eight nucleotide matches, one nucleotide mismatch, and three gaps. Thus $a = 8$, $b = 1$, and $c = 3$ for the coefficients in this

```
A-ACTTGG--AT  (#4)
ACACTTCGGTAT  (#6)
```

FIGURE 6.4 Final alignment of sequences 4 and 6 from Fig. 6.3.

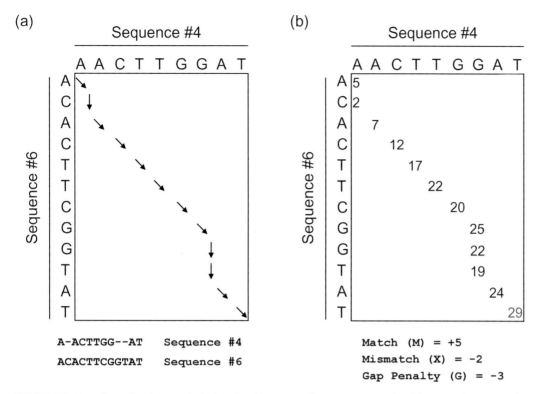

FIGURE 6.5 Two-dimensional array depicting the alignment of sequences 4 and 6. (a) Diagonal arrows indicate a match or mismatch between the nucleotides in the sequences being compared, while vertical arrows denote gaps in one sequence or the other. (b) Scores for matches (M), mismatches (X), or gaps (G) tallied for each move depicted in panel a. The nominal score corresponds to the number in red at the lower right corner of the array. Illustration by Craig Herbold.

alignment. If the values for awards and penalties are set at $M = +5$, $X = -2$, and $G = -3$, the nominal alignment score (N) for the alignment of sequences 4 and 6 is equal to 29 $[(8 \times +5) + (1 \times -2) + (3 \times -3) = 40 - 2 - 9 = 29]$.

The nominal score (N) is determined based on the values selected for each match award (M), mismatch penalty (X), and gap penalty (G). It is useful to adjust these values so that they are appropriate for the two sequences being compared (Baxevanis, 2005). However, if the values used for one alignment are different from those used for a second alignment, the nominal scores calculated for these two alignments cannot be directly compared.

If it has been empirically determined that gaps are more or less likely to exist between two sequences, the gap penalty should be adjusted likewise. It is generally thought that opening a gap should carry a greater penalty than extending an already existing gap. Therefore, an adjustment to the gap penalty has been introduced to some alignment algorithms that assigns two gap penalties. Termed an affine gap penalty, this adjustment assigns both a gap opening penalty and a gap extension penalty. While a thorough discussion of the mathematics and computational implementation of using an affine gap penalty is beyond the scope of this text, it is prudent to be aware that the use of such penalties has become standard practice and that these two gap penalties should be adjusted based on your knowledge of the sequence as well as your best judgment. As a reasonable rule of thumb, no more than one gap per 20 bp should be introduced into an alignment.

The nominal score provides the basis for determining the best pairwise alignment. If one were to enumerate all possible alignments between two sequences, one (or more) of these alignments would yield a maximum nominal score. The alignment that returns a maximum score is interpreted as the one with the highest probability of being the true alignment. Methods have been developed to help determine the best alignment without enumerating all possible alignments.

Global alignment (Needleman-Wunsch algorithm)

One method that is useful for the alignment of two sequences that are about the same size and that one believes should be aligned globally, or across the entire length of the sequences, is the Needleman-Wunsch algorithm (Needleman and Wunsch, 1970). While an extensive formal treatment of the Needleman-Wunsch algorithm is beyond the scope of this text, an introduction to the method should be useful for understanding different alignment strategies and tools available in this course.

For setting up the Needleman-Wunsch algorithm, the two sequences to be aligned are placed onto the axes of a two-dimensional array similar to what is shown in Fig. 6.5. Any alignment of the two sequences can be represented as a path through the array. One starts in the top left corner, making choices whether to introduce a gap or to align matched or mismatched characters. Depicted as arrows on the array, diagonal moves correspond to matches and mismatches while vertical and (although not shown in this example) horizontal moves correspond to gaps (−). The alignment given in Fig. 6.5a is depicted as a path through the two-dimensional array. First, there is a diagonal "align" move, followed by a vertical "gap" move. This gap move corresponds to the dash placed in sequence 4. The vertical move is followed by six "align" moves which in turn is followed by two more "gap" moves and finally with two more "align" moves.

Each move also affects the nominal score for the alignment. Each "align" move can be classified as a match (M) or a mismatch (X) and scored accordingly. Likewise, each horizontal or vertical move corresponds to a gap penalty (G). In Fig. 6.5b, the score is tallied for the pathway depicted in Fig. 6.5a. Just like in the previous calculation, the nominal score for the alignment is 29.

The power of the Needleman-Wunsch algorithm is that it can be used to find a single pathway that maximizes the nominal score. The alignment that produces the maximum nominal score is assumed to be the one most likely to be the true alignment. The computational process of finding the optimum alignment from all possible alignments is illustrated in Fig. 6.6. Two arrays are shown, one for keeping track of pathway choices (Fig. 6.6a) and one for scorekeeping (Fig. 6.6b). An additional row and an additional column are found in these arrays so that one or more gaps may be allowed in the first positions of the alignment. Once the entire array is filled out in accordance with the rules in the Needleman-Wunsch algorithm, the maximum nominal score is contained in the bottom right corner and the pathway choices that led to the score in that box correspond to the maximum scoring alignment. To prepare for the calculation, a zero is placed into the top left corner. The first row and the first column are then filled in as a series of successive "gap" moves. Each gap move carries a penalty so that the effect of several successive gaps is an extreme penalty. The score for any one of these boxes is simply the score of the box next to it plus an additional gap penalty.

For all other boxes, there are three ways to move into a given box from immediately adjacent boxes. One could move diagonally from the upper left, vertically from the top,

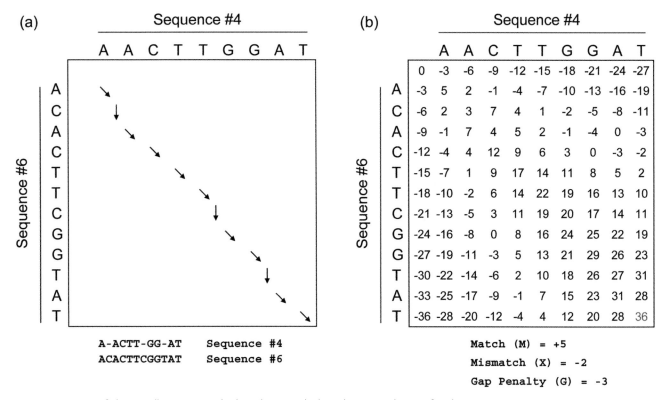

FIGURE 6.6 Use of the Needleman-Wunsch algorithm to calculate the nominal score for the maximum scoring alignment. See the text for a detailed explanation. Illustration by Craig Herbold.

or horizontally from the left. For each box, three scores corresponding to those moves are calculated. The Needleman-Wunsch algorithm makes the calculation easy by asserting that the only value chosen for a box is the maximum possible score of these three move choices. In Fig. 6.7, for example, the three possible moves into the lower right-hand box from the other three boxes are shown. One possible value for the lower right-hand box is the value of the box to its left with a gap penalty assessed ($10 - 3 = 7$). Another possible value is the value of the box above it with a gap penalty assessed ($14 - 3 = 11$). The third possible value is obtained with a diagonal move. In this case, the nucleotide character corresponding to that position in the alignment must be examined. If the two

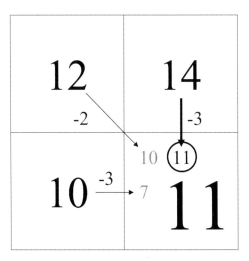

FIGURE 6.7 Three ways to move into the lower right-hand box of a two-dimensional array. The only move recorded is the move that maximizes the value of the lower right-hand box. Illustration by Craig Herbold.

nucleotides are the same a "match" score is added, and if they are different the mismatch penalty is assessed. In this instance, a mismatch was scored ($12 - 2 = 10$). Whatever move maximizes the score in the lower right-hand box is recorded, while the other two moves are ignored. For this example, a vertical move produced the highest score (e.g., 11 versus 7 or 10) and is thus the only move that is recorded.

The same type of calculation produced the results observed in Fig. 6.6. The lower right-hand box has a score of 36, which is the maximum of three potential values. The score is obtained by adding a match award for aligning two T characters (both the row and the column contain a T character) to the value found in the box to the upper left ($31 + 5 = 36$). This operation corresponds to a diagonal move and produces a match. One of the other two possible values is found by adding a gap penalty to the value found in the box above our box of interest [$28 + (-3) = 25$], which corresponds to a vertical move. The vertical move does not produce a score higher than the diagonal move, and thus the move and the score calculated by the vertical move are ignored. The third value is calculated similarly by adding a gap penalty to the value to the left of our box of interest [also $28 + (-3) = 25$ for this example]. This final possibility corresponds to a horizontal move. Again, this move also does not produce a maximum score, so the score corresponding to the horizontal move is not recorded. Thus, the value in the lower right corner of the two-dimensional array is maximized with a score of 36 and only the diagonal move is recorded. The boxes to the left, to the top, and to the upper left were similarly maximized when they were calculated. For instance, the score of 31 in the box to the upper left was the maximum of the three possible moves into that box and was also obtained with a diagonal move.

The process of filling in a two-dimensional array in the manner depicted in Fig. 6.6b is known as dynamic programming because the maximum value that can be calculated for a given box is dependent on earlier calculations that similarly maximized those box values. The moves that produced a maximum score are recorded as a pathway through the two-dimensional array. For clarity, only the path that produced the maximum nominal score is shown in Fig. 6.6a. The alignment that corresponds to this path (the maximum scoring alignment) is displayed at the bottom of the array. The nominal score can be double-checked by counting up matches (M), mismatches (X), and gaps (G). Since there are nine match positions and three gaps, the nominal score is $N = 45 - 9 = 36$, which is precisely the value obtained with the Needleman-Wunsch algorithm.

It may come as a surprise that the maximum scoring alignment in Fig. 6.6 differs from the true alignment depicted in Fig. 6.4 and 6.5. The true alignment has a score of 29, while the maximum scoring alignment has a score of 36. Unfortunately, there is very little that can be done to overcome such discrepancies, but it is good practice to think about how a computationally derived, theoretically optimized alignment may differ from the true alignment and how such differences may affect analyses that use the alignment. In the case presented here, the transversion that occurred in the evolution of sequence 4 is not recovered in the maximum scoring alignment and one gap is misplaced; however, it also is important to note that 9 of the 12 positions were accurately aligned by the Needleman-Wunsch algorithm.

Local alignment (Smith-Waterman algorithm)

The Needleman-Wunsch algorithm aligns two sequences across their entire length. With global alignments, it is assumed that the first and last sequence characters of both sequences should be aligned to one another. Furthermore, a gap carries a penalty whether it occurs at the beginning, at the end, or in the middle of the alignment. For the example in Fig. 6.6, the first and last nucleotides of both sequences can be aligned because they are identical. Sometimes the use of different primers in sequencing a particular gene can produce sequences that do not start and stop at the same position in the gene sequence. As depicted in Fig. 6.8, perhaps one sequencing reaction was carried out using primer 519R to obtain sequence positions 1 to 518 of the 16S rRNA gene from one organism, while another sequencing reaction was carried out using primer 27F to obtain sequence positions 28 to 750 of the same gene from a different organism. Only nucleotides corresponding to positions 28 to 518 in both sequences could be aligned (overlap highlighted in yellow), and it would not be appropriate in such a case to penalize all the gaps that would need to be placed at the ends of these two sequences obtained with different primers. In instances such as this, a local alignment, which allows a smaller sequence to be aligned to a portion of a larger sequence, is preferred over a global alignment.

The modified Needleman-Wunsch algorithm that allows a local alignment to be calculated is known as the Smith-Waterman algorithm (Smith and Waterman, 1981). The two algorithms are surprisingly similar, and if one understands one, understanding the other one is quite straightforward. Again, the affine gap penalty is ignored for this text. There are four major differences between the Smith-Waterman algorithm and the Needleman-Wunsch algorithm. First, the successive gap penalties that can be observed in the first row and first column of Fig. 6.6 are all set to 0. Therefore, there is no penalty at all for gaps placed at the beginning or end of a sequence. Second, box scores are never allowed to be negative. If all three moves produce a score that is less than 0, the score in that box is set to 0. Third, the highest score observed in the entire two-dimensional array (not necessarily the score in the lower right-hand corner) is where the pathway ends and is traced back from. Similarly, the path starts where it can be traced back to a 0 score, rather than being constrained to a path that leads to the top left corner. The fourth difference is that any unaligned, leftover sequence is not reported in the alignment. These differences are illustrated in Fig. 6.9. In this example, the last three nucleotides

FIGURE 6.8 Necessity of using local sequence alignments, rather than global alignments, when only portions of the sequences can be aligned. The black plus strand was derived from one bacterial isolate, while the orange minus strand was obtained from another bacterial isolate. Different primers (519R and 27F) were used to sequence the 16S rDNA PCR products. Lines are not to scale. Illustration by Erin Sanders.

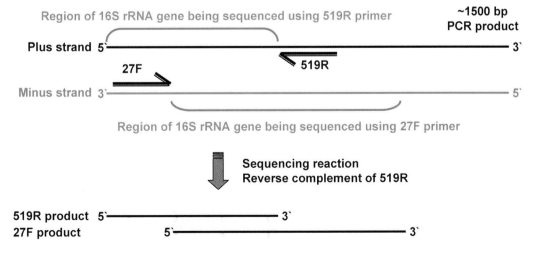

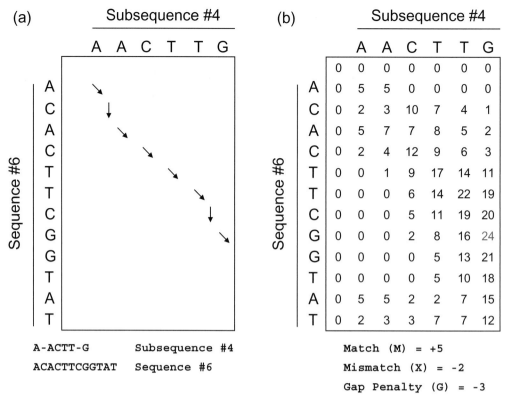

FIGURE 6.9 Computation of the nominal score from local sequence alignments, using the Smith-Waterman algorithm. See the text for a detailed explanation. Illustration by Craig Herbold.

are removed from sequence 4 and the remaining nucleotide subsequence is aligned to sequence 6. Since a portion of sequence 4 is being aligned to sequence 6, a global alignment would unnecessarily penalize the gaps that are required to align a 6-nucleotide sequence to a 12-nucleotide sequence.

In Fig. 6.9, it should first be noted that all values in the first row and first column are set to 0, this being the first major difference between the Needleman-Wunsch algorithm and the Smith-Waterman algorithm. The scores are then filled in using the same basic approach as is used with the Needleman-Wunsch algorithm; however, negative scores are not allowed. The maximum score observable in the two-dimensional array is 24, so this is where the alignment pathway must end. It is traced back to a zero value, which in this case happens to correspond to the top left corner, but this is not always the situation. Finally, the local alignment is reported as an alignment between the shortened subsequence 4 and the aligned portion of sequence 6 (Fig. 6.10). As shown in Fig. 6.10, the last four nucleotides of sequence 6 (GTAT) are not reported because they are not aligned to any part of the other sequence. With respect to nucleotide sequences, local alignments are more useful than global alignments in part because they allow one to identify undesirable sequence data often found at the beginning and ends of sequencing reads. Specifically, local alignment methods identify the regions of the sequence that

A-ACTT-G (#4)
ACACTTCG (#6)

FIGURE 6.10 Final local alignment of sequences 4 and 6 from Fig. 6.9.

are readily aligned and those that are poorly aligned. Perhaps poorly aligned sections of the sequences are of low quality and should be excluded from further analysis. One should always be careful when declaring sequence data unusable, though. It is best to decide on rigorous rules for declaring data usable or unusable and maintain those standards throughout the analysis. All sequences should be evaluated by the same criteria. Before a global alignment can be performed, however, all sequences being compared must be about the same size and aligned reasonably well using a local alignment tool.

KEY TERMS

Affine gap penalty A method for scoring gap penalties in which a fixed deduction is made for introducing a gap and then an additional deduction is made that is proportional to the length of the gap. Because the gap opening penalty is greater than the gap extension penalty, lengthening existing gaps is favored over creating new gaps.

Bootstrapping A statistical method by which the frequencies of particular observations are used to construct a new data set. In alignment bootstrapping, new alignments are created by sampling columns of the original alignment. Each column has an equally likely chance of being sampled, and each column may be sampled more than once. The new alignment is usually the same length as the original alignment.

Deletion Removal of nucleotides from a sequence, sometimes resulting in a frameshift mutation.

Dynamic programming The process of filling in a two-dimensional array such that the value obtained for any position in the array is dependent upon calculations done previously to fill other positions in the array.

Edit distance The number of nucleotide changes that distinguish any two sequences being compared in a pairwise alignment (e.g., number of mismatches and gaps).

Extant Not extinct; still alive today.

Extinct No longer living.

Gaps Insertion or deletion events represented as dashes in sequence alignments.

Identical by descent Originating from a common ancestor.

Indel Insertion and/or deletion in an alignment.

Insertions Additions of nucleotides to a sequence, sometimes resulting in a frameshift mutation.

Maximum scoring alignment The pathway that leads to the highest possible nominal score in the Needleman-Wunsch algorithm.

Needleman-Wunsch algorithm A global alignment method for comparing two sequences.

Nominal score A calculated metric that essentially represents the inverse of the edit distance in that it reflects the number of nucleotides that match when two sequences are compared in a pairwise alignment while taking the number of mismatches and gaps into account.

Robustness The credible strength and reliability of a particular data set.

Smith-Waterman algorithm A local alignment method for comparing two sequences.

Substitution A change in primary sequence that may result in missense, nonsense, or silent mutations.

Transition A change from a purine to another purine (A to G or G to A), or a pyrimidine to another pyrimidine (C to T or T to C).

Transversion A change from a purine (A or G) to a pyrimidine (C or T) and vice versa.

True alignment Alignment between two existing sequences that results from the actual evolutionary path taken by ancestral sequences; may or may not be the same as the maximum scoring alignment if nucleotide changes that took place in the ancestral state of either sequence are not obvious.

REFERENCES

Altschul, S. F., T. L. Madden, A. A. Schäffer, J. Zhang, Z. Zhang, W. Miller, and D. J. Lipman. 1997. Gapped BLAST and PSI-BLAST: a new generation of protein database search programs. *Nucleic Acids Res.* **25:**3389–3402.

Baxevanis, A. D. 2005. Assessing sequence similarity: BLAST and FASTA, p. 295–324. *In* A. D. Baxevanis and B. F. F. Ouellette (ed.), *Bioinformatics: a Practical Guide to the Analysis of Genes and Proteins,* 3rd ed. John Wiley & Sons, Inc., New York, NY.

Dunbrack, R. L., Jr. 2006. Sequence comparison and protein structure prediction. *Curr. Opin. Struct. Biol.* **16:**374–384.

Jones, N. C., and P. A. Pevzner. 2004. *An Introduction to Bioinformatics Algorithms.* MIT Press, Cambridge, MA.

Needleman, S., and C. Wunsch. 1970. A general method applicable to the search for similarities in the amino acid sequence of two proteins. *J. Mol. Biol.* **48:**443–453.

Smith, T. F., and M. S. Waterman. 1981. Identification of common molecular subsequences. *J. Mol. Biol.* **147:**195–197.

SECTION 6.2
DNA SEQUENCE ALIGNMENTS IN A BLAST SEARCH

Introduction to the NCBI nucleotide sequence database

One application for local DNA sequence alignments is in database searches. BLAST (Altschul et al., 1997) is a widely used search tool that is available for searching query sequences against massive genetic databases such as GenBank (Benson et al., 2009). Representing a primary sequence database, or one in which the sequence data have not necessarily been reviewed and verified experimentally before submission, GenBank is one of three sequence archives run as part of an international collaboration between data collection centers including the National Center for Biotechnology Information (NCBI), the European Molecular Biology Laboratory (EMBL), and the DNA Databank of Japan (DDBJ) (Fig. 6.11). Together, these agencies have amassed more than 50 billion base pairs of nucleotide sequence data and more than 45 million individual sequences. Sequence submissions are independently updated every 24 hours at all three centers, and records are distributed in a common format among all three databases. Thus, a query to any one of the databases will produce results derived from submissions made at all three centers.

Conducting a database search

The power of the NCBI-BLAST algorithm lies in its organization. A short query sequence is submitted to GenBank, and the most similar sequences in the database are found and returned to the user (Fig. 6.12). By default, NCBI-BLAST displays each of the returned sequences along with the query sequence as a pairwise local alignment. NCBI-BLAST also reports several statistics that assist in assessing the level of relatedness between the query and database sequences. We will return to a discussion of output of BLAST results following dissection of the algorithm itself.

NCBI-BLAST does not actually conduct an exhaustive search through GenBank (Fig. 6.13). Instead, a secondary database linked to all the sequence information found in

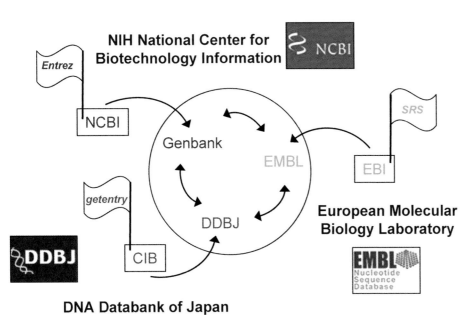

FIGURE 6.11 International Nucleotide Sequence Database (INSD) collaboration. Data flow among the three data collection centers (NCBI, EMBL, and DDBJ). Illustration by Erin Sanders.

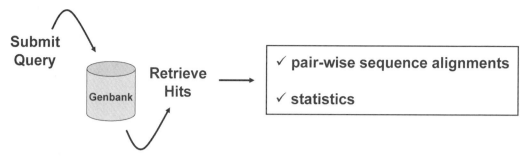

FIGURE 6.12 Overview of BLAST search process and results output. Illustration by Erin Sanders.

GenBank is queried during a BLAST search. This secondary database is organized by chopping the sequence of each GenBank record into smaller "words" (Baxevanis, 2005). BLAST also takes the query sequence, chops it up into all possible words of the user-defined size, and compares these to the GenBank database words. For nucleotide database searches, words may be as small as 7 nucleotides (blastn default = 11) or as large as 64 nucleotides (megablast default = 28). Ultimately, decreasing the word size relaxes the stringency imposed on the search, allowing detection of more distant relationships between sequences. On the other hand, increasing the word size tends to speed up a BLAST search by requiring more exact matches between the query and other sequences in the database; however, this situation runs the risk of overlooking potentially informative biological relationships.

The BLAST program begins by seeding a search of the secondary database with a query word, in which exact (100% identity) words accumulate as potential search hits that are investigated further by the algorithm. As shown in Fig. 6.14, each GenBank sequence that contains the query word is then aligned to the query sequence, in which the neighborhood is extended from the query word in both directions. The score for the

FIGURE 6.13 Effect of database organization on BLAST search process. Illustration by Craig Herbold.

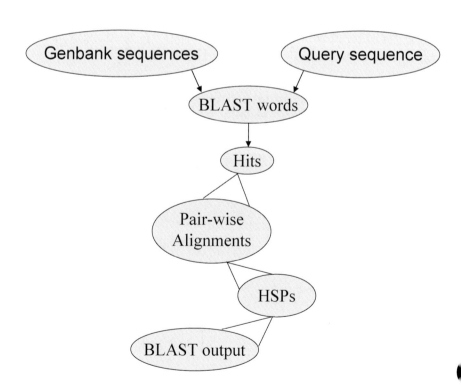

Query: GTAGCGGTAAATTAACGTCA**GGTACTTCTAG**GACGTAGGGATGCGCAATCCACTGCTA

GTA--GG-A-ATT-A--TC-**GGTACTTCTAG**GA--TAG-GAT-CG-AATC-A-T-CTA

Sbjct: GTACTGGCAGATTCATATCC**GGTACTTCTAG**GATCTAGAGATCCGTAATCGATTCCTA

FIGURE 6.14 BLAST search query and neighborhood extension in blastn. Illustration by Erin Sanders.

alignment is calculated, continuing until a maximal local alignment length is achieved. Any search hit that can be used to construct a pairwise alignment with a high enough score is returned to the user.

Interpreting the results of a nucleotide BLAST (blastn) search based on user-defined search parameters

As portrayed in Fig. 6.15, a search hit that results from an optimal local alignment of maximal length is called a high-scoring segment pair (HSP), which can be graphically represented if the number of aligned nucleotides (e.g., length of extension) is plotted against the cumulative score obtained as an alignment is extended along its length. The plot begins with the alignment of the original query word with one word from a GenBank sequence. The threshold (T) which defines the word size used to seed the search is set by the user (e.g., blastn default $T = 11$). As extension ensues, the score is calculated and continues to increase provided that matches outnumber mismatches and gaps. BLAST output to the user in the form of search hits actually starts once the score exceeds another threshold (S), which is discussed in detail below. At some point during the extension, the number of mismatches and gaps begins to exceed the number of matches, causing the score to decrease. Note how the curve begins to decay as extension proceeds beyond the length of the HSP. If the two sequences being compared by the BLAST algorithm were plotted on a two-dimensional array, the HSP would correspond to the maximum observable score, denoting where the alignment terminates.

Identifying an HSP that is biologically significant is highly dependent on scoring statistics. Users specify an "Expect" parameter, or E-value, before a BLAST search is initiated. This parameter specifies the statistical significance threshold ascribed to a database match, indicating whether the alignment portrays a biological relationship or whether the observed sequence similarity is simply due to chance. As shown in Fig. 6.16, the default E-value threshold is set to 10 in the blastn implementation, meaning that one would expect about 10 matches to be found purely by chance that return the minimal

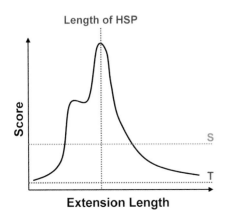

FIGURE 6.15 Cumulative score as the extension of pairwise alignment increases, resulting in production of an HSP when the maximum optimal length is achieved. Illustration by Erin Sanders.

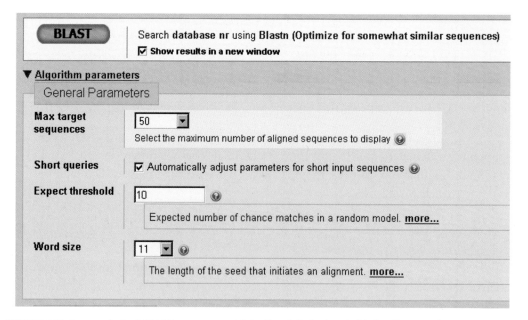

FIGURE 6.16 Screen shot of BLAST query page. Note the default thresholds for Expect values set at 10 and Word size (T) set at 11. (From http://blast.ncbi.nlm.nih.gov/Blast.cgi.)

acceptable nominal score. Only alignments that have Expect scores less than or equal to the Expect threshold are reported. The lower the threshold, the fewer the hits that are conveyed to the user. Although the HSPs with the lowest scores may simply be random junk, it is a good idea to set the Expect threshold high enough and collect a little junk so that no biologically significant, albeit more distant, relationships are missed.

The user-defined E-value defines a threshold score (S) above which a pairwise alignment must score to be designated an HSP (refer again to Fig. 6.15). Once the starting and ending locations of a particular alignment are known, whose details are discussed below, BLAST calculates the nominal score for the entire alignment, which includes extensions in both directions from the original query word used to seed the search. The maximum nominal score, which is matrix dependent, is then normalized into a value called the maximum bit score, allowing results from different BLAST searches with the same query sequence or multiple HSPs of different lengths from a single search to be compared since the matrix scoring strategy itself is taken into account. If an alignment was calculated with a match award of +4 and another was calculated with a match award of +7, the higher match award will yield a higher nominal score but will return a similar bit score. The bit score (B) is calculated from a nominal score (N) by using the following equation:

$$B = (\lambda N - \ln K)/\ln 2$$

where λ and K are constants that are specific to the combination of match, mismatch, and gap penalties used in the BLAST search. This normalization makes the bit score more comparable between searches than the nominal score. Consequently, it is actually a threshold bit score, not the threshold nominal score, against which local alignments are judged before being called an HSP (e.g., S in Fig. 6.15). The threshold bit score (S) is that which an alignment must score above to be an HSP and is calculated according to the following equation:

$$S = \ln\ (mn/E)$$

where E is the user-defined E-value, m is the length of the query sequence, and n is the number of nucleotides contained in a database. Since sequences are being deposited to GenBank all the time, n increases from one day to the next. Any local alignment scoring higher than the value of S is reported as a search hit to the user. In the most recent version of NCBI-BLAST, the maximum bit score for a particular alignment is reported as "Max score" (Fig. 6.17).

The total score is another value that NCBI-BLAST reports to the user. The total score is calculated as the sum of the bit scores (Max scores) of all the HSPs found in a particular database entry. For example, in Fig. 6.17 the top HSP, using a query encoding the 16S rRNA gene, is to a genome sequence of *Bacillus halodurans*. While the Max score is the value of the highest-scoring HSP, eight different HSPs corresponding to eight ribosomal operons in the genome of *B. halodurans* were actually found in GenBank. The total score for the genome sequence entry for *B. halodurans* is simply the summation of the maximum bit scores found for eight high-scoring local alignments found within its genome. This score is much less informative for the individual entries for each of the eight operons, since in most cases there is only one HSP so the total score and Max score are the same. The total score does not provide additional information for organismal classification, so for the purposes of this text, this score is not discussed further.

After conversion of the nominal score to the Max score (bit score [B]), the E-value is calculated for each HSP so that short HSPs and long HSPs can be compared more effectively (see column 6 in Fig. 6.17). The equation used to compute the E-values is similar to that used to determine the threshold bit score (S):

$$E = mn/2^B$$

where m is the length of the query, n is the number of nucleotides in the database, and B is the Max score calculated for the HSP. Any HSP that meets the Expect threshold is then returned to the user in pairwise alignment form. Nominal scores, Max scores, and E-values are associated with each HSP and allow one to interpret the quality of the database match (see Fig. 6.17 and 6.18). As an indicator of how good the alignment is, the higher the Max score, the better the alignment. As an indicator of the biological significance of the alignment, the lower the E-value, the more significant the HSP.

FIGURE 6.17 One-line descriptions in the BLAST report. The Max score is the maximum possible bit score. (From http://blast.ncbi.nlm.nih.gov/Blast.cgi.)

Sequences producing significant alignments:
(Click headers to sort columns)

Accession	Description	Max score	Total score	Query coverage	E value	Max ident
BA000004.3	Bacillus halodurans C-125 DNA, complete genome	2800	2.228e+04	100%	0.0	100%
AB031209.1	Bacillus halodurans orf1, orf2, rrnA-16S, rrnA-23S, rrnA-5S, yaaC, guaB genes, partial and complete cds	2800	2800	100%	0.0	100%
AB031214.1	Bacillus halodurans yvgQ, rrnG-16S, trn-Ile, trn-Ala, rrnG-23S, trn-5S, ykfC, nasA genes, partial and complete cds	2785	2785	100%	0.0	99%
AB031212.1	Bacillus halodurans orf1, kbaA, ybaN, rrnE-16S, rrnE-23S, rrnE-5S, trn-Asn, trn-Thr, rrnE-23S, rrnE-5S, trn-Asn, tr	2785	2785	100%	0.0	99%
AB031210.1	Bacillus halodurans dnaX, yaaK, recR, yaaL, bofA, rrnB-16S, rrnB-23S, rrnB-5S, orf1, csfB genes, partial and com	2785	2785	100%	0.0	99%
AB031213.1	Bacillus halodurans ydcI, ydcK, trn-Asn, trn-Ser, trn-Glu, trn-Val, trn-Asp, trn-leu1, trn-leu2, trn-Arg, trn-Gly, trn-I	2782	2782	100%	0.0	99%
AB031211.1	Bacillus halodurans lysS, rrnC-16S, trn-Ile, trn-Ala, rrnC-23S, rrnC-5S, trnJ-Val, trnJ-Thr, trnJ-Lys, trnJ-Leu1, trnJ-	2782	5562	100%	0.0	99%
AB031215.1	Bacillus halodurans ygaG, orf, ygxA, rrnH-16S, rrnH-23S, rrnH-5S, trnD-Asn, trnD-Ser, trnD-Glu, trnD-Val, trnD-M	2780	2780	100%	0.0	99%
AB013373.1	Bacillus halodurans C-125 23S, 5S and 16S rRNA genes and tRNA-Ala, Arg, Gly, Leu, Lys, Pro, Thr and Val genes i	2780	2780	100%	0.0	99%
AB002661.1	Bacillus sp. gene for 16S rRNA, complete sequence	2780	2780	100%	0.0	99%
AB043971.1	Bacillus halodurans gene for 16S rRNA, complete sequence	2780	2780	100%	0.0	99%
AB027713.1	Bacillus halodurans gene for 16S rRNA, strain:AH-101	2731	2731	98%	0.0	99%
AB043856.1	Bacillus sp. A-59 gene for 16S rRNA	2720	2720	97%	0.0	99%
AB043844.1	Bacillus sp. 202-1 gene for 16S rRNA	2715	2715	97%	0.0	99%
AJ302709.1	Bacillus halodurans 16S rRNA gene, strain DSM 497T	2708	2708	97%	0.0	99%
AB043847.1	Bacillus sp. C-3 gene for 16S rRNA	2704	2704	97%	0.0	99%

Closer inspection of BLAST pairwise sequence alignments

In practice, the DNA sequence alignments generated by BLAST are used to identify database sequences that share characteristics with the query sequence. Therefore, the similarity of the query sequence to the sequence identified in the database (the subject sequence) is of particular interest to the user. Several statistics are presented in the results of a nucleotide BLAST search that allow a hypothesis to be formed about the evolutionary history of a particular gene or the identity of an organism from which the gene was derived. These sorts of hypotheses can be tested later using phylogenetic analysis (to be covered in Unit 7). For example, the goal for the "I, Microbiologist" project is to analyze 16S rDNA sequences to determine microbial community composition. A nucleotide BLAST search can be used to generate plausible hypotheses about the identity of the isolates or clones that can then be examined further by appropriate phylogenetic methods.

In Fig. 6.18, one pairwise alignment resulting from an NCBI-BLAST search is presented. The length of the query was 231 nucleotides (data not shown). The first few lines report the identity of the sequence that BLAST was able to find in the database (referred to as the Subject [Sbjct] sequence in the pairwise alignment). In this case, a sequence belonging to *M. fervidus,* a type of archaeon, was found to possess significant similarity to the query sequence. The length of this database entry is reported as 3,365 nucleotides. BLAST was able to find a region of local similarity between a query of only 231 nucleotides and this much longer database entry for *M. fervidus.* The local alignment of these two sequences comprises an HSP. The length of the match to the query sequence, termed the Query coverage, should be examined for all HSPs reported. In the example in Fig. 6.18, 63% of the query (145 of 231 nucleotides) could be aligned to the sequence for *M. fervidus.* Up to 100% of the query may align to database sequences. Note that alignment of gene fragments may generate hits with artificially low E-values.

Continuing with the examination of Fig. 6.18, the Max score (in bits) is presented with the nominal score in parentheses, followed by the E-value (Expect). Based on the

FIGURE 6.18 Pairwise sequence alignment for one HSP generated from a BLAST report. (From http://blast.ncbi.nlm.nih.gov/Blast.cgi.)

```
> gb|M32222.1|MEFTGSRNA  M.fervidus 7S RNA, Ser-tRNA, 16S rRNA and Ala-tRNA genes
Length=3365


 Score =  131 bits (70),  Expect = 1e-27
 Identities = 120/145 (82%), Gaps = 0/145 (0%)
 Strand=Plus/Plus


Query  86    GGCTTTTGGGGAGTGTAAGTAGCTCCCCGAATAAGCGGTGGGCAAGAGGGGTGGCAGCCG  145
             |||||||  ||||||||| |||||| ||||||| | ||||||||| |||| |||||||
Sbjct  1977  GGCTTTTCCGGAGTGTAAAAAGCTCCGGGAATAAGGGCTGGGCAAGACCGGTGCCAGCCG  2036


Query  146   CCGCGGGAACACCCCCACCGCGAGCGGTGGCCGTGATTATTGGGCCTAAAGGGGCCGTAG  205
             |||||| ||||||  || | |||| |||||||| | ||||||||||||||||| | ||||||
Sbjct  2037  CCGCGGTAACACCGGCAGCCCGAGTGGTGGCCGCGTTTATTGGGCCTAAAGCGTCCGTAG  2096


Query  206   CCGGGCCGGTGTGGCTCCGGTGAAA  230
             |||| |||||| | |||||||||||
Sbjct  2097  CCGGTCCGGTAAGTCTCCGGTGAAA  2121
```

database size and the query length, a score like the one presented would be expected to occur only 1e−27 (10^{-27}) times by chance. In other words, it would be nearly impossible to accidentally find this sequence by randomly jumbling the query sequence. Since E-values represent the expected number of times a BLAST result would be expected to happen by chance, lower E-values may be interpreted to mean that an HSP is more biologically relevant. In general, it is recommended to use an E-value of about 1e−6 (10^{-6}) as the upper limit for a 16S rDNA search when attempting to identify an isolate by BLAST (Baxevanis, 2005). Any E-value higher than 10^{-6} for a nucleotide BLAST search will probably not help form a hypothesis about the taxonomic identity of the query sequence.

Next, BLAST reports the number of identities and gaps. This particular alignment is 145 positions long, so the finding of 120 matches means that the two sequences are 82% identical over the length of the alignment. In addition, there happen to be no gaps in this particular alignment (0/145 = 0%). An important consideration when interpreting the BLAST results is that short alignments tend to have a higher percent identity. Therefore, the length of the alignment should also be taken into account when assessing the percent identity. In general, it is best to assume that different strains of the same species have >99% sequence identity and different species of the same genus have >95% sequence identity. However, these are only general guidelines and any taxonomic assignments should be considered preliminary, pending a thorough phylogenetic analysis. If the query sequence finds no database entries with higher than about 70% identity, the query sequence either is a low-quality sequence or may represent a whole new branch of life hitherto never observed.

BLAST provides clues to the orientation of the query sequence relative to another sequence in the database. Notice in Fig. 6.18 that the orientation of the query and the database sequence is given as "Strand = Plus/Plus." In addition, the orders of the base count origin numbers for the Query and Sbjct sequences are both ascending (86 to 230 and 1977 to 2121, respectively). Considering the conventions involved in expressing DNA strandedness (e.g., the 5′-to-3′ directionality and antiparallel nature of the strands in a DNA duplex), these observations mean that the query sequence and the database sequence are both in the same orientation. Therefore, if the database sequence is in the correct orientation, which by convention would be the plus strand written 5′ to 3′, so is the query. On the other hand, if the query is actually the reverse complement of the gene in the database, the strand specifications would instead state Strand = Plus/Minus because BLAST automatically reports the order of the Sbjct sequence in the same orientation as the query (see Fig. 6.21 for an example).

It is possible that the published sequence in GenBank actually is the reverse complement of the query sequence because the gene is encoded by the minus strand. For instance, genome sequences often contain ribosomal operons in one or both orientations. To ensure proper and consistent orientation of all sequences being evaluated in the BLAST search, it is best to visit the Report page for the database sequence, which is available in the Fig. 6.18 example by clicking on the database identifier in the definition line (gb|M32222.1|MEFTGSRNA), and determine the orientation of the gene sequence. In general, if the entry asserts that the only gene present is the 16S rRNA gene, it is usually prudent to assume that the database entry is in the correct orientation. Otherwise, a quick search through the Report page will identify the gene as being coded as "complement" if the database entry contains the reverse complement of the gene and was annotated correctly (Fig. 6.19).

☐ 1: AY959328. Reports Bacillus halodura...[gi:62005122]

Features Sequence

```
LOCUS       AY959328              1500 bp   DNA     linear   BCT 04-APR-2005
DEFINITION  Bacillus halodurans strain XJU-5 16S ribosomal RNA gene, partial
            sequence.
ACCESSION   AY959328
VERSION     AY959328.1  GI:62005122
KEYWORDS    .
SOURCE      Bacillus halodurans
  ORGANISM  Bacillus halodurans
            Bacteria; Firmicutes; Bacillales; Bacillaceae; Bacillus.
REFERENCE   1  (bases 1 to 1500)
  AUTHORS   Deng,A., Rahman,E. and Yi,X.
  TITLE     Isolation and Identification of Several Strains of Bacillus
            halodurans from Xinjiang in China
  JOURNAL   Unpublished
REFERENCE   2  (bases 1 to 1500)
  AUTHORS   Rahnan,E., Deng,A., Yi,X., Yang,Y. and Wang,N.
  TITLE     Direct Submission
  JOURNAL   Submitted (10-MAR-2005) Life Science and Technology, Xinjiang
            University, Shengli Road 14#, Wulumuqi, Xinjiang 830046, China
FEATURES            Location/Qualifiers
     source         1..1500
                    /organism="Bacillus halodurans"
                    /mol_type="genomic DNA"
                    /strain="XJU-5"
                    /db_xref="taxon:86665"
     rRNA           complement(<1..>1500)
                    /product="16S ribosomal RNA"
ORIGIN
        1 cttccgatac ggctaccttg ttacgacttc accccaatca tctgtcccac cttaggcggc
       61 tggctccaaa aggttacccc accgacttcg ggtgttacaa actctcgtgg tgtgacgggc
      121 ggtgtgtaca aggcccggga acgtattcac cgcggcatgc tgatccgcga ttactagcaa
      181 ttccggcttc atgcaggcga gttgcagcct gcaatccgaa ctgagaatgg ctttctggga
      241 ttggcttcac ctcgcgggtt cgcaacccctt tgtaccatcc attgtagcac gtgtgtagcc
```

FIGURE 6.19 Screen shot of the Report page from a blastn search in GenBank. (From http://blast.ncbi.nlm.nih.gov/Blast.cgi.)

A word of caution about database entries

Several issues with GenBank and BLAST must be kept in mind when a rigorous bioinformatics analysis of DNA sequence data is carried out. The first is quality control of the database entries. No system is perfect, and any database (such as GenBank) that allows users to upload sequences carries a risk of containing inaccurate, unsubstantiated, or poorly curated information. The sequence records in GenBank reflect only what has been contributed by people, and being naturally fallible, people make mistakes all the time. Annotation of gene sequences should not be trusted blindly, and one should remain skeptical of any taxonomic classifications asserted if no supportive experimental evidence exists. It is especially important to be skeptical when the title of a sequence starts with the word "uncultured" or "uncultivated" (Fig. 6.20). The Report page always should be checked in these cases so that the taxonomic assignment of the submitted sequence may be verified. If the scientist who submitted the sequence published the data as part of a

FIGURE 6.20 One-line descriptions in the BLAST report. Note that hits 1 and 3 are from uncultured sources. (From http://blast.ncbi.nlm.nih.gov/Blast.cgi.)

Sequences producing significant alignments:
(Click headers to sort columns)

Accession	Description	Max score	Total score	Query coverage	▲ E value	Max ident
DQ828401.1	Uncultured firmicute clone DOK_CONFYM_clone098 16S ribosomal RNA gene, partial sequence	713	713	58%	0.0	92%
AY936911.1	Marine sediment bacterium ISA-3100 16S ribosomal RNA gene, partial sequence	713	713	58%	0.0	92%
AY694435.1	Uncultured Bacillus sp. clone JAB SHC 05 16S ribosomal RNA gene, partial sequence	711	711	58%	0.0	92%
EU257452.1	Bacillus subtilis isolate C14-1 16S ribosomal RNA gene, partial sequence	710	884	92%	0.0	92%
EF428248.2	Bacillus megaterium strain HDYM-24 16S ribosomal RNA gene, partial sequence	710	884	92%	0.0	92%

rigorous phylogenetic analysis, there will be a citation for the relevant publication. Without a publication, there is no way to evaluate and verify the taxonomic assignment of sequences belonging to uncultured or uncultivated organisms.

Another important issue is the way in which BLAST finds the optimum alignment. It is possible that the BLAST alignment made during a search is longer than what is reported in the results; however, if extension of the alignment results in a lower score, BLAST does not report the extended alignment to the user. Therefore, any local alignment returned by BLAST may represent only a portion of the true alignment, which may very well extend well beyond the edges that BLAST reports. This particular limitation with respect to the BLAST algorithm results in shorter alignments with very high sequence identity and longer alignments with lower sequence identity. It is quite likely that if those high-scoring small alignments were extended to encompass the same coverage as their lower-scoring counterparts, the sequence identity could decrease significantly.

In Fig. 6.21, two results from a blastn search are depicted. The first database entry has 84% identity to the query, while the second exhibits only 75% identity. A closer examination of the pairwise sequence alignments shows that a large indel is present in the second HSP that is missing in the first. The effect of this indel was to truncate the first HSP in favor of choosing the maximum score possible. Both HSP subjects align to the query sequence positions 91 to 284. Over this range, the first HSP subject has 88% sequence identity to the query and the second HSP has 85% identity. Extension of the HSP beyond this range had an extensive impact on the second HSP, and its score was decreased accordingly.

A typical BLAST search returns several hundred HSPs. The key is to look at several of the top-scoring HSPs and use every bit of available data to formulate a hypothesis about the identity of the query sequence. A pitfall comes with the size of the database and number of nearly identical entries it contains. For instance, as exemplified in Fig. 6.17, several of the top HSPs may belong to one prokaryotic species and may just indicate that there are multiple entries of a nearly identical sequence. The exact rank of an HSP is much less important than the overall statistics that BLAST returns. Therefore, it is wise to search through several of the top HSPs from obviously different sources and compare the BLAST statistics for each HSP with one another. It is easy to see how the mindless BLAST system can result in the production of erroneous interpretations of statistics and scores, so caution must be used with every step of a BLAST analysis. Remember that skepticism is an ally in the bioinformatics business, not an enemy.

Nucleotide BLAST program selection

The previous segments focused mainly on a discussion of blastn (nucleotide BLAST). Two additional options are available through the NCBI suite of search programs. Megablast is designed to quickly compare a query sequence to closely related sequences in the database, efficiently finding alignments between highly similar sequences. A comparison of the default parameters for Megablast (Fig. 6.22) and blastn (Fig. 6.23) reveals differences in the default match scores, mismatch scores, and gap penalties for the two search programs. Furthermore, Megablast extends the exact word size from 11 to 28 nucleotides. Because the length of the initial search word governs the sensitivity of the BLAST search, Megablast (although faster) is less sensitive than blastn because it fails to align more distantly related sequences.

A third option is Discontiguous Megablast, which is designed to find database sequences that are similar, but not identical, to the search query (Ma et al., 2002). Rather

```
>gb|EF430681.1|   Unidentified archaeon clone D1003F08.x 16S ribosomal RNA gene,
partial sequence
Length=370

 Score =  251 bits (278),  Expect = 1e-65
 Identities = 198/233 (84%), Gaps = 3/233 (1%)
 Strand=Plus/Minus

Query  55   CATTGTCCTTCCTGGCAACAGAGTTTTACGATCCGAAAG-CCTTCATCACTCACGCGGCG  113
            |||| || ||  ||||||| || |||||||  || ||||| |||||||||||||||||||
Sbjct  332  CATTATCATCCCGGGTAAAAGAGCTTTACAAGCCCTAAGGCCTTCATCACTCACGCGGCA  273

Query  114  TTGCTCCG-TCAGACTTTCGCCCATTGCGGAAGATTCCCTACTGCTGCCTCCCGTAGGAG  172
            ||||| || ||||| |||||||||||| | ||| ||||||||||||||||||||||||||
Sbjct  272  TTGCTTGGATCAGGCTTTCGCCCATTGTCCAATATTCCCACTGCTGCCTCCCGTAGGAG  213

Query  173  TCTGGGCCGTGTCTCAGTCCCAGTGTGGCCGATCACCCTCT-CAGGTCGGCTATGCATCG  231
            |||||||||||||||||||||||||||||| |||||| | | || ||||||||||| |||
Sbjct  212  TCTGGGCCGTGTCTCAGTCCCAGTGTGGCTGATCATCCTCTACAGACCAGCTAAGGATCG  153

Query  232  TCGCCTTGGTGAGCCGTTACCTCACCAACTAGCTAATGCGCCGCGGGCCCATC  284
            ||||||||||||||| ||||||||||||||||||||||| |  |||| | |||
Sbjct  152  TCGCCTTGGTGAGCCTTTACCTCACCAACTAGCTAATCCTACGCGGGCTCATC  100

>gb|EF430682.1|   Unidentified archaeon clone D1003F10.x 16S ribosomal RNA gene,
partial sequence
Length=359

 Score =  221 bits (244),  Expect = 2e-56
 Identities = 241/319 (75%), Gaps = 26/319 (8%)
 Strand=Plus/Minus

Query  91   AAGCCTTCATCACTCACGCGGCGTTGCTCCGTCAGACTTTCGCCCATTGCGGAAGATTCC  150
            ||| ||||||||||||||||||| |||||  ||||| |||||||||||| | ||| ||||
Sbjct  295  AAGGCTTCATCACTCACGCGGCATTGCTGGATCAGGCTTTCGCCCATTGTCCAATATTCC  236

Query  151  CTACTGCTGCCTCCCGTAGGAGTCTGGGCCGTGTCT-CAGT-CCCAGTGTGGCCGATCAC  208
            | ||||||||||||||||||||||||||||| |||| |||| ||||||||||| ||||| 
Sbjct  235  CCACTGCTGCCTCCCGTAGGAGTCTGGGCCGCGTCTCCAGTCCCCAGTGTGGCTGATCAT  176

Query  209  CCTCTCAGGTCGGCTATGCATCGTCGCCTTGGTGAGCCGTTACCTCACCAACTAGCTAAT  268
            |||||||| ||||||||| | |||  || |||||||| ||||| |||||||||||||||| 
Sbjct  175  CCTCTCAGACCAGCTAAAGATCGTAGCCTTGGTAGGCCTTTACCCCACCAACTAGCTAAT  116

Query  269  GCGCCGCGGGCCCATCTGCAAGTGATAGATCGATCCGTCTTTCATTCTTCCCCCATGAAG  328
             | ||||||||||||||              |||  ||||| ||| || ||||   ||| 
Sbjct  115  CAGACGCGGGCCGATC---------------CTTCGGCAATTAATCTTTCCCCA--AAG  74

Query  329  GAGAAGATCCTATCCGGTATTAGCTCCGGTTTCCCGAAGTTATCCCAGTCTTGCGGGCAG  388
             | |     |||||||||||||||||||| ||||||||| || ||| || ||  |||   
Sbjct  73   GGG-----CGTATCCGGTATTAGCTCCAGTTTCCCGGAGTTGTCCCGAACCAAAGGGTAC  19

Query  389  GTTGCCCACGTGTTACTCA  407
            ||| ||||||| |||||||
Sbjct  18   GTT-CCCACGCGTTACTCA  1
```

FIGURE 6.21 Comparison of two pairwise sequence alignments from a BLAST search. (From http://blast.ncbi.nlm.nih.gov/Blast.cgi.)

than requiring an exact word match to seed the search, Discontiguous Megablast uses a word size of 11 or 12 nucleotides but allows mismatches in certain positions. The BLAST program selected for a particular search depends upon the nature and size of the query sequence. The goal of the "I, Microbiologist" project is to use the nucleotide BLAST output from a 16S rDNA query sequence to make a preliminary taxonomic assignment that can be verified by phylogenetic analysis. Sequences derived from cultivated bacterial isolates are more likely to resemble existing sequences in GenBank than are 16S rRNA genes cloned directly from soil samples, suggesting that Megablast may be a more appropriate search program for the isolate sequences while blastn (with a word size of only 7) may be a good place to start for the cloned sequences. Regardless of which program is used, it is important to maintain an accurate record of the search parameters used to generate the BLAST results.

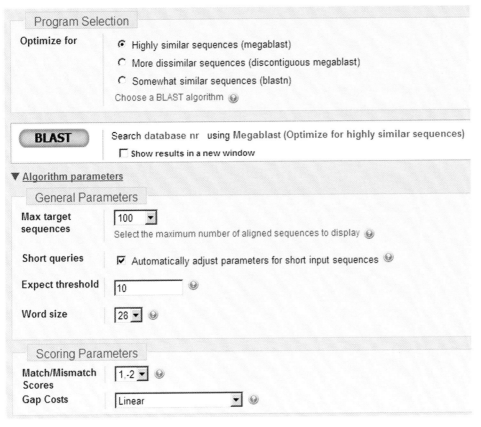

FIGURE 6.22 Megablast search parameters. Note that the word size is 28, the match score is +1, and the mismatch score is −2; also note the linear gap penalty (i.e., deduction for gap opening, irrespective of the length of the gap). (From http://blast.ncbi.nlm.nih.gov/Blast.cgi.)

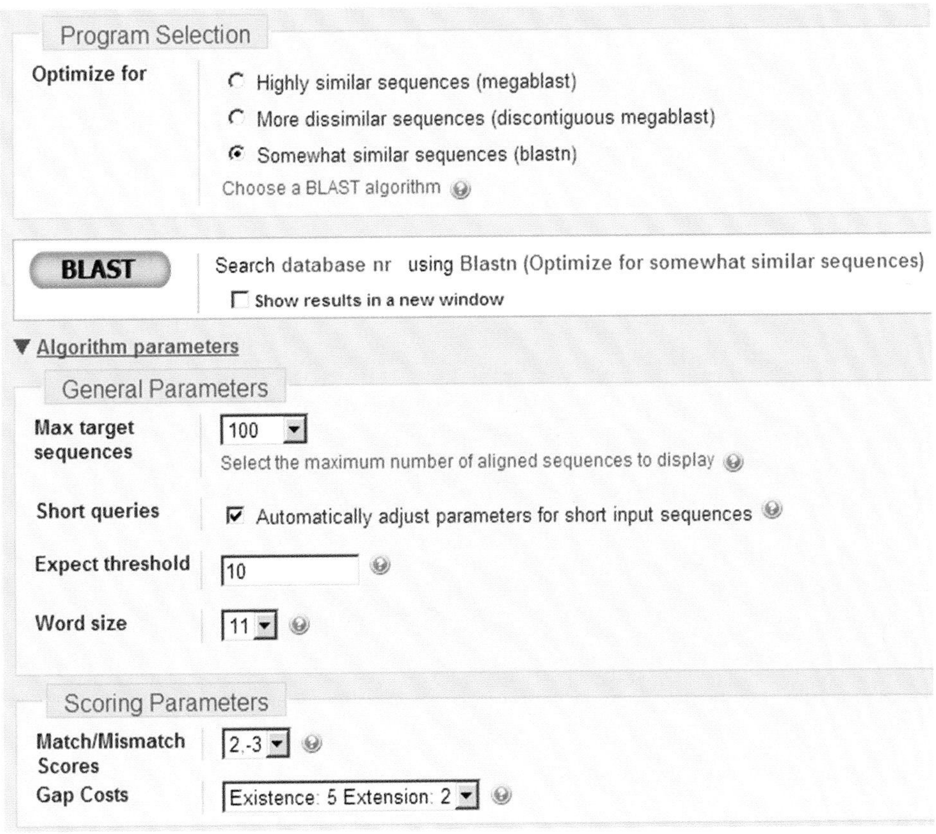

FIGURE 6.23 Blastn search parameters. Note how they differ from the default Megablast parameters depicted in Figure 6.22, with a word size of 11, a match score of +2, a mismatch score of −3, and an affine gap penalty [i.e., deduction for gap opening, or existence, (+5) and gap extension (+2)]. (From http://blast.ncbi.nlm.nih.gov/Blast.cgi.)

BOX 6.1 Understanding a Nucleotide BLAST Search as a Two-Dimensional Array

Constructing pairwise alignments in BLAST, which includes determination of the nucleotide positions at which the alignment starts and stops, proceeds using a two-dimensional array, just like the Smith-Waterman algorithm for determining local alignment. However, there is one major exception. The only alignment paths calculated traverse through the exact word match. In the first figure, a two-dimensional alignment array is represented.

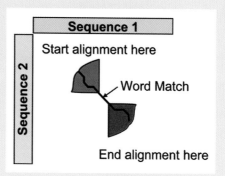

Depiction of a BLAST search using a two-dimensional array. Illustration by Craig Herbold.

The path extends in each direction from the word match as long as the nominal score can be increased. Many locations in the green area represent potential starting points for an alignment, and many locations in the red area represent potential ending points. One such path is traced using a black line. All alignments, regardless of where exactly they begin in the green area and regardless of where they end in the red area, traverse the word match.

Due to some nifty properties of the alignment arrays, the sequence locations where the alignment begins and where it ends can be calculated independently of one another. In the next figure, the two-dimensional alignment array is shown again but without all the potential alignment ending locations.

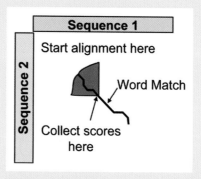

Generating a nominal score during BLAST. Illustration by Craig Herbold.

BOX 6.1 Understanding a Nucleotide BLAST Search as a Two-Dimensional Array (continued)

If an arbitrary start point was chosen within the green area, and a path ends at the word match, then the nominal score at the location of the word match is the score of one candidate alignment starting location. All the possible scores, representing all the possible starting locations, could be collected and made into a graph like that shown in Fig. 6.15, where it can be observed that there is an optimum length to the alignment that coincides with the point at which the nominal score reaches a maximum (i.e., the HSP). Therefore, identification of the HSP specifies an optimum starting location for the alignment. There also is an optimum ending location, found within the red region of the first figure in this box. These scores are calculated in analogous fashion, except that the alignment must begin with the hit and end somewhere in the red area. Those scores also can be plotted against extension length, with the HSP stipulating the optimum end location for that particular alignment.

KEY TERMS

Bit score The normalized maximum nominal score for an alignment calculated by BLAST.

E-value A BLAST parameter that specifies the statistical significance threshold attributed to a match in the database. This value gives an indication of whether the alignment portrays a biological relationship; the lower the E-value, the less likely it is that the aligned sequence similarity is due to chance.

High-scoring segment pair (HSP) A BLAST search hit with maximum observable score resulting from an optimal local alignment of maximal length.

Max score The maximum bit score for an alignment reported by BLAST; this score gives an indication of the quality of the alignment (the higher the Max score, the better the alignment).

Primary sequence database An archival database that contains experimental results that are not necessarily curated (e.g., reviewed and verified); it contains hypothetical annotations of sequence data.

Query A nucleotide or amino acid sequence specified by the user.

Total score The score calculated by BLAST as the sum of the Max scores of all HSPs found in a single database entry; an informative parameter for genome entries in which more than one HSP may be present (e.g., multiple copies of rRNA operon in single genome sequence).

REFERENCES

Altschul, S. F., T. L. Madden, A. A. Schäffer, J. Zhang, Z. Zhang, W. Miller, and D. J. Lipman. 1997. Gapped BLAST and PSI-BLAST: a new generation of protein database search programs. *Nucleic Acids Res.* **25**:3389–3402.

Baxevanis, A. D. 2005. Assessing sequence similarity: BLAST and FASTA, p. 295–324. *In* A. D. Baxevanis and B. F. F. Ouellette (ed.), *Bioinformatics: a Practical Guide to the Analysis of Genes and Proteins*, 3rd ed. John Wiley & Sons, Inc., New York, NY.

Benson, D. A., I. Karsch-Mizrachi, D. J. Lipman, J. Ostell, and E. W. Sayers. 2009. GenBank. *Nucleic Acids Res.* **37**(Database Issue):D26–D31. doi:10.1093/nar/gkn723.

Ma, B., J. Tromp, and M. Li. 2002. PatternHunter: faster and more sensitive homology search. *Bioinformatics* **18**:440–445.

Web Resources

NCBI-BLAST http://www.ncbi.nlm.nih.gov/BLAST/

NCBI Education Resources, *BLAST Tutorial* http://www.ncbi.nlm.nih.gov/Education/

The NCBI Handbook Chapter 16: "The BLAST Sequence Analysis Tool" by Tom Madden http://www.ncbi.nlm.nih.gov/books/bv.fcgi?rid=handbook.chapter.610

SECTION 6.3
MODEL-BASED ALIGNMENT TOOLS FOR 16S rRNA

The Ribosomal RNA Database Project

Another massive database that is geared specifically toward the analysis of ribosomal DNA (rDNA) sequences is the Ribosomal Database Project (RDP-II) (Cole et al., 2005). This database contains hundreds of thousands of ribosomal sequences, submitted by various researchers from all over the world. As exemplified in Fig. 6.24, the RDP-II acquires bacterial rDNA sequences from the International Nucleotide Sequence Database (INSD) every month. To assist analyses, several online tools are available; these are constantly being updated to meet the needs of researchers who focus on rDNA sequences.

One of the most valuable resources at the RDP-II is the aligned database of 16S rDNA sequences. In this database, sequences are prealigned to a secondary-structure model. rRNA, an essential component of ribosomes, forms extensive and predictable secondary structure (Fig. 6.25) (Cannone et al., 2002; CRW Site). It is thought that secondary structure may be more highly conserved than the nucleotide sequences based on its ties to translation, the biological function of the ribosomes. In stem-loop structures, a nucleotide change in one location may be easily tolerated if there is an additional change in its base-pairing partner. Quickly evolving regions may be difficult to align by traditional pairwise alignment schemes since they use only nucleotide data and not secondary-structure data. The secondary structure does not evolve as quickly as the primary sequence, provided that function is preserved, so an alignment can still be performed if it can be shown that the "unalignable" regions occupy the same position in the structure.

Remember that the goal of sequence alignment tends to be the maximization of a nominal alignment score. This score has very little to do with whether a molecule is functional. Structure-based alignment has a different goal altogether and one that is

FIGURE 6.24 Data flow from the INSD to RDP-II. Illustration by Erin Sanders.

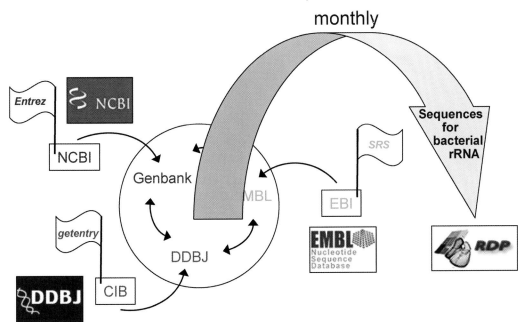

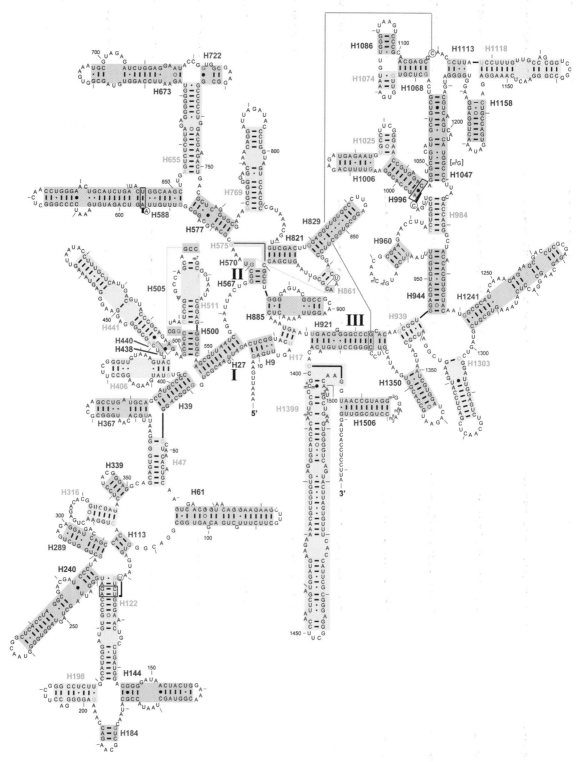

FIGURE 6.25 Primary and secondary structure of 16S rRNA from *E. coli* (J01695). Each base pair is numbered (5′:3′), so that the first base pair in this molecule is (9:25). Each helix is numbered with the 5′ nucleotide of the initial base pair; the first helix in this molecule starts with pair (9:25) and is numbered H9. Helices are shaded (in light red, green, and blue) and labeled with the "helical element identifier" in the same color. The 5′ nucleotide of the initial base pair is colored as well. Reprinted from Cannone et al. (2002) with permission (http://www.rna.ccbb.utexas.edu/); model version, November 1999; numbering diagram version, July 2001.

intuitively based more completely on the evolution of the gene. Due to the overall structural conservation, every nucleotide occupies a specific location. It does not matter if a nucleotide is an A or a G; instead, it matters whether that particular nucleotide occupies position 456 or 833 in the secondary structure (based on *E. coli* numbering in Fig. 6.25). This increase in specificity allows the construction of an alignment that is far more evolutionarily relevant. The structure dependence of the alignment should place a heavier constraint on the alignment, increasing the chances that two nucleotides that are aligned to one another truly are related by descent from a common ancestor. Thus, structure-based alignments are potentially more accurate than sequence-based alignments and may be more useful for conducting a meaningful phylogenetic analysis.

The secondary-structure model of Cannone et al. is the starting point for structure-based alignment at the RDP-II (Cannone et al., 2002). These researchers took on the bold goal of defining the secondary structure of rDNA sequences in an evolutionary context. Secondary structures, which are predicted based on the probability of loop and helix formation by canonical base-pairing rules, were calculated for a broad sampling of rDNA sequences. These secondary structures were aligned to one another, and the nucleotide sequences were then remapped onto the secondary-structure alignment. Predictable patterns of nucleotide use in specific secondary-structure locations became clear. These patterns could then be used to build an all-encompassing model of the evolution of secondary structure. Specific nucleotides are more likely to exist at specific locations, and these may act as anchors for the alignment. The nucleotides between these anchors may then be aligned to one another if they occupy the same location within the secondary structure. It is this secondary-structure model that new sequences are aligned to, not necessarily any one specific sequence contained in the database. Nor is it necessary to recalculate the secondary structure for most newly deposited sequences, since very similar sequences usually exist within the database already. The sequences contained within the RDP-II are ordered within a phylogenetic framework, which can be rapidly explored by using the Hierarchy Browser tool (Cole et al., 2005) (Fig. 6.26). The browser also provides a means for the user to select individual sequences for later download.

Using a modified version of an rRNA aligner called RNACAD, each new sequence is aligned to the sequence to which it is most closely related according to an internal model that directly incorporates secondary-structure and primary-sequence information (Cole

FIGURE 6.26 Modified screen shot from RDP-II, highlighting the functions of each analysis tool: myRDP, Hierarchy Browser, Classifier, and SeqMatch. (Reprinted from Cole et al. [2005] with permission; http://rdp.cme.msu.edu/.)

RIBOSOMAL DATABASE PROJECT

Explore our online analysis tools:

myRDP	→ Align and classify your 16S rRNA sequences. Use the RDP Pipeline to process sequence libraries from raw sequencer output to analysis.
BROWSERS	→ **Hierarchy Browser** – Browse a phylogenetic hierarchy and compile a list of 16S rRNA sequences for download or use. Also search by publication or genome.
CLASSIFIER	→ Assign 16S rRNA sequences to our taxonomical hierarchy.
SEQ MATCH	→ Upload your sequence and search for its nearest neighbors.

et al., 2005; Brown, 2000). Then, this preexisting sequence is used as a guide to align the new sequence to the rest of the alignment. Occasionally, the model is updated and some sequences are reassigned and/or realigned, depending on any new information gained since the last update. The alignment tool can be accessed by creating an account at myRDP (Fig. 6.26).

The secondary-structure-based alignments found at the RDP-II provide a basis for a meaningful comparative analysis of new sequences with the database. The SeqMatch tool at the RDP-II can be used to find the nearest neighbors, which are the closest matching sequences within the database to a query sequence (Fig. 6.26) (Cole et al., 2005). Seq-Match resembles BLAST in its database structure, using nucleotide "words" to search a query against the database. However, the resemblance stops there since no alignment is performed. SeqMatch looks for the database entries that share the largest number of words with the query. As shown in Fig. 6.27, the highest-scoring sequences are then returned to the user, along with several scoring statistics. SeqMatch reports the number of "unique common oligomers," which is the total number of words shared between the

FIGURE 6.27 Screen shots from RDP-II, evaluating search hits obtained using SeqMatch. (Reprinted from Cole et al. [2005] with permission; http://rdp.cme.msu.edu/.)

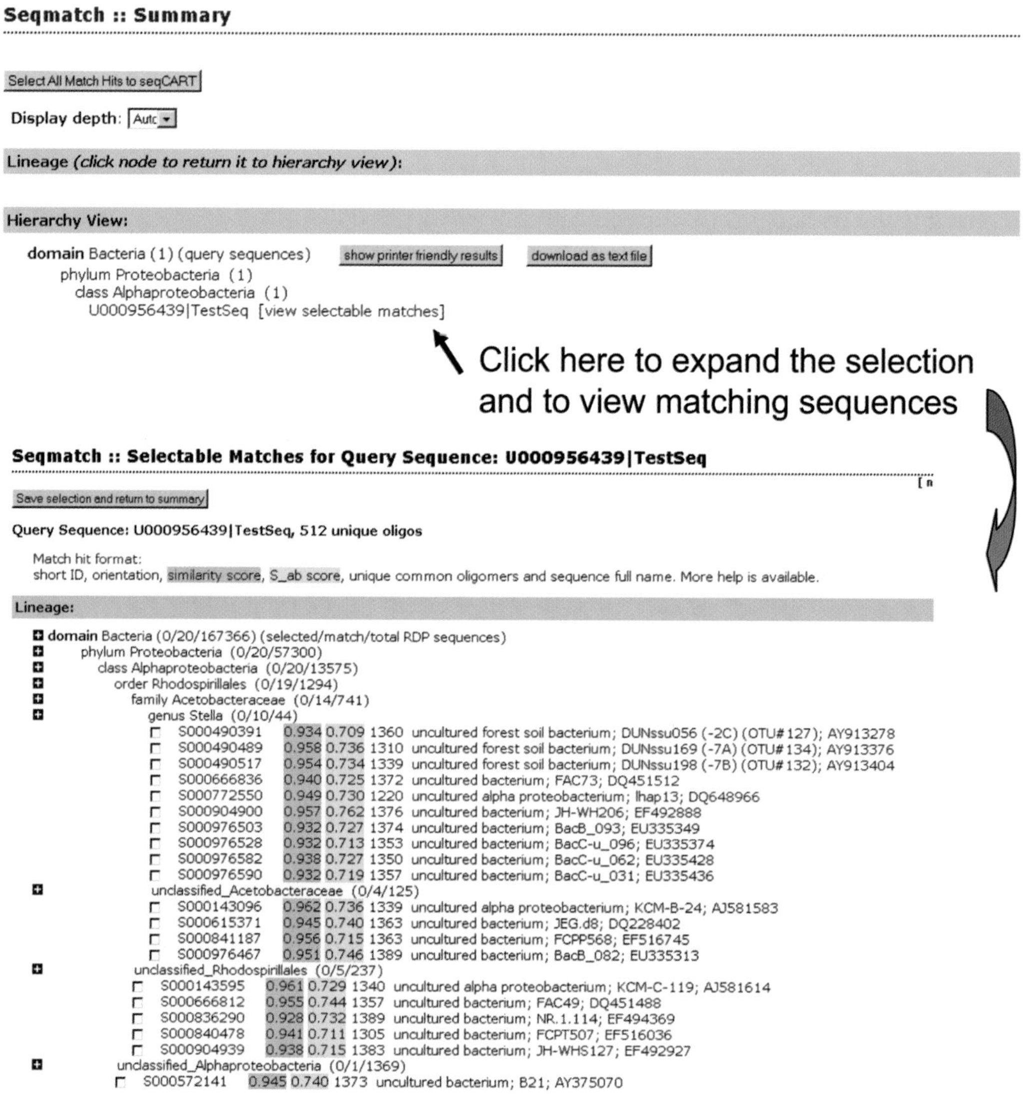

Table 6.1 Performance of an rRNA search using SeqMatch and BLAST[a]

| Program[b] | % of 16S rRNA queries[c] returning the most similar sequence[d] among the highest-scoring N results | | |
	N = 1	N = 10	N = 20
SeqMatch	65	92	95
BLAST	39	53	55

[a]Reprinted from Cole et al. (2005) with permission.

[b]For both programs, the data set consisted of 37,456 near-full-length (≥1,200 bases) rRNA sequences from the RDP release 9.20 alignment database.

[c]1,000 query sequences were selected randomly from the data set.

[d]The most similar sequence to each query was determined by exhaustive pairwise similarity comparison of each query against the data set. In cases of a tie, only one similar sequence was returned by the program.

query sequence and the database entry, as well as the SeqMatch score (S_ab score, highlighted in orange), which is the ratio of shared words to unique words. The similarity score (highlighted in pink), which reflects the percent identity of the query sequence to the SeqMatch sequences, also is reported but only if the query sequence has been aligned successfully to the RDP-II model. The plain-text files in FASTA format for these high-scoring sequences can be retrieved easily from this page.

A recent comparison of SeqMatch to BLAST suggests that the former may be more accurate than the latter at finding closely related rRNA gene sequences (Table 6.1) (Cole

FIGURE 6.28 Screen shots from RDP-II, evaluating search hits obtained using Classifier. (Reprinted from Cole et al. [2005] with permission; http://rdp.cme.msu.edu/.)

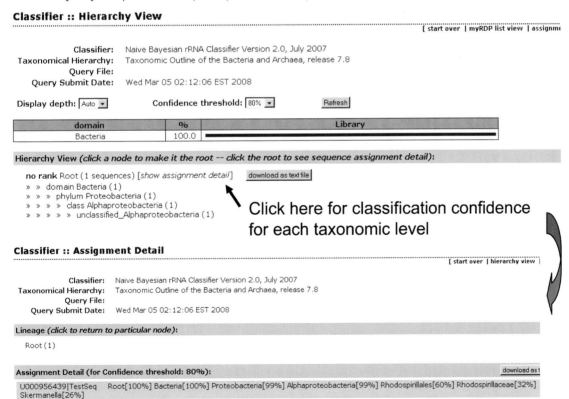

et al., 2005); however, the results produced by either method should be viewed strictly as a hypothesis awaiting a more detailed phylogenetic analysis.

Classifier, another tool at RDP-II, allows one to classify query sequences at different taxonomic levels (Fig. 6.26) (Wang et al., 2007). This tool uses "words" as well, but in a rather different way. Using the NCBI database as a source and the classification scheme in *Bergey's Manual* (Garrity et al., 2004; *Bergey's Manual* Online), all database words are assigned a probability of being observed in a particular taxonomic group. Each hierarchical grouping has its own set of word probabilities. Given a set of words for a query sequence, one can (with sufficient knowledge of probability and statistics) calculate the probability of its belonging to a particular taxonomic group. RDP-II makes this easy by doing all the necessary calculations and reporting confidence estimates for classification of the query sequence. An example is presented in Fig. 6.28. Assignment of this sequence as a member of the *Proteobacteria* and of the *Alphaproteobacteria* is quite robust (99% for both assignments). The sequence is most likely a member of the *Rhodospirillales,* but the confidence is not significant enough for us to make the assignment (60%). This feature of the RDP-II is quite useful and statistically reliable; however, it is also based on proper genus-level annotation in the NCBI database. Any human errors in annotation have a poisoning effect on this analysis, and skepticism should again be the rule of thumb.

KEY TERMS

FASTA format Text-based format representing DNA, RNA, or protein sequences in which the nucleotides or amino acids are depicted as single-letter codes; the sequence begins with a description line preceded by a greater-than symbol (>).

Nearest neighbors Highly similar sequences in a database that exhibit the greatest number of nucleotide matches to a query sequence.

REFERENCES

Brown, M. P. S. 2000. Small subunit ribosomal RNA modeling using stochastic context-free grammar, p. 57–66. *In Proceedings of the 8th International Conference on Intelligent Systems for Molecular Biology (ISBM 2000),* San Diego, CA.

Cannone, J. J., S. Subramanian, M. N. Schnare, J. R. Collett, L. M. D'Souza, Y. Du, B. Feng, N. Lin, L. V. Madabusi, K. M. Muller, N. Pande, Z. Shang, N. Yu, and R. R. Gutell. 2002. The Comparative RNA Web (CRW) Site: an online database of comparative sequence and structure information for ribosomal, intron, and other RNAs. *BioMed. Central Bioinformatics* **3:**2. (Correction, **3:** 15.) http://www.rna.ccbb.utexas.edu/.

Cole, J. R., B. Chai, R. J. Farris, Q. Wang, S. A. Kulam, D. M. McGarrell, G. M. Garrity, and J. M. Tiedje. 2005. The Ribosomal Database Project (RDP-II): sequences and tools for high-throughput rRNA analysis. *Nucleic Acids Res.* **33:**D294–D296. doi: 10.1093/nar/gki038

Garrity, G. M., J. A. Bell, and T. G. Lilburn. 2004. Taxonomic outline of the prokaryotes. *Bergey's Manual of Systematic Bacteriology,* 2nd ed., Release 5.0. Springer-Verlag, New York, NY.

Wang, Q., G. M. Garrity, J. M. Tiedje, and J. R. Cole. 2007. Naive Bayesian classifier for rapid assignment of rRNA sequences into the new bacterial taxonomy. *Appl. Environ. Microbiol.* **73:**5261–5267.

Web Resources

Bergey's Manual Online http://dx.doi.org/10.1007/bergeysoutline

The Comparative RNA Web (CRW) Site http://www.rna.ccbb.utexas.edu/

The Ribosomal Database Project (RDP-II) http://rdp.cme.msu.edu/

SECTION 6.4
MULTIPLE SEQUENCE ALIGNMENTS

Aligning three or more DNA sequences

Algorithms for aligning pairs of DNA sequences have been discussed in previous sections; these methodologies facilitate fast and efficient nucleotide database searches. Most advanced applications of alignments, such as phylogenetic reconstruction, require the use of a multiple alignment. Multiple alignments contain several sequences, all aligned to one another. The ideal multiple-alignment method simultaneously aligns all sequences together. The Needleman-Wunsch algorithm for global alignment and the Smith-Waterman algorithm for local alignment have both been generalized to allow for such a simultaneous comparison of three or more sequences; however, the computational requirements for calculating a multiple alignment by these methods become cumbersome with the addition of more sequences. To overcome these limitations, two unique approaches have been developed that utilize the easily calculated pairwise alignments. These methods are termed sequential alignments (also called progressive alignments) and star alignments (Lake, 1991).

Both sequential and star alignments begin with a calculation of several pairwise alignments and follow this by combining the different pairwise alignments. The two approaches differ in the exact steps in which the simpler pairwise alignments are combined to make the more complex multiple alignment. These two methods are graphically summarized in Fig. 6.29. In the star alignment method (Fig. 6.29a), all sequences are aligned to a common sequence, in this case sequence 1. The pairwise alignment of sequences 1 and 2 is combined with the pairwise alignment of sequences 1 and 3 and the pairwise alignment of sequences 1 and 4 by using sequence 1 as a guide. Any nucleotides in sequence 2, 3, or 4 or any other sequence that are aligned to a particular nucleotide in sequence 1 are automatically aligned to one another. In the figure, the arrows point to sequence 1 because each of the sequences is aligned to sequence 1. It becomes visually apparent why this method is termed the star alignment, since sequence 1 acts as a "star center," with each of the other sequences acting as the tips of the star.

Sequential alignments differ from star alignments in that not all sequences are necessarily aligned to a single star center. In the sequential alignment depicted in Fig. 6.29b, sequence 1 is aligned to sequence 2, sequence 2 is aligned to sequence 3, sequence 3 is aligned to sequence 4, etc. To combine these sequences into a larger alignment, the pairwise alignment of sequences 1 and 2 is combined with the pairwise alignment of sequences 2 and 3, using sequence 2 as a guide. This is essentially the same procedure as is used in the star alignment, except that sequence 2 is acting as a star center and the

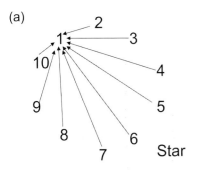

(a) Star

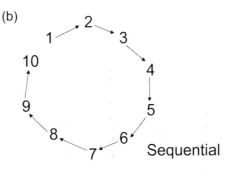

(b) Sequential

FIGURE 6.29 Overview of multiple sequence alignment methods. (a) Star alignment; (b) sequential, or progressive, alignment. Illustration by Craig Herbold.

multiple alignment at this step contains sequences 1, 2, and 3. The next step, though, is very different from the star alignment method in that the star center changes. The next pairwise alignment that is incorporated into the multiple alignment is the pairwise alignment of sequences 3 and 4. The common sequence between the multiple alignment and the pairwise alignment is sequence 3. Therefore, sequence 3 acts as a temporary star center, just long enough to add sequence 4 into the multiple alignment. Many iterations of this procedure occur until all sequences are aligned.

Star alignments and sequential alignments both have advantages and disadvantages. Each method has been developed to decrease computational costs. There is no guarantee that the answer that emerges is accurate or unbiased. In fact, it has been clearly shown that the order in which pairwise alignments are combined will tend to bias any phylogenetic analysis performed with that alignment (Lake, 1991). Sequences that are aligned to one another have a tendency to group together in the subsequently calculated tree, regardless of the phylogenetic method employed in the analysis. In Fig. 6.29, the sequential-alignment method therefore would be likely to produce a tree that groups sequences 1 and 2 together, regardless of whether they actually belong together as a reflection of their ancestral origins. The star alignment does not suffer from this problem because all sequences are aligned through sequence 1. Any introduced bias would be expected to be distributed evenly among all the other sequences. This does not mean that the star alignment method is ideal by any means. Actually, the choice of star center probably does have an effect on the subsequent phylogeny. Thus, when opting to use the star method, it would be wise to calculate several star alignments, using different sequences as the star center. Each of these alignments can then be used to calculate an independent phylogenetic tree. Comparison of the phylogenetic trees obtained in such a manner might then reveal whether any particular star center leads to a particularly biased tree.

Creating a multiple sequence alignment in CLUSTAL

There has been a push to overcome the biases associated with sequential alignments and to develop user-friendly software. One particular method of constructing multiple alignments has addressed both these concerns to the satisfaction of the general scientific community and is one of the most widely used multiple-alignment methods available. The CLUSTAL algorithm, which is available as a stand-alone program or packaged into other alignment programs such as MEGA4, utilizes components of the sequential-alignment methodology but uses a guide tree to determine the order in which sequential alignment is carried out (Higgins and Sharp, 1988; Thompson et al., 2003; Tamura et al., 2007). CLUSTAL begins by doing pairwise comparisons between each combination of two sequences. It then uses distances calculated from the pairwise alignments to estimate a phylogenetic tree, called a guide tree since it is used as a guide to combine pairwise alignments to make a multiple sequence alignment. One justification for the preferred use of this method is that an alignment between closely related sequences is probably more accurate than alignments between distantly related sequences. Under this assumption, the accuracy of a multiple alignment is maximized by aligning sequences that are closely related before attempting to align those that are more distantly related.

A portion of a CLUSTAL alignment is presented in Fig. 6.30. The guide tree in the figure reveals the order of alignment that is followed to make a multiple alignment. Initially, sequence 2 is aligned to sequence 6 and this alignment is used to create a

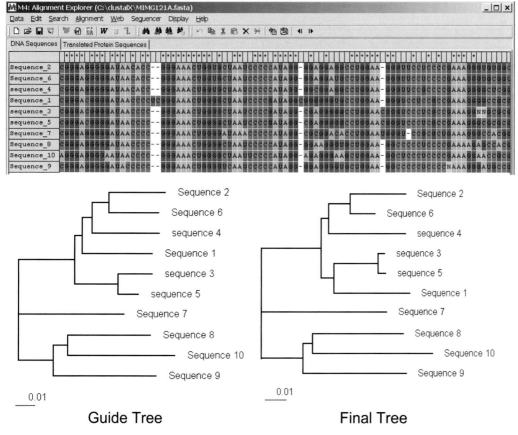

FIGURE 6.30 Creating a multiple sequence alignment in CLUSTAL that is used to construct a phylogenetic tree. The order of alignment is determined based on the guide tree, while the eventual product of the analysis is the final tree. (Reprinted from Tamura et al. [2007] with permission; http://www.megasoftware.net/.)

consensus sequence, which refers to a sequence that summarizes the information in an alignment. The consensus sequence is the same length as the alignment and contains nucleotides that are conserved at a particular position. If both sequences 2 and 6 contain the same nucleotide at a particular alignment position, the consensus sequence also contains that nucleotide at the same position. Sequence 4 is then aligned to the consensus sequence obtained from the alignment of sequences 2 and 6, since it is the next related sequence according to the guide tree. The consensus sequence then acts as a star center to align sequence 4 to sequences 2 and 6. Sequence 1 is then added to the alignment of sequences 2, 4, and 6 by a similar procedure. The next combination step requires the consensus sequence from the pairwise alignment of sequences 3 and 5 to be aligned to the consensus sequence of the alignment of sequences 1, 2, 4, and 6. This method of combining groups proceeds until all sequences are contained in a multiple alignment. This sequential method of alignment is extremely fast and produces visually pleasing alignments. For these reasons, it currently reigns as the most popular method for calculating alignments.

Model-based methods used to create multiple sequence alignments

Additional methods exist for aligning sequences, but the inner workings of such methods are beyond the scope of this text. Still, it is appropriate to be introduced to the conceptual bases for such methods. Structure-based alignment was introduced in Section 6.3

and is increasingly perceived as more appropriate than CLUSTAL for sequences for which adequate structure is known (like 16S rRNA). Structure-based alignments can be called "model based" since they rely on model information rather than primary sequence to align data sets. Since sequences are aligned to a common structural model, they are automatically aligned to one another. Conceptually, this is similar to star alignments, except that in this approach the structure model acts as the star center. The model itself is constructed using precalculated structures. This innovation sidesteps the computational problems associated with constructing multiple alignments, but the model itself requires a great deal of expertise and time to develop.

Other model-based methods for constructing multiple alignments utilize a hidden Markov model (HMM) to align sequences. HMMs can be based on secondary structure, sequence motifs, or any preexisting alignment that has been examined thoroughly. HMMs use a carefully assembled alignment as a set of training data to build a "hidden" model of nucleotide alignment based on probability. Specifically, the carefully assembled training alignment is used to calculate a site-specific probability value for observing a particular nucleotide or a gap at a given position in an alignment. This model is then used to calculate the most likely path of any sequence through the model space. This approach allows a likelihood value to be calculated for any possible alignment of a particular sequence. Again, a multiple sequence alignment is achieved easily because all sequences aligned to the model are automatically aligned to one another. An added benefit is that uncertainty in the alignment can be assessed. Thus, decisions on whether to retain a portion of poorly aligned sequence can become much more objective. One HMM-based phylogenetic method that explores the uncertainty in an alignment is Bali-Phy (Suchard and Redelings, 2006). This program explores different high-likelihood alignments of the same set of sequences with a view to determining a phylogeny that is robust to the specific alignment used. Although a thorough discussion of this method is beyond the scope of this text, the curious student is encouraged to investigate this topic further by reading additional literature (Wong et al., 2008).

Troubleshooting multiple sequence alignments

Several problems can arise when one constructs alignments of three or more nucleotide sequences. Even when automated alignment methods are used, the alignment must be examined by eye to find any obvious problems. It is generally useful to conduct a preliminary alignment to troubleshoot data. Sequences that are in the wrong orientation will be poorly aligned to other sequences in the multiple alignment. Sometimes more than one sequence is in the wrong orientation. If CLUSTAL is used, these sequences will be well aligned to one another but poorly aligned to everything else. To help troubleshoot this problem, it is always a good idea to include a sequence in the analysis that is known to be in the proper orientation and which spans the entire sequence region that should be contained in the alignment.

Figure 6.31 displays the same sequences as in Fig. 6.30, except that the reverse complements of sequences 2 and 10 (minus strands) were "accidentally" used instead of the properly oriented sequence (plus strands). The alignment between sequences 2 and 10 is good, but these sequences are not aligned well to the other sequences. The resulting tree from this poorly annotated alignment is shown below the multiple alignment. Sequences 2 and 10 cluster together and are separated from the other sequences by an extremely long branch. While these patterns in the alignment and phylogeny can certainly

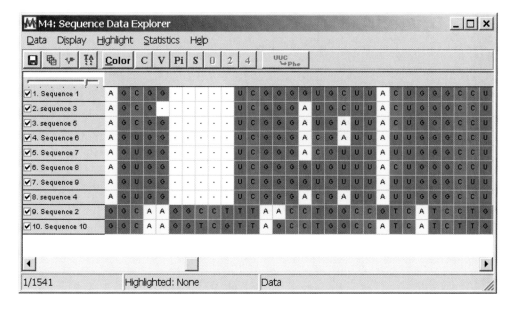

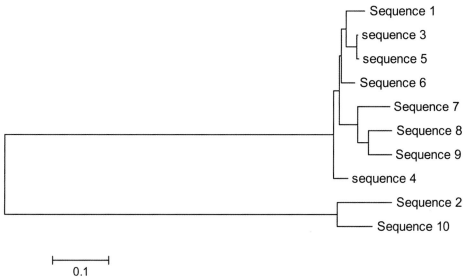

FIGURE 6.31 Effect on alignment and phylogenetic tree when two of the sequences (2 and 10) are in the opposite orientation relative to the other sequences in the alignment. (Reprinted from Tamura et al. [2007] with permission; http://www.megasoftware.net/.)

arise from the evolutionary process, they should raise suspicion regarding alignment quality. If this pattern is encountered, sequences should be double-checked for proper orientation. Knowing the orientation of at least one sequence in the alignment will assist this effort by indicating which set of sequences is in the correct orientation.

The "edges" of an alignment are another hot spot of problems. Most alignment algorithms do quite poorly near the beginning and end of the alignment. The only way of troubleshooting this problem is by visual examination followed by manual adjustments. Initially, long sequences tend to define the length of the alignment, and consequently shorter sequences tend to be stretched out in an attempt by the alignment algorithm to maximize the nominal alignment score. The introduction of several gaps can be mediated by forcing nucleotides to match one another. One such example is illustrated in the top panel of Fig. 6.32. Here the beginning of an alignment is shown in

```
CTGGACACATGGAACCTGACCTG    Sequence #1
C----C---TG---C---AC-TG    Sequence #2
C------C-TGCA-C------TG    Sequence #3
-----C-C--------TGCACTG    Sequence #4
-----C-C-TG--------CTG    Sequence #5
```

```
CTGGACACATGGAACCTGACCTG    Sequence #1
---------------CCTGCACTG   Sequence #2
---------------CCTGCACTG   Sequence #3
---------------CCTGCACTG   Sequence #4
---------------CCTG--CTG   Sequence #5
```

FIGURE 6.32 Alignment problems at sequence edges. Illustration by Craig Herbold.

which the first sequence is much longer than the others. Assume that the alignment is very good to the right of the portion shown. The four shorter sequences have been stretched out so that more nucleotides match. A closer examination reveals that all the shorter sequences should be aligned to the highlighted region in the long sequence, as done in the lower panel of Fig. 6.32. A similar effect can be found in alignment regions containing large inserts in one or a few sequences. Sometimes the effect of long sequences or long inserts on the alignment is so disastrous that those troublesome sequences must be edited before an alignment can be constructed. For example, some *Archaea* contain self-splicing introns within the 16S rDNA sequence (Nomura et al., 2002). These introns are present in the sequence but should not be included in a phylogenetic analysis. In this case, it is prudent to excise them from the sequence so that the effect on the sequence alignment is lessened.

Methods of alignment also provide ample opportunity for problems. Remember that CLUSTAL uses a rudimentary guide tree to align sequences to one another. It should not be overlooked that phylogenies calculated using CLUSTAL alignments usually match the CLUSTAL guide tree regardless of the phylogenetic method employed (i.e., compare the guide tree and the final tree in Fig. 6.30). By basing the alignment on a rudimentary tree, the alignment is biased toward that tree. As discussed in Unit 7, the artifact known as long-branch attraction (LBA) is the bane of phylogenetic analysis. The theory that explains how LBA arises also implicates the CLUSTAL method as exacerbating the problem. Thus, alignments suffer from several drawbacks, only one of which is human error. A rigorous analysis requires alignment troubleshooting to remove as much of the human error component as possible; it is also necessary to consider the biases that are propagated onward into the phylogenetic stage of analysis.

BOX 6.2 Construction of Consensus Sequence for Contig Assembly Based on Alignment of Overlapping DNA Ends

Current DNA sequencing technology using the Sanger capillary method limits reads to approximately 1,000 bases or fewer (Madabhushi, 1998; Hall, 2007). Thus, no single end-to-end sequencing reaction can be performed that will cover any stretch of sequence space larger than 1,000 bp, including the 16S rRNA gene, which is approximately 1,500 bp. Recall from Unit 5 that the 16S rRNA gene is cloned into vectors and that two sequencing primers (T7 and M13R), which anneal to regions on the vector flanking the DNA insert, are used to amplify the intervening gene segment (see Fig. 3.8). Because the polymerase enzyme is expected to extend both primers to produce sequencing reads approaching 1 kb in size (usually more like 700 to 850 bases), a central region of the 16S rRNA gene (typically no more than 150 bases) will overlap, or share a sequence of bases. These two independent sequencing reads can be combined to form a single, contiguous DNA sequence called a contig. The process involves performing a pairwise comparison (also known as alignment) of the DNA ends for the two sequencing reads. Because one primer amplifies the plus strand and the second primer amplifies the minus strand, the reverse complement of one of the two products, whichever encodes the minus strand, first must be obtained before the alignment can be done (see Unit 3 and Experiment 6.1). Using computer software, the region of overlap between the two sequence reads is merged to form a contig, and a consensus sequence for the entire DNA fragment, now approximately 1,500 bases in length for the 16S rDNA clones, is generated.

There are a suite of programs available to perform this task; however, NCBI-BLAST has a special function, called BLAST2Sequences, which allows one to quickly perform a pairwise, local alignment of two sequences, assessing whether or not there is any region of overlap between them (Tatusova and Madden, 1999). Like the other BLAST tools, BLAST2Sequences results provide numerous statistics that help the user determine the length, quality, and orientation of the overlap region. In the following example, two sequences were uploaded to BLAST2Sequences, one representing the sequence read from the T7 primer and the other from the M13R primer.

(a)

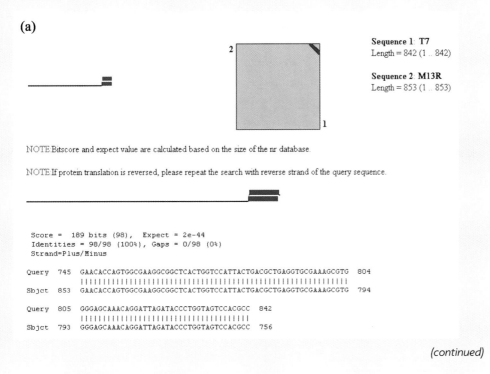

Sequence 1: T7
Length = 842 (1 .. 842)

Sequence 2: M13R
Length = 853 (1 .. 853)

NOTE Bitscore and expect value are calculated based on the size of the nr database.

NOTE If protein translation is reversed, please repeat the search with reverse strand of the query sequence.

```
Score =  189 bits (98),  Expect = 2e-44
Identities = 98/98 (100%), Gaps = 0/98 (0%)
Strand=Plus/Minus

Query  745  GAACACCAGTGGCGAAGGCGGCTCACTGGTCCATTACTGACGCTGAGGTGCGAAAGCGTG  804
            |||||||||||||||||||||||||||||||||||||||||||||||||||||||||||
Sbjct  853  GAACACCAGTGGCGAAGGCGGCTCACTGGTCCATTACTGACGCTGAGGTGCGAAAGCGTG  794

Query  805  GGGAGCAAACAGGATTAGATACCCTGGTAGTCCACGCC  842
            ||||||||||||||||||||||||||||||||||||||
Sbjct  793  GGGAGCAAACAGGATTAGATACCCTGGTAGTCCACGCC  756
```

(continued)

BOX 6.2 Construction of Consensus Sequence for Contig Assembly Based on Alignment of Overlapping DNA Ends (continued)

(b)

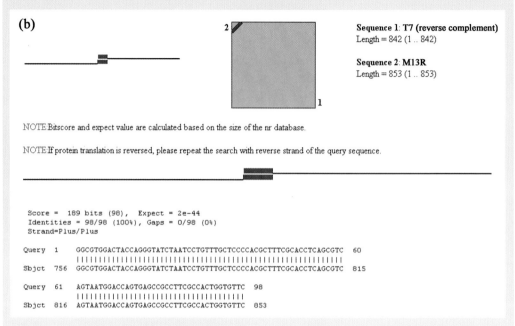

Sequence 1: **T7 (reverse complement)**
Length = 842 (1 .. 842)

Sequence 2: **M13R**
Length = 853 (1 .. 853)

NOTE:Bitscore and expect value are calculated based on the size of the nr database.

NOTE:If protein translation is reversed, please repeat the search with reverse strand of the query sequence.

```
 Score =  189 bits (98),  Expect = 2e-44
 Identities = 98/98 (100%), Gaps = 0/98 (0%)
 Strand=Plus/Plus

Query  1    GGCGTGGACTACCAGGGTATCTAATCCTGTTTGCTCCCCACGCTTTCGCACCTCAGCGTC  60
            ||||||||||||||||||||||||||||||||||||||||||||||||||||||||||||
Sbjct  756  GGCGTGGACTACCAGGGTATCTAATCCTGTTTGCTCCCCACGCTTTCGCACCTCAGCGTC  815

Query  61   AGTAATGGACCAGTGAGCCGCCTTCGCCACTGGTGTTC  98
            ||||||||||||||||||||||||||||||||||||||
Sbjct  816  AGTAATGGACCAGTGAGCCGCCTTCGCCACTGGTGTTC  853
```

Assembly of contigs from two sequences based on an alignment of overlapping DNA ends. (a) Aligned using raw sequence T7 and M13R reads; (b) aligned using reverse complement of T7 sequence but un-modified read of M13R. (From http://blast.ncbi.nlm.nih.gov/bl2seq/wblast2.cgi.)

In the first scenario, the user forgot to create the reverse complement of one of the two sequencing reads, thus producing a reversed segment of overlap (Strand = Plus/ Minus in panel a). In the second scenario, where the user submitted the reverse complement of the second sequence (in this case corresponding to the T7 read), the region of overlap was found between two DNA reads in the same orientation (Strand = Plus/Plus in panel b). If the two sequences were aligned across their entire length, then the blue diagonal in the graphical representation of the alignment would span from the lower left corner to the upper right corner. However, the diagonal length is proportional to the length of the overlap. Since only a portion of the two sequences aligns in our example, the diagonal is shorter, reflecting the small region of overlap relative to the queried sequence space. Inspection of the statistics for the aligned ends provides information about the quality of the DNA sequence reads. Particular attention should be paid to identities and gaps, which ideally should be 100% and 0%, respectively. However, deviations from these results suggest that one of the two sequence reads was of questionable quality, requiring the user to visually inspect the raw sequence files (e.g., chromatograms) for individual reads in this region. When a consensus sequence is generated, some judgment is needed to manually resolve the base call or gap characters for all ambiguous sites. If the raw data are not sufficiently reliable to make such decisions, then an N character should be assigned to that site until the region can be resequenced.

Consensus sequence A hypothetical nucleotide or protein sequence that is constructed from an alignment. Nucleotides or amino acids at a particular position in the consensus sequence reflect agreement among the aligned sequences about which nucleotide or amino acid predominates at a particular position.

Contig A single, contiguous nucleic acid or protein sequence derived from the merging of two independent sequences with a common series of nucleotides or amino acids.

Guide tree A distance-based phylogenetic tree used to direct the sequential alignment of sequence pairs during the creation of a multiple sequence alignment in CLUSTAL.

Likelihood Similar to probability. Given a model and data, the likelihood and probability values are the same. Probability refers to the distribution of expected values for data given a model (expected data are variable), while likelihood refers to the distribution of model parameters given a set of data (model parameters are variable).

Long-branch attraction A problem encountered in phylogenetic analysis of nucleic acid and protein sequences whereby long branches in a tree cluster with other long branches regardless of the true evolutionary history of either branch.

Multiple alignment An alignment that contains three or more sequences all aligned to one another.

Sequential (progressive) alignment A multiple-alignment method in which pairwise alignments are sequentially combined. Any pairwise alignment can be combined into the multiple alignment, provided that the multiple alignment and the pairwise alignment have one sequence in common.

Star alignment A multiple-alignment method in which all pairwise alignments contain a common sequence. The sequence common to all the pairwise alignments is used to facilitate alignment.

REFERENCES

Hall, N. 2007. Advanced sequencing technologies and their wider impact in microbiology. *J. Exp. Biol.* **209:**1518–1525.

Higgins, D. G., and P. M. Sharp. 1988. CLUSTAL: a package for performing multiple sequence alignment on a microcomputer. *Gene* **73:**237–244.

Lake, J. A. 1991. The order of sequence alignment can bias the selection of tree topology. *Mol. Biol. Evol.* **8:**378–385.

Madabhushi, R. S. 1998. Separation of 4-color DNA sequencing extension products in noncovalently coated capillaries using low viscosity polymer solutions. *Electrophoresis* **19:**224–230.

Nomura, N., Y. Morinaga, T. Kogishi, E. J. Kim, Y. Sako, and A. Uchida. 2002. Heterogeneous yet similar introns reside in identical positions of the rRNA genes in natural isolates of the archaeon *Aeropyrum pernix*. *Gene* **295:**43–50.

Suchard, M. A., and B. D. Redelings. 2006. BAli-Phy: simultaneous Bayesian inference of alignment and phylogeny. *Bioinformatics* **22:**2047–2048.

Tamura, K., J. Dudley, M. Nei, and S. Kumar. 2007. MEGA4: Molecular Evolutionary Genetics Analysis (MEGA) software version 4.0. *Mol. Biol. Evol.* **24:**1596–1599.

Tatusova, T. A., and T. L. Madden. 1999. BLAST2Sequences, a new tool for comparing protein and nucleotide sequences. *FEMS Microbiol. Lett.* **174:**248–250.

Thompson, J. D., T. J. Gibson, and D. G. Higgins. 2003. Multiple sequence alignment using ClustalW and ClustalX, p. 2.3.1–2.3.22. *In Current Protocols in Bioinformatics.* John Wiley & Sons, Inc., Hoboken, NJ.

Wong, K. M., M. A. Suchard, and J. P. Huelsenbeck. 2008. Alignment uncertainty and genomic analysis. *Science* **319:**473–476.

Web Resource

Molecular Evolutionary Genetics Analysis software, version 4.0 (MEGA4) http://www.megasoftware.net/

READING ASSESSMENT

1. Given the multiple alignment below (length = 44 positions), write out the four original DNA sequences used to construct the alignment.

   Sequence 1 `--ACCGGTTGATCCTGCC-GACCCGACC-CTATCG---TAGATG`

   Sequence 2 `AATCCG-TT--TCCTGCCGGACC-GACTGCTATCGGATTGAGCA`

   Sequence 3 `---CCGGTT--TCCT-CCGGACC-GACCGC---CGGGGTAGATA`

   Sequence 4 `--ACCG-TTGA-CCTGCC-GACCCGA-GCTATCGGGG-TAGACG`

   Sequence 1: _____

   Sequence 2: _____

   Sequence 3: _____

   Sequence 4: _____

2. Complete the construction of a distance matrix for the sequences in question 1 using "Edit Distance." Show your work for full credit.

	1	2	3	4
Sequence 1				
Sequence 2				
Sequence 3				
Sequence 4				

Hint: Realign sequences pairwise, calculating the edit distance for each pair. See the examples below.

Sequence 1 --**ACCGGTTGA**TCCTGCC-GACC**C**GAC**C**-CTATCG---**TAGATG**

Sequence 2 **AAT**CCG-TT--TCCTGCC**GG**ACC-GAC**TGC**TATCG**GATT**GAGCA

Sequence 1 **A**CCGGTT**GA**TCCT**G**CC-GACC**C**GACC-C**TAT**CG---TAGAT**G**

Sequence 3 -CCGGTT--TCCT-CC**G**GACC-GACC**G**C---C**G**G**GG**TAGAT**A**

Sequence 1 ACCG**G**TTGA**T**CCTGCCGACCCGA**CC**-CTATCG---TAGA**T**G

Sequence 4 ACCG-TTGA-CCTGCCGACCCGA--**G**CTATCG**GGG**TAGACG

3. Complete the construction of a similarity matrix for the sequences in question 1 using a match score of +5, a mismatch penalty of −2, and a gap penalty of −3. Show your work for full credit.

	1	2	3	4
Sequence 1				
Sequence 2				
Sequence 3				
Sequence 4				

Hint: Calculate the nominal score for each pairwise alignment.

Sequence 1 --**ACCGGTTGA**TCCTGCC-GACC**C**GAC**C**-CTATCG---**TAGATG**

Sequence 2 **AAT**CCG-TT--TCCTGCC**GG**ACC-GAC**TGC**TATCG**GATT**GAGCA

$N = aM + bX + cG = 26\,(5) + 7(-2) + 11(-3) = 130 - 14 - 33 = 83$

4. Complete the construction of a similarity matrix for the alignment in question 1 using a match score of +7, a mismatch penalty of −5, and a gap penalty of −5. Show your work for full credit.

	1	2	3	4
Sequence 1				
Sequence 2				
Sequence 3				
Sequence 4				

Hint: Calculate the nominal score for each pairwise alignment as done in question 3, but use the new values for awards and penalties.

Sequence 1 `--A`**CCGGTTGA**`TCCTGCC-GACC`**C**`GAC`**C**`-CTATCG---`**TAGATG**

Sequence 2 **AAT**`CCG-TT--TCCTGCC`**G**`GACC-GAC`**TG**`CTATCG`**GATT**`GAGCA`

$$N = aM + bX + cG = 26\ (7) + 7(-5) + 11(-5) = 182 - 35 - 55 = 92$$

5. Align the following two sequences using the Needleman-Wunsch algorithm (*Hint:* used to create global alignments). Report the nominal score for the resulting alignment. Note that there may be more than one alignment that will return the same nominal score.

   Sequence 1 ACCGGTTGATCC

   Sequence 2 ACCGTTGACCTG

   Match (M) = +5

   Mismatch (X) = −2

   Gap (G) = −3

6. Align the following two sequences using the Smith-Waterman algorithm (*Hint:* used to create local alignments). Report the nominal score for the resulting alignment. Note that more than one alignment may return the same nominal score.

   Sequence 1 CGGGATAAA

   Sequence 2 GCTATCAGGGTAGACG

   Match (M) = +5

   Mismatch (X) = −2

   Gap (G) = −3

7. If the same scoring strategy is applied to the alignment of the same two sequences, first using the Needleman-Wunsch algorithm and next using the Smith-Waterman algorithm, why might each algorithm return a different nominal score?

8. In community analyses, it is common to sequence the first 500 nucleotides of the 16S rRNA gene rather than the whole gene, which is approximately 1,500 bp. How would you go about aligning the 500-base sequencing products to a known 16S rRNA gene if global alignment is the only method available?

9. Given the following query word and database sequences, which database sequences would be evaluated as a potential HSP in a BLAST search? In other words, which sequences would be expected to return a match to the query word?

   Word = AGCCTCCA

   Sequence 1 CCCGGGAGGGCTACGCTAGCTTCGATAGCTC

   Sequence 2 CGATTCGCTAGAGCCTCCATCGATACGCTCA

   Sequence 3 GATAGCTCGCATATATCGCTCTAGACTCGAT

   Sequence 4 TAGAGAGCCTCGCCTCTCGCGCGCTAGAGTC

   Sequence 5 TCGCGCGATATTATATCGCAGCCTCCAAGAA

   Sequence 6 GTCGCTCGCATGCGCTAGCTCTCGCTACGCC

10. When comparing multiple BLAST searches of the same query sequence (i.e., using BLAST searches each with a different match award and gap penalty), which BLAST statistics can be directly compared? Why?

11. In a BLAST search, if the nominal score for a particular hit is 47, under which scenario(s) below would no match be returned to the user? Assume that the same match/mismatch/gap scoring scheme was used for each search and that $\lambda = 0.63$ and $K = 2.4$ for this particular scoring scheme.

$$B = (\lambda N - \ln K)/\ln 2$$

Scenario	No. of nucleotides in database	Expect cutoff	Bit score (S)
1	10^6	10	23.25
2	10^6	0.01	33.22
3	10^6	0.000001	46.51
4	10^9	10	33.22
5	10^9	0.01	43.19
6	10^9	0.000001	56.47

12. Which BLAST program is most appropriate to use if the query sequence is derived from metagenomically derived clones containing 16S rRNA genes? Which is most appropriate if the query is PCR amplified from a bacterial isolate? Why?

13. For rRNA molecules, why might structure-based alignments be considered more reliable than sequence-based alignments? What is one argument against this reasoning?

UNIT 6

EXPERIMENTAL OVERVIEW

In Experiments 3.6 and 5.9, students verified the DNA sequence information obtained for the 16S rRNA genes submitted for their isolates and clones, respectively. A high-quality DNA sequence should be used for all subsequent bioinformatics experiments (described in this section) and phylogenetic analyses (Unit 7), meaning the ambiguous base calls (N's) at the ends of the sequences have been removed and their identity within internal stretches of sequence has been resolved upon manual inspection of the chromatograms. If DNA sequences are not already in FASTA format, they will be converted to FASTA format so they are ready for analysis.

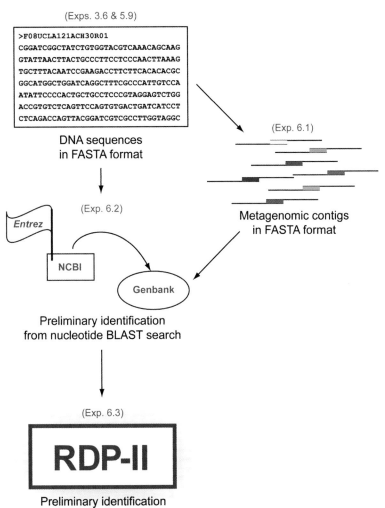

(Exps. 3.6 & 5.9)

```
>F08UCLA121ACH30R01
CGGATCGGCTATCTGTGGTACGTCAAACAGCAAG
GTATTAACTTACTGCCCTTCCTCCCAACTTAAAG
TGCTTTACAATCCGAAGACCTTCTTCACACACGC
GGCATGGCTGGATCAGGCTTTCGCCCATTGTCCA
ATATTCCCCACTGCTGCCTCCCGTAGGAGTCTGG
ACCGTGTCTCAGTTCCAGTGTGACTGATCATCCT
CTCAGACCAGTTACGGATCGTCGCCTTGGTAGGC
```

**DNA sequences
in FASTA format**

(Exp. 6.1)

**Metagenomic contigs
in FASTA format**

(Exp. 6.2)

Entrez

NCBI

Genbank

**Preliminary identification
from nucleotide BLAST search**

(Exp. 6.3)

RDP-II

**Preliminary identification
from SeqMatch & Classifier**

Before students working on the cultivation-independent project can begin their analyses, the DNA sequence reads for each clone from the T7 and M13R primers must be combined into a single, contiguous sequence (contig) in FASTA format as directed in Experiment 6.1.

In Experiment 6.2, all students will perform a nucleotide BLAST (blastn) search with their DNA sequences, examining the pairwise sequence alignments and scoring statistics for evidence that can be used to assign a preliminary taxonomic identification for their isolate or clone. In Experiment 6.3, all students will create a myRDP account, upload their DNA sequences to RDP-II, and perform searches with SeqMatch and Classifier. The results will be explored for additional evidence in support of a taxonomic identification for their isolate or clone. Students should compare the output from each method (blastn, SeqMatch, and Classifier), collecting sequences for nearest neighbors that may be used to construct a multiple sequence alignment and phylogenetic trees in Unit 7.

EXPERIMENT 6.1 Assembly of Consensus DNA Sequences

Using the DNA sequences obtained from the T7 and M13R primer pair of the same sample, students will need to find the region where they overlap (*hint*: the reverse complement of one of the two sequences will need to be made) and combine the two sequences into a single, contiguous sequence (contig). The region of overlap is comprised of a consensus sequence generated from the data produced for this region by the forward and reverse primers.

BLAST2Sequences first will be used to quickly confirm that the sequence reads from the T7 and M13R primers generated from the same clone as the DNA template are long enough to generate a region of overlap between them. Next, the T7 and M13R DNA sequence reads will be assembled into a contig by using an alignment program such as AlignIR, Lasergene SeqMan, DNAStar, MacVector, or Sequencher. A site license is required for these programs. The instructions provided below are specific for AlignIR but the principles may be applied to any alignment program.

Note: Open all websites and links for data output in new tabs or windows rather than in the same window, because you will need to scroll back and forth between screens.

METHODS

Part I BLAST2Sequences

1. Compile DNA sequence data into a FASTA file. Open both T7 and M13R primer plain-text DNA sequencing files generated in Experiment 5.9. These should be in FASTA format. To create a new FASTA file with both DNA sequences, perform the following simple steps.

    a. Open a text editor such as Notepad (go to the Start menu, and find Notepad under Accessories).

    b. Copy your DNA sequences and paste them sequentially into Notepad, with a hard return after each DNA sequence.

    c. Add or modify the description line for each sequence. Precede the identifier with a greater-than symbol (>) and press Enter after the name. Use the sample ID given to your clone in Unit 5, adding the primer name used to generate the sequence (e.g., >F08UCLA121ACH01 T7 primer). The nucleotide sequence will begin on the second line.

    d. Save the new FASTA text file (e.g., contig seq1.txt).

2. Prepare DNA sequences for contig assembly using BLAST. For each T7/M13R sequence pair, generate the reverse complement for one of the two sequences (T7 or M13R primer sequence; it does not matter which one at this point). Use any of the following websites to determine the reverse complement: the Sequence Manipulation Suite (http://www.bioinformatics.org/SMS/rev_comp.html), Baylor College of Medicine (BCM) Human Genome Sequencing Center (HGSC) (http://searchlauncher.bcm.tmc.edu/seq-util/Options/revcomp.html), or the Bio-Web Python CGI Scripts for Molecular Biology & Bioinformatics (http://www.cellbiol.com/scripts/complement/reverse_complement_sequence.html). Copy and paste the reverse complement into the FASTA text file (e.g., contig seq1.txt), then save as a new FASTA text file (e.g., contig seq2.txt).

Experiment continues

Note: Do not assume that the M13R reverse primer gives the sequence of the minus/template strand or that the T7 primer gives the sequence of the plus/nontemplate strand. The 16S rDNA PCR product may have been cloned in either orientation as diagrammed in the figure. The orientation of the sequence can be resolved based on the results of a BLAST search in Experiment 6.2.

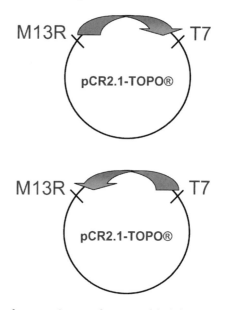

3. Use BLAST to confirm that contigs can be assembled from T7/M13R DNA sequence reads for each clone. Open your Internet browser, and navigate to the NCBI BLAST homepage (www.ncbi.nih.gov/BLAST). Select "nucleotide blast" under the Basic BLAST heading. Select the box labeled "Align two or more sequences" at the bottom of the section Enter Query Sequence, which will automatically modify the BLAST page, allowing you to enter two sequences, one as the Query and the other as the Subject.

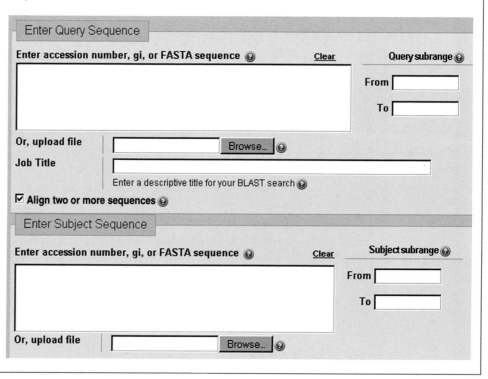

For each clone, copy and paste one of the sequences from the contig seq2.txt file into the Query box, excluding anything that is not nucleotide sequence (e.g., terminal N's or ">description line"). In the Subject box, copy and paste the nucleotide sequence for the reverse complement of the other sequence corresponding to the same clone sample pasted in the Query box. Click the BLAST button. The default parameters are sufficient for our purposes.

The results for BLAST2Sequences are presented in a format similar to that produced for a nucleotide BLAST (blastn) search, except that the Subject sequence has been defined by the user (rather than being retrieved from GenBank database). If there is any overlap between the two uploaded sequences, BLAST2Sequences will produce a pairwise sequence alignment of the overlap region with additional statistics about the quality of the alignment (e.g., length of overlap region, presence of mismatches or gaps, or relative orientation of the two strands). BLAST2Sequences also produces a dot matrix diagram similar to that shown below. For instance, a single diagonal line from the lower left corner to the upper right corner reflects an overlap region between two plus-strand sequences.

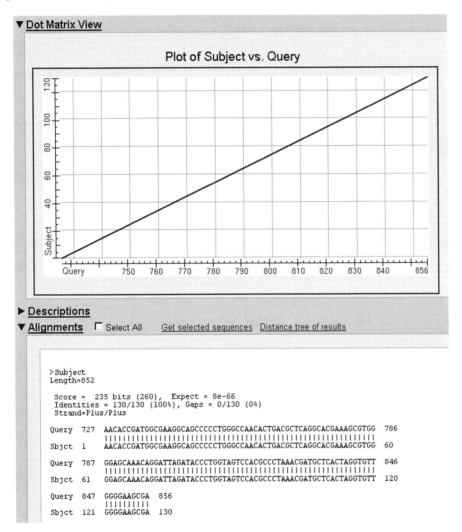

▼ **Dot Matrix View**

Plot of Subject vs. Query

▶ **Descriptions**

▼ **Alignments** ☐ Select All <u>Get selected sequences</u> <u>Distance tree of results</u>

```
>Subject
Length=852

 Score =  235 bits (260),  Expect = 8e-66
 Identities = 130/130 (100%), Gaps = 0/130 (0%)
 Strand=Plus/Plus

Query  727  AACACCGATGGCGAAGGCAGCCCCCTGGGCCAACACTGACGCTCAGGCACGAAAGCGTGG  786
            |||||||||||||||||||||||||||||||||||||||||||||||||||||||||||||
Sbjct  1    AACACCGATGGCGAAGGCAGCCCCCTGGGCCAACACTGACGCTCAGGCACGAAAGCGTGG  60

Query  787  GGAGCAAACAGGATTAGATACCCTGGTAGTCCACGCCCTAAACGATGCTCACTAGGTGTT  846
            |||||||||||||||||||||||||||||||||||||||||||||||||||||||||||||
Sbjct  61   GGAGCAAACAGGATTAGATACCCTGGTAGTCCACGCCCTAAACGATGCTCACTAGGTGTT  120

Query  847  GGGGAAGCGA  856
            ||||||||||
Sbjct  121  GGGGAAGCGA  130
```

Experiment continues

Part II Contig Assembly with AlignIR

1. Compilation of DNA sequence data into FASTA file. Retrieve the DNA sequences in FASTA format from the contig seq2.txt files for each clone. As long as each sequence is in FASTA format with a greater-than symbol (>) at the beginning of the header line, you may create a new FASTA file compiling all DNA sequences into a single document.

    a. Copy and paste the T7/M13R sequences used to build contigs in BLAST2Sequences into a text editor such as Notepad, one clone at a time. Use sequences which are in the proper orientation for contig assembly as confirmed in step 3 above.

    b. SAVE as a new FASTA text file (e.g., all contigs seq2.txt).

2. Using AlignIR to assemble consensus sequences using properly oriented contigs. Open the program AlignIR. Under File, select New. Enter a project name in the appropriate directory. Click Next. Locate your FASTA text file (e.g., all contigs seq2.txt), and add the sequences by selecting Add. Then click Finish. Your DNA sequences will be uploaded to the AlignIR main window as shown:

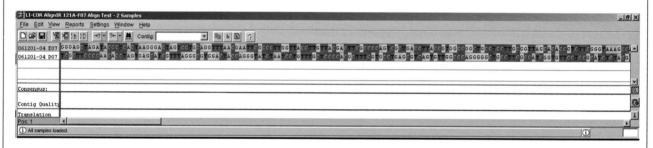

Under File, select Assemble and then Start on the Sequence Assembler Console. The assembled contig will appear at the bottom of the main window screen:

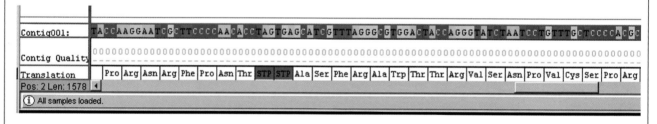

Go to the View menu and click Overview to see contig with region of overlap. Use the horizontal scroll bars to view entire length of contig:

If you return to the main window, use the horizontal scroll bar to view the actual sequence of the overlap region so you can scan for any mismatched bases. If you find any, you will need to return to the Chromatograms for both the T7 and M13R primer sequences to determine the correct base assignment and then repeat the sequence assembly process up to this point.

When you have verified the overlap region, report the results by using the Report menu and clicking Consensus. Select LI-COR for Report Type. For Report Destination, select File and hit Browse to find the folder you created for the project. Change the file name to something meaningful using the HTML extension (.htm).

Open your Internet browser, and select Open File from the File menu. Locate the HTML file of your consensus sequence, and open it within the browser. Print the report and include a copy in your laboratory notebook. Note the total number of bases in the contig (No. of Bases).

To copy and paste the consensus sequence in FASTA format into Notepad, you will need to return to the main window. With the cursor, select the line corresponding to the consensus sequence at the bottom of the page (e.g., Contig001), which will in turn become highlighted in blue. Next, choose Selected sample from the Report menu (or press Alt + F2). Save the file of the consensus sequence in FASTA format to the appropriate directory.

You will need to add or revise the description line for the consensus sequence (e.g., >F08UCLA121ACH01 Contig). Be certain that the identifier is preceded by a greater-than symbol (>).

Highlight the nucleotide sequence, and select Word Count under the Tools menu. The character count (no spaces) should be equal to the base count as shown if all data were saved and transferred properly.

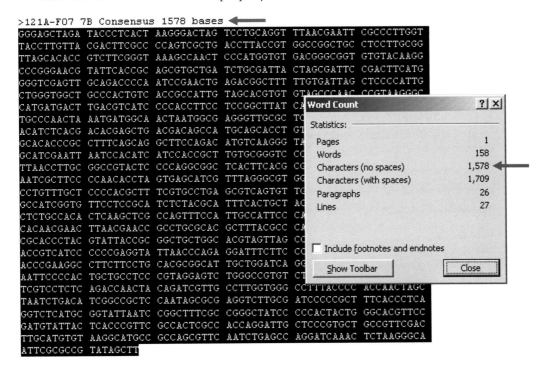

Experiment continues

Once you finish assembling the consensus sequences for each of your cloned 16S rRNA genes, compile them into a single Word document (.doc) for electronic submission to your instructor. Then proceed to Experiment 6.2.

BLAST is a Web-based program used to calculate sequence similarity by comparing a nucleotide query sequence against nucleotide sequences in databases such as GenBank. In this experiment, BLAST is used to make a preliminary identification for the microorganisms from which your 16S rRNA genes were obtained.

To identify your microorganisms using a BLAST search, perform the following series of steps.

Note: Open all websites and links for data output in new tabs or windows, rather than in the same window, because you will need to scroll back and forth between screens.

METHODS

1. Access the NCBI BLAST homepage (http://www.ncbi.nlm.nih.gov/BLAST/).

2. Under the Basic BLAST heading, click "nucleotide blast."

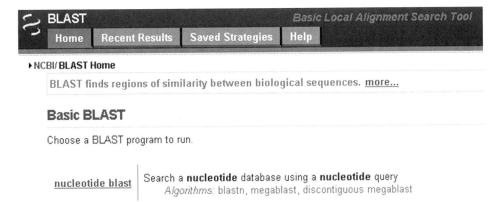

3. Under the heading Enter Query Sequence, paste the DNA sequence for one of your isolates (i.e., reverse complement of 519R) or clones (i.e., contig made from T7/M13R sequences). These sequences should have few or no ambiguities (i.e., N's).

Experiment continues

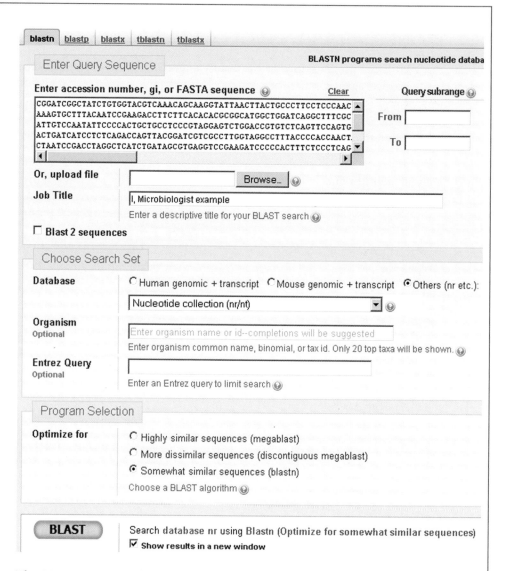

The DNA sequences should be in FASTA format as shown in the text box below. Include the entire sequence (highlighted section in text box), but exclude the description line information.

```
>060823-01_A01_HK_1-10_pmol_per_ul_519R.ab1   853   0   853 ABI
CGGATCGGCTATCTGTGGTACGTCAAACAGCAAGGTATTAACTTACTGCCCTTCCTCCCAACTTAAAGTGCT
TTACAATCCGAAGACCTTCTTCACACACGCGGCATGGCTGGATCAGGCTTTCGCCCATTGTCCAATATTCCC
CACTGCTGCCTCCCGTAGGAGTCTGGACCGTGTCTCAGTTCCAGTGTGACTGATCATCCTCTCAGACCAGTT
ACGGATCGTCGCCTTGGTAGGCCTTTACCCCACCAACTAGCTAATCCGACCTAGGCTCATCTGATAGCGTGA
GGTCCGAAGATCCCCCACTTTCTCCCTCAGGACGTATGCGGTATTAGCGCCCGTTTCCGGACGTTATCCCCC
ACTACCAGGCAGATTCCTAGGCATTACTCACCCGTCCGCCGCTGAATCCAGGAGCAAGCTCCCTTCATCCGC
TCGACTTGCATGTGTTAGGCCTGCCGCCAGCGTTCAATCTGAGCCATGACAAAACTCTAAA
```

Assign a title to your search in the Job Title box. Alternatively, if you include the description line when you paste the DNA sequence into the text box, BLAST will use the description as the job title. Note, however, that the description line must contain the greater-than character (>) and must fit on one line.

4. Under the Choose Search Set heading, click the "Others" Database button. From the drop-down menu, select Nucleotide collection (nr/nt). Under the Program

Selection heading, choose Somewhat Similar Sequences (blastn). For now, keep all other settings at default. Click the blue <BLAST> button at the lower left corner of the page.

5. A page will come up during the search with your Job Title and a request ID number. This series of numbers can be used to retrieve your BLAST results again later. You should record this number in your notebook.

6. When the search finishes, you will be forwarded to a new page with a graphic summary of the results at the top.

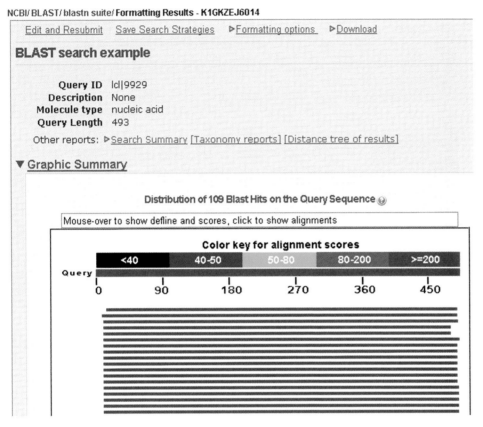

NCBI/ BLAST/ blastn suite/ **Formatting Results - K1GKZEJ6014**

Edit and Resubmit   Save Search Strategies   ▷Formatting options   ▷Download

BLAST search example

Query ID lcl|9929
Description None
Molecule type nucleic acid
Query Length 493

Other reports: ▷Search Summary [Taxonomy reports] [Distance tree of results]

▼ Graphic Summary

Distribution of 109 Blast Hits on the Query Sequence ☺

Mouse-over to show defline and scores, click to show alignments

Color key for alignment scores
<40   40-50   50-80   80-200   >=200

7. Scroll down to "Sequences producing significant alignments" under the Descriptions heading. Below should be listed several matches to your query (search hits), in table form, with the following column headings for each hit: Accession, Description, Max score, Total score, Query coverage, E value, and Max ident (identity). If you scroll through the entire list, notice that the sequence hits are sorted in ascending order according to E-values, with the first hit having the lowest E-value.

Sequences producing significant alignments:
(Click headers to sort columns)

Accession	Description	Max score	Total score	Query coverage	△ E value	Max ident
EU670073.1	Uncultured bacterium clone FIU_KM_MD_021 16S ribosomal RNA gene, p	843	843	96%	0.0	99%
EU670075.1	Uncultured bacterium clone FIU_KM_MD_023 16S ribosomal RNA gene, p	839	839	97%	0.0	98%
EU037096.1	Pseudomonas aeruginosa strain CMG860 16S ribosomal RNA gene, parti	838	838	97%	0.0	98%
EF427781.1	Pseudomonas aeruginosa strain 1S280 16S ribosomal RNA gene, partial	838	838	95%	0.0	99%

Homology between your query sequence and the sequences in the database is generally based on three parameters (although Query coverage also should be examined):

Experiment continues

a. the Max score, which specifies the similarity between the two sequences. The higher the Max score, the better the match.

b. the E-value, which denotes the probability of a random match between the two sequences being compared. The lower the E-value, the less likely that the two sequences are a random match.

c. The Max Ident, which indicates the percentage of identical nucleotides at aligned positions between the query and the database sequence. The higher the Max Ident, the better the match.

8. Click on the Accession link for the first search hit, which corresponds to the hyperlink in the first column for the one-line description. The next page that comes up (called the Report) contains information about the matching organism, including the full nucleotide sequence associated with the database entry and the relative orientation of your search hit with homology to your query sequence, the scientific name of the source organism or environmental isolate from which the hit sequence is obtained, detailed information about the identity of the gene that was queried (*hint:* it should be 16S rRNA), a features table, and bibliographic references.

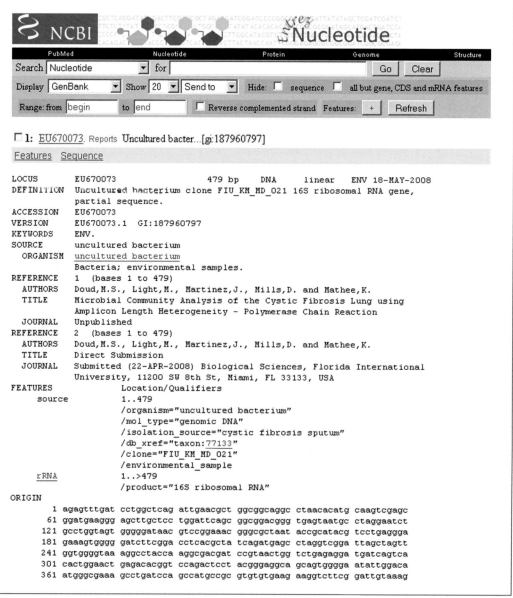

```
LOCUS       EU670073                 479 bp    DNA     linear   ENV 18-MAY-2008
DEFINITION  Uncultured bacterium clone FIU_KM_MD_021 16S ribosomal RNA gene,
            partial sequence.
ACCESSION   EU670073
VERSION     EU670073.1  GI:187960797
KEYWORDS    ENV.
SOURCE      uncultured bacterium
  ORGANISM  uncultured bacterium
            Bacteria; environmental samples.
REFERENCE   1  (bases 1 to 479)
  AUTHORS   Doud,M.S., Light,M., Martinez,J., Mills,D. and Mathee,K.
  TITLE     Microbial Community Analysis of the Cystic Fibrosis Lung using
            Amplicon Length Heterogeneity - Polymerase Chain Reaction
  JOURNAL   Unpublished
REFERENCE   2  (bases 1 to 479)
  AUTHORS   Doud,M.S., Light,M., Martinez,J., Mills,D. and Mathee,K.
  TITLE     Direct Submission
  JOURNAL   Submitted (22-APR-2008) Biological Sciences, Florida International
            University, 11200 SW 8th St, Miami, FL 33133, USA
FEATURES             Location/Qualifiers
     source          1..479
                     /organism="uncultured bacterium"
                     /mol_type="genomic DNA"
                     /isolation_source="cystic fibrosis sputum"
                     /db_xref="taxon:77133"
                     /clone="FIU_KM_MD_021"
                     /environmental_sample
     rRNA            1..>479
                     /product="16S ribosomal RNA"
ORIGIN
        1 agagtttgat cctggctcag attgaacgct ggcggcaggc ctaacacatg caagtcgagc
       61 ggatgaaggg agcttgctcc tggattcagc ggcggacggg tgagtaatgc ctaggaatct
      121 gcctggtagt gggggataac gtccggaaac gggcgctaat accgcatacg tcctgaggga
      181 gaaagtgggg gatcttcgga cctcacgcta tcagatgagc ctaggtcgga ttagctagtt
      241 ggtggggtaa aggcctacca aggcgacgat ccgtaactgg tctgagagga tgatcagtca
      301 cactggaact gagacacggt ccagactcct acgggaggca gcagtgggga atattggaca
      361 atgggcgaaa gcctgatcca gccatgccgc gtgtgtgaag aaggtcttcg gattgtaaag
```

If you select FASTA from the drop-down menu next to Display at the top left corner of the page, the full nucleotide sequence (in FASTA format) for the search hit will be provided. You can copy and paste the sequence with description line into a new document for later use.

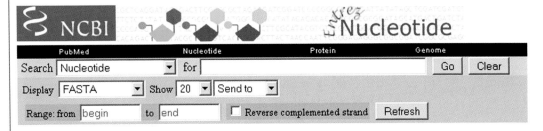

☐ 1: EU670073. Reports Uncultured bacter...[gi:187960797]
>gi|187960797|gb|EU670073.1| Uncultured bacterium clone FIU_KM_MD_021 16S ribosomal
AGAGTTTGATCCTGGCTCAGATTGAACGCTGGCGGCAGGCCTAACACATGCAAGTCGAGCGGATGAAGGG
AGCTTGCTCCTGGATTCAGCGGCGGACGGGTGAGTAATGCCTAGGAATCTGCCTGGTAGTGGGGGATAAC
GTCCGGAAACGGGCGCTAATACCGCATACGTCCTGAGGGAGAAAGTGGGGGATCTTCGGACCTCACGCTA
TCAGATGAGCCTAGGTCGGATTAGCTAGTTGGTGGGGTAAAGGCCTACCAAGGCGACGATCCGTAACTGG
TCTGAGAGGATGATCAGTCACACTGGAACTGAGACACGGTCCAGACTCCTACGGGAGGCAGCAGTGGGGA
ATATTGGACAATGGGCGAAAGCCTGATCCAGCCATGCCGCGTGTGTGAAGAAGGTCTTCGGATTGTAAAG
CACTTTAAGTTGGGAGGAAGGGCAGTAAGTTAATACCTTGCTGTTTGACGTACCAACAG

9. Return to the summary page, and find the sequence you previously selected in the list of search hits. This time click on the Max score link to the right of the description. A pairwise sequence alignment should appear. For example:

>☐gb|EU670073.1| Uncultured bacterium clone FIU_KM_MD_021 16S ribosomal RNA gene,
partial sequence
Length=479

 Score = 843 bits (934), Expect = 0.0
 Identities = 472/475 (99%), Gaps = 0/475 (0%)
 Strand=Plus/Minus

Query 16   TGGTACGTCAAACAGCAAGGTATTAACTTACTGCCCTTCCTCCCAACTTAAAGTGCTTTA 75
            ||||||||||||||||||||||||||||||||||||||||||||||||||||||||||||
Sbjct 475 TGGTACGTCAAACAGCAAGGTATTAACTTACTGCCCTTCCTCCCAACTTAAAGTGCTTTA 416

Query 76   CAATCCGAAGACCTTCTTCACACACGCGGCATGGCTGGATCAGGCTTTCGCCCATTGTCC 135
            ||||||||||||||||||||||||||||||||||||||||||||||||||||||||||||
Sbjct 415 CAATCCGAAGACCTTCTTCACACACGCGGCATGGCTGGATCAGGCTTTCGCCCATTGTCC 356

This section of the BLAST results provides detailed information about the quality of the alignment between the query sequence (Query) and the database sequence (Sbjct). The number of matches and mismatches that exist between the two sequences is reported [e.g., Identities = 472/475 (99%) meaning 472 matches and 3 mismatches]. In general, an identity of ≤99% indicates that the query and subject sequences were obtained from distinct species and an identity of ≤95% indicates that the query and subject sequences were acquired from distinct genera. This section also indicates whether there were any differences between the two sequences that resulted in extra or missing nucleotides (gaps) in the sequence [e.g., Gaps = 0/475 (0%)].

Experiment continues

Each sequence in the alignment is numbered. The subject sequence (Sbjct) is numbered according to the system used for the full sequence found on the Report page. Locate the Base Count Origin number for the query sequence (16 in the above example) and for the subject sequence (475 in the above example).

Notice that the numbers for the query are ascending (e.g., 16–75 and 76–135) whereas the numbers for the subject sequence are descending (e.g., 475–416 and 415–356). Think about what these observations tell you about the DNA strand orientation of the original sequence (Query) relative to the search hit (Sbjct). Consider 5′-to-3′ directionality and the antiparallel nature of a single DNA strand. Furthermore, notice that the Strand specifications state "Strand = Plus/Minus," reflecting the conventions and nomenclature involved in expressing DNA strandedness.

Recall that a reverse primer was used in the sequencing reaction of the query 16S rRNA gene (e.g., 519R, T7, or M13R depending on the clone direction in the sequencing vector). If the proper orientation of the query sequence was not obtained prior to initiating the BLAST search, an alignment will result between the query sequence and the reverse complement of the search hit (Sbjct) sequence in the database. This is because the BLAST algorithm presumes that the Query sequence is the plus strand and chooses whichever strand in the database (plus or minus) produces the best alignment.

Assuming that the database sequences are in the proper orientation (one must check the Report page for evidence, as demonstrated in Fig. 6.19), the pairwise sequence alignment provides clues to whether the query sequence is in the proper orientation (i.e., that it actually is the plus-strand sequence). If so, the results should give the strand specifications for the Query and Sbjct as "Strand=Plus/Plus," with the base count origin numbers both in ascending order.

```
>  gb|EU670073.1|  Uncultured bacterium clone FIU_KM_MD_021 16S ribosomal RNA gene,
partial sequence
Length=479

 Score =  843 bits (934),  Expect = 0.0
 Identities = 472/475 (99%),  Gaps = 0/475 (0%)
 Strand=Plus/Plus

Query  4     AGAGTTTTGTCATGGCTCAGATTGAACGCTGGCGGCAGGCCTAACACATGCAAGTCGAGC  63
             ||||||| || |||||||||||||||||||||||||||||||||||||||||||||||||
Sbjct  1     AGAGTTTGATCCTGGCTCAGATTGAACGCTGGCGGCAGGCCTAACACATGCAAGTCGAGC  60

Query  64    GGATGAAGGGAGCTTGCTCCTGGATTCAGCGGCGGACGGGTGAGTAATGCCTAGGAATCT  123
             ||||||||||||||||||||||||||||||||||||||||||||||||||||||||||||
Sbjct  61    GGATGAAGGGAGCTTGCTCCTGGATTCAGCGGCGGACGGGTGAGTAATGCCTAGGAATCT  120
```

Notice in the above example that none of the scoring statistics change, demonstrating how the results of a BLAST search essentially are independent of strand orientation. Instead, strand orientation becomes a user-defined parameter. The programs used in subsequent bioinformatics experiments to generate DNA sequence alignments and phylogenetic trees require the DNA sequences to be in the proper orientation. Thus, students should use the BLAST results to confirm that all query sequences produce alignments with strand specifications as "Strand=Plus/Plus," provided that the gene is in the plus orientation on the report page, before going on to Experiment 6.3 and Unit 7 experiments.

10. Prepare a table summarizing the data generated from your BLAST search, recording the accession number, description, Max score, total score, query coverage, E-value, and Max ident (identity) for the best search hit(s). For the description, be sure to include the gene name (indicating whether it is a partial or complete sequence or derived from complete genome sequence) as well as the organism information (phylum, class, order, family, genus, and species). Include as much of the hierarchy as is known (see the BLAST Results page for a summary). You should include more than one hit for each sequence. For example, the top hit may be a database sequence derived from an uncharacterized or uncultured organism, which is not terribly informative if you are trying to identify the organism from which your query sequence was obtained. Instead, you should scroll down the search hit list (see BLAST "one-line-descriptions") and find a hit in which the sequence (Sbjct) was derived from a known organism. It is best if there is a reference associated with that sequence, indicating that the sequence and organism information has been verified experimentally (i.e., the orientation of the sequence is reliably reported as plus or minus strand; the sequence is not predicted from the automatic annotation of a genome).

The sequences for the top hits from your BLAST search should be included in the phylogenetic tree produced in Experiment 7.2.

For each isolate or clone sequenced and initially identified using a nucleotide BLAST search, you will use the analysis tools available at the RDP-II website to search for nearest neighbors and generate a phylogenetic lineage for each of your organisms. For isolates, use the reverse complement of the sequence as your query; and for clones, use the full-length consensus sequence as your query. These sequences should have few or no ambiguities (e.g., N's) and should be in the proper orientation (verified by BLAST search).

METHODS

Part I Upload FASTA sequence files to RDP-II

1. Access the RDP homepage: http://rdp.cme.msu.edu/.

2. Click on myRDP to start.

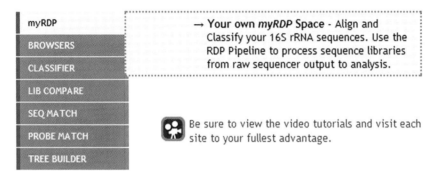

3. Sign up for a new account by creating a login ID and password (or log in to an existing account). The account is free and easy to create. Registration requires a valid e-mail account.

    Note: There are short video tutorials available to explain various features and tools of RDP-II. We recommend that the first-time user view these tutorials before proceeding.

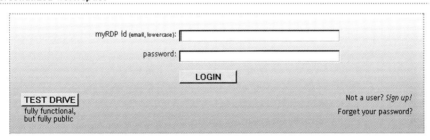

Experiment continues

4. After logging in, click on the Account Info link next to the welcome message. Enter the e-mail addresses of your partner or team members to add them as "super buddies" on your RDP-II account. Any sequences that are uploaded by you or your group can be viewed as shared data by clicking on the Overview link.

5. Click the Upload link to upload your 16S rDNA sequences to myRDP. The sequences must be in FASTA format. In addition, it will be easier to manipulate the sequences in RDP-II if all the DNA sequences are in the same file. To create a new FASTA file compiling your DNA sequences, perform the following simple steps.

a. Open a text editor such as Notepad (go to the Start menu, and find Notepad under Accessories).

b. Copy your DNA sequences and paste them sequentially into Notepad, with a hard return after each DNA sequence. Use the text files for either the 519R reverse complement DNA sequences or contigs of the T7 and M13R sequence reads that are in the proper orientation and that have few or no ambiguities (e.g., N's).

c. Add a description line to each sequence preceded by a greater-than symbol (>), then press Enter after the name. For this file, use only the sample ID given to your isolate in Unit 3 or clone in Unit 5.

d. Save the new FASTA text file as a .txt file.

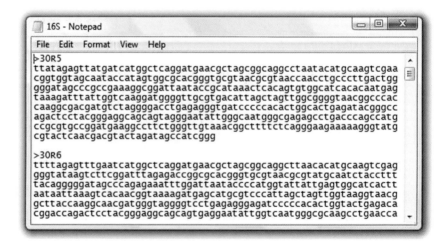

6. Select Bacteria 16S rRNA from the drop-down menu for "Choose a gene for aligner," and assign a group name. Click on the Browse button, select your FASTA file, and press the Upload button to upload your sequences to myRDP.

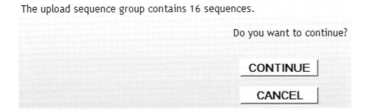

7. RDP-II will indicate how many sequences your FASTA file contains. Confirm that the number reported is the same as the number of sequences in your file. If the number in RDP-II is wrong, press Cancel and then open the FASTA file again in Notepad to make sure that all the sequences are in the proper format. If the number is correct, click Continue.

The upload sequence group contains 16 sequences.

Do you want to continue?

CONTINUE

CANCEL

8. Uploaded sequences that meet the minimal criteria for processing by RDP-II are aligned to the database reference sequences by using a sequence alignment program that takes RNA secondary structure, as well as primary sequence information, into consideration. This process takes time; the length of time depends on the number and length of sequences uploaded, as well as how busy the server is at the time. Short sequences, low-quality sequences, and any sequence that is not 16S rDNA usually fail to align. However, processing should never take longer than 1 to 2 days.

Click the Overview link to check the alignment status of your sequences. RDP-II will display how many sequences are aligned and how many are pending.

Clicking ⊞ or ⊞ Selects (Adds) · Clicking ⊟ Deselects (Removes) for download and analysis

Aligned · Failed · Unaligned

group name (selected)	submitter id	date	project	total	pending	A	F	U
⊞ Cultivation-Dependent (0)	jbruin@ucla.edu	25 Aug, 08		16	0	16	0	0

9. Once all your sequences are aligned (0 in the Pending column), go on to Part II.

Experiment continues

Part II Sequence Match

Using Sequence Match (SeqMatch), you will generate a hierarchical view of the taxonomic lineage for each of your isolates or contigs.

1. Go to myRDP, and select the Overview link to view aligned sequences. Click the box next to the group name to select all the sequences. When all the sequences are selected, the box will change from grey to red.

➕ Cultivation-Dependent (0) ➡ ➖ Cultivation-Dependent (16)

2. Click on the SEQMATCH link in the orange bar at the top of the page.

3. Note that several options can be selected or deselected to filter the results obtained with SeqMatch. These filters apply to other RDP-II tools as well.

16 sequences selected from myRDP account.

Do Seqmatch with Selected Sequences

Choose a file to upload: Browse...

Cut and paste sequence(s) (in Fasta, GenBank, or EMBL format):

Strain: ○ Type	○ Non Type	◉ Both
Source: ○ Uncultured	◉ Isolates	○ Both
Size: ◉ ≥1200	○ <1200	○ Both
Quality: ◉ Good	○ Suspect	○ Both
Taxonomy: ◉ Nomenclatural ○ NCBI		
KNN matches: 20 ∨		

Submit   Reset

Strain Type strain information is provided by bacterial taxonomy. *Hint:* Type strains link taxonomy with phylogeny. Include type strain sequences in your analysis to provide documented landmarks.

Source View only environmental (uncultured) sequences, only sequences from individual isolates, or both. Source classification is based on sequence annotation and the NCBI taxonomy.

Size View only nearly full-length sequences ($\geq$1,200 bases), only short partial sequences ($<$1,200 bases), or both.

Quality View only good-quality sequences, only suspect-quality sequences, or both. Sequences were flagged (*) as suspect quality using Pintail (Ashelford et al., 2005).

Taxonomy View sequences placed into a new phylogenetically consistent higher-order bacterial taxonomy overlaid on the 16S rRNA classification. For the nomenclatural taxonomy, a set of well characterized (vetted) sequences was provided by Garrity et al. (2007). Other sequences were placed into this scheme using the RDP Naïve Bayesian classifier (Wang et al., 2007).

KNN matches The number of matches displayed per sequence; also, the number used to classify queries by unanimous vote.

4. For your first SeqMatch search, change the Source from Both to Isolates. By doing this, you will exclude uncultured bacteria from the results. After changing the setting, press the Do SeqMatch with Selected Sequences button.

5. The results of the SeqMatch search are presented as a hierarchical view in which a taxonomic category can be expanded by clicking View Selectable Matches.

Hierarchy View:

no rank Root (16) (query sequences) | show printer friendly results | | download as text file |

   domain Bacteria (16)
     phylum Proteobacteria (4)
      class Betaproteobacteria (1)   **Click here to expand the selection**
       order Burkholderiales (1)   **& to view matching sequences**
        family Oxalobacteraceae (1)
          U001706176|F08UCLA121ACK30I12 [view selectable matches]
      class Gammaproteobacteria (3)
       order Pseudomonadales (1)
        family Pseudomonadaceae (1)
         genus Pseudomonas (1)
           U001706171|F08UCLA121ACK30N07 [view selectable matches]
       order Enterobacteriales (2)
        family Enterobacteriaceae (2)
         U001706168|F08UCLA121ACK30M04 [view selectable matches]
         U001706180|F08UCLA121ACK30V16 [view selectable matches]

| Save selection and return to summary |

Expanding the selection will give you the top hits with the best similarity scores, S_ab scores, and unique common oligomers. The full sequence name is provided for each match. Print the lineage information and note the top hits for each of your isolates or clones, as these sequences will be retrieved later for use in the multiple sequence alignment in Experiment 7.1.

Experiment continues

Query Sequence: U001706171|F08UCLA121ACK30N07, 473 unique oligos

Match hit format:
short ID, orientation, similarity score, S_ab score, unique common oligomers and sequence full name.

Lineage:

➕ **no rank** Root (0/20/245557) (selected/match/total RDP sequences)
➕   domain Bacteria (0/20/240078)
➕   phylum Proteobacteria (0/20/78391)
➕   class Gammaproteobacteria (0/20/38100)
➕   order Pseudomonadales (0/20/13194)
➕   family Pseudomonadaceae (0/20/10579)
➕   genus Pseudomonas (0/20/10362)

	Short ID	Sim	S_ab	Oligos	Name
☐	S000019748	0.983	0.913	1409	uncultured eubacterium WD259; AJ292672
☐	S000361081	0.981	0.899	1420	Pseudomonas sp. ARCTIC-P37; AY573031
☐	S000372274	0.981	0.901	1386	Pseudomonas sp. KBOS 17; AY653222
☐	S000392868	0.983	0.913	1409	Pseudomonas sp. NZ099; AF388207
☐	S000401660	0.983	0.899	1428	Pseudomonas jessenii; PS06; AY206685
☐	S000413062	0.983	0.899	1441	Pseudomonas sp. A-13; AY556391
☐	S000426393	0.985	0.905	1392	Pseudomonas sp. Fa2; AY747590
☐	S000434402	0.983	0.907	1371	Pseudomonas sp. NZ011; AY014803
☐	S000546343	0.985	0.928	1399	Pseudomonas sp. Fa2; AY131214
☐	S000558124	0.981	0.901	1416	Pseudomonas sp. CSS-1; DQ084462
☐	S000607495	0.983	0.903	1306	Pseudomonas sp. PH8F; AY835583
☐	S000607497	0.986	0.903	1270	Pseudomonas sp. PH20A; leaf spot; AY835585
☐	S000651634	0.983	0.909	1319	Pseudomonas sp. 6A; DQ417331
☐	S000806475	0.983	0.915	1409	Pseudomonas sp.BSi20664; EF382726
☐	S000859343	0.983	0.899	1441	Pseudomonas sp. OS3; EF491953
☐	S000860078	0.985	0.907	1374	Pseudomonas sp. 1/4_O_3; EF540467
☐	S000966635	0.983	0.920	1442	Pseudomonas sp. PSB1; EU184081
☐	S000980833	0.983	0.907	1411	Pseudomonas sp. 0704CCBr; EU335083
☐	S000980834	0.985	0.915	1409	Pseudomonas sp. 0704CCL; EU335084
☐	S001155393	0.978	0.928	1383	Pseudomonas sp. ABc2; EU862560

Part III Classifier

Using Classifier, you will obtain confidence estimates for the taxonomic assignment given to your query sequence by RDP-II. Classifier assigns 16S rRNA gene sequences to the new phylogenetically consistent higher-order bacterial taxonomy proposed by Garrity et al. (2007). Hierarchical taxa are based on a naïve Bayesian rRNA classifier, which allows classification of both bacterial and archaeal 16S rRNA sequences (Wang et al., 2007).

1.   Go to myRDP, and select the overview link to view aligned sequences. Click the box next to the group name to select all the sequences. Click the Classifier link in the orange bar at the top of the page.

2.   Click Do Classification with Selected Sequences. There are no parameters to change for Classifier.

3. The results of the Classifier search are presented as a hierarchical view displaying taxonomic ranks with a confidence level equal to or greater than the threshold set by the user (the default is 80%). To examine the confidence score for all taxonomic levels, click on Show Assignment Detail.

Classifier:	Naive Bayesian rRNA Classifier Version 2.0, July 2007
Taxonomical Hierarchy:	Taxonomic Outline of the Bacteria and Archaea, release 7.8
Query File:	
Query Submit Date:	Fri Sep 26 05:01:53 EDT 2008

Display depth: Auto ⌄   **Confidence threshold:** 80% ⌄   [Refresh]

domain	%	Library
Bacteria	100.0	▬▬▬▬▬▬▬▬▬▬▬▬▬▬▬▬▬▬▬▬▬▬▬

Hierarchy View *(click a node to make it the root -- click the root to see sequence assignment detail)*:

no rank Root (16 sequences) *[show assignment detail]*   [download as text file]
» » domain Bacteria (16)
» » » phylum Actinobacteria (6)
» » » » class Actinobacteria (6)
» » » » » subclass Actinobacteridae (6)
» » » » » » order Actinomycetales (6)
» » » phylum Proteobacteria (4)
» » » » class Betaproteobacteria (1)
» » » » » order Burkholderiales (1)
» » » » » » family Oxalobacteraceae (1)
» » » » class Gammaproteobacteria (3)
» » » » » order Pseudomonadales (1)
» » » » » » family Pseudomonadaceae (1)

Click here for classification confidence for each taxonomic level

Lineage *(click to return to particular node)*:

Root (16)

Expanding the selection gives a confidence score for those taxonomic ranks in which the confidence level is equal to or greater than the threshold set by the user as well as those below the threshold. Print the assignment detail information and/or download the information as a text file. These results will be referenced when choosing sequences to include in the multiple sequence alignment in Experiment 7.1.

Assignment Detail (for Confidence threshold: 80%):   [download as text file]

U001706165|F08UCLA121ACK30R01   Root[100%] Bacteria[100%] Bacteroidetes[100%] Sphingobacteria[100%] Sphingobacteriales[100%] Flexibacteraceae[100%] Hymenobacter[100%]

U001706166|F08UCLA121ACK30N02   Root[100%] Bacteria[100%] Bacteroidetes[100%] Flavobacteria[100%] Flavobacteriales[100%] Flavobacteriaceae[100%] Flavobacterium[100%]

U001706167|F08UCLA121ACK30M03   Root[100%] Bacteria[100%] Firmicutes[100%] "Bacilli"[100%] Bacillales[100%] Bacillaceae[100%] "Bacillaceae 1"[100%] Bacillus[100%] Bacillus c[100%]

U001706168|F08UCLA121ACK30M04   Root[100%] Bacteria[100%] Proteobacteria[100%] Gammaproteobacteria[100%] Enterobacteriales[100%] Enterobacteriaceae[100%] Pectobacterium[28%]

U001706169|F08UCLA121ACK30V05   Root[100%] Bacteria[100%] Actinobacteria[100%] Actinobacteria[100%] Actinobacteridae[100%] Actinomycetales[100%] Micrococcineae[99%] Micrococcaceae[96%] Arthrobacter[80%]

Experiment continues

REFERENCES

Ashelford, K. E., N. A. Chuzhanova, J. C. Fry, A. J. Jones, and A. J. Weightman. 2005. At least 1 in 20 16S rRNA sequence records currently held in public repositories is estimated to contain substantial anomalies. *Appl. Environ. Microbiol.* **71:**7724–7736.

Garrity, G. M., T. G. Lilburn, J. R. Cole, S. H. Harrison, J. Euzeby, and B. J. Tindall. 2007. *The Taxonomic Outline of Bacteria and Archaea.* TOBA Release 7.7, March 2007. Michigan State University Board of Trustees, East Lansing. http://www.taxonomicoutline.org/.

Wang, Q., G. M. Garrity, J. M. Tiedje, and J. R. Cole. 2007. Naïve Bayesian classifier for rapid assignment of rRNA sequences into the new bacterial taxonomy. *Appl. Environ. Microbiol.* **73:**5261–5267.

UNIT 7

Molecular Evolution: Phylogenetic Analysis of 16S rRNA Genes*

SECTION 7.1
DEPICTING EVOLUTIONARY RELATIONSHIPS WITH PHYLOGENETIC TREES

In Unit 6, methods for constructing pairwise and multiple sequence alignments were introduced. While pairwise alignments are currently exploited in searches to identify and retrieve homologous sequences, whether they are orthologs or paralogs, from public databases (e.g., GenBank), multiple sequence alignments are useful when building phylogenetic trees to depict evolutionary relationships among and between genes from different organisms. It should be reemphasized that construction of the alignment is the most important step in building phylogenetic trees because the alignment contains all the information used to infer the evolutionary history, or ancestral origin, of a gene. Simply speaking, the various phylogenetic methods available today interpret the data in different ways. The alignment consists of raw data, while any phylogenetic tree is merely one of many possible interpretations of those data.

If each phylogenetic reconstruction method can produce a unique interpretation for a given alignment, can any interpretation be considered reliable? There are numerous ways to address this question, but some gene sequences simply do not contain enough information for us to deduce an accurate history or to answer a specific question about the history of that gene. With a little knowledge about how to build and interpret phylogenetic trees, one can learn to recognize such situations and make informed decisions about the reliability of that tree.

The information in the first section of Unit 7 introduces important characteristics of phylogenetic trees, dissecting their parts and explaining the criteria by which taxonomic classifications can be made based on the pattern and timing of events depicted in a tree. These data can be used to formulate specific hypotheses about evolutionary relationships among organisms in a tree. The second section describes tools that can be utilized for

*This unit was coauthored with Craig Herbold.

325

tree building and that enable phylogeneticists to confidently reject or accept their phylogenetic hypotheses by using the statistical support for specific features of the tree. As mentioned above, there are several different methods for building phylogenetic trees, each of which distinctively interprets the multiple sequence alignment data. The third section provides an overview of a few methods commonly used for inferring phylogeny from nucleotide sequence alignments, explaining some of the models that can be incorporated into an analysis depending on the nature of the gene under study. Finally, the fourth section offers a historical examination of the role that phylogenetic analysis has played in the development of a taxonomic system that begins to bring order to the immense diversity found in the prokaryotic realm.

Essential features of unrooted phylogenetic trees

A phylogenetic tree, also referred to as a phylogenetic reconstruction or simply a phylogeny, is a graphical summary of evolutionary history. The term topology is often used to refer to the shape of a phylogenetic tree. While evolutionary distance, which is a measure of the differences that distinguish organisms, is an important component of phylogenetic trees, indeed is commonly used to build phylogenetic trees, the topology of a tree defines the relationships among organisms or genes. As illustrated in Fig. 7.1, the simplest possible topology is a single branch connecting two terminal nodes, here labeled Monkey and Banana. Each terminal node represents a gene from an extant, or existing, organism, and is also referred to as a taxon. The branch length between the two nodes is proportional to the number of changes that have occurred in the branch as a function of time. The branch connecting the two taxa represents the intermediate sequence of the same gene from now-extinct organisms that existed during the course of evolution. One of these intermediates is the most recent common ancestor of the monkey and the banana. The point at which that intermediate exists along the branch is where this tree may be rooted. In practice, no standard phylogenetic method used in sequence analysis can infer the root of a phylogenetic tree. Roots are inferred from data other than sequences or are based on assumptions. Therefore, a discussion of rooting will be postponed for now. Instead, the focus initially will be on the unrooted tree, which specifies the relationships among taxa, but does not say anything about the order of evolutionary events leading to the creation of each lineage, nor does it specify the position of the common ancestor.

The phylogenetic tree in Fig. 7.1 indicates that there is a relationship between the monkey and the banana, but more taxa are needed before a tree emerges that is informative as to the nature of that relationship. To add another taxon to the two-taxon tree, an additional branch must be introduced. Additional branches may be added to a pre-existing tree somewhere along the length of an existing branch, as shown in Fig. 7.2. The addition of a third branch to the two-taxon tree creates an important feature of phylogenetic trees known as an internal node. Again, although some root surely exists and is located on one of the three branches represented in Fig. 7.2, the discussion will

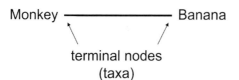

FIGURE 7.1 Monkeys and bananas may be related through a two-taxon tree. Monkeys and bananas are represented by terminal nodes, or taxa, and the evolutionary history separating them is represented as a single line. Illustration by Craig Herbold.

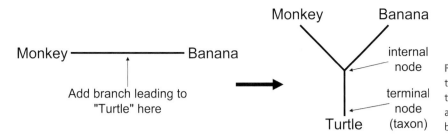

FIGURE 7.2 The addition of a third taxon to the single branch of the two-taxon tree that relates monkeys and bananas to one another creates an internal node. Illustration by Craig Herbold.

stay focused on the unrooted phylogenetic tree. Internal nodes, regardless of the location of the root, represent divergence in which a single parental (ancestral) branch on one side of the internal node splits into two daughter branches (lineages) on the other side of the internal node. The creation of this type of branch point, called a bifurcating event, gives rise to sister taxa that have a common ancestor. In studies of eukaryotes, these nodes (internal branch points) are often interpreted as speciation events; however, due to the lack of a solid framework for defining bacterial or archaeal species, this term is not entirely applicable to phylogenies that relate these organisms. A more thorough discussion of this issue is provided in Section 7.4.

The addition of a fourth taxon to the three-taxon tree is slightly more complicated. Since branches may be added to any existing branch and there are three branches in the three-taxon tree, there are three possible unrooted, four-taxon trees. Starting with the three-taxon tree of Monkey, Banana and Turtle, a branch leading to a fourth taxon, Badger, may be added to any one of the three existing branches. In Fig. 7.3, all of the possible four-taxon trees that relate Monkey, Banana, Turtle, and Badger are shown. Badger could be added to the branch leading to Monkey to make the four-taxon tree shown in blue. This tree has Monkey and Badger grouped on one side of an internal branch, called a bipartition, with Banana and Turtle grouped on the other side. Likewise, adding Badger to the branch leading to Banana results in the tree shown in red and adding Badger to the branch leading to Turtle produces the tree shown in green. Comparative sequence analysis, which entails the construction of a multiple sequence align-

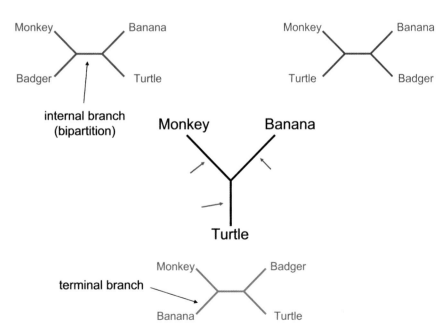

FIGURE 7.3 The addition of a fourth taxon to a three-taxon tree may be accomplished in three different ways; thus, there are three unique four-taxon trees that may relate the four taxa. These trees are the simplest trees that contain an internal branch, or bipartition, and allow the mutually exclusive grouping of two groups of taxa. Illustration by Craig Herbold.

ment with a gene common to all four taxa, would show only one of these three possible trees to be the correct tree. In this example, it is the tree that properly places mammals together (e.g., Monkey and Badger)—the blue tree.

At this point, four features of phylogenetic trees have been introduced. These features may be classified as nodes (internal and terminal nodes) and branches (internal branches, or bipartitions, and terminal branches). All phylogenetic trees are made up of these four simple features. Four-taxon trees hold a special place in the study of phylogenetics because they are the simplest trees that contain all four fundamental features, and all four fundamental features are required to define a phylogenetic relationship. The placement of taxa has no effect on two-taxon and three-taxon trees. As seen in Fig. 7.1, any two taxa are related through a single branch. It does not matter whether a taxon is placed on one end or the other because the relationship between any two taxa is always a single branch connecting the two. Also, as seen in Fig. 7.2, any three taxa are related through a single internal node and three terminal branches. These phylogenetic relationships are uninformative and offer no help in sorting taxa into proper phylogenetic groups. Bipartitions, or internal branches, are the most important feature of phylogenetic trees because they categorize the set of taxa being examined into two subsets of taxa. Each subset must contain at least two taxa (external branches cannot be considered a bipartition). Since a four-taxon tree is the simplest tree that contains a bipartition, the four-taxon tree is the fundamental unit of informative phylogenetic trees.

Phylogenetic lineages created by unrooted trees

All phylogenetic trees are composed of internal and terminal branches and nodes. Likewise, any four taxa contained in a phylogenetic tree may be grouped into two subsets of taxa that are separated by a bipartition. From these humble fundamental units, complex trees may emerge with properties that require a specific vocabulary. In Fig. 7.4, clans are introduced as a way to describe groups separated by a bipartition in a phylogenetic tree (Wilkinson et al., 2007). In Fig. 7.4a, the phylogenetically correct four-taxon tree from Fig. 7.3 is reproduced, showing the two clans that are implied by the bipartition. Monkey, Badger, and their adjacent internal node form a clan that is highlighted with a light blue oval, while Banana, Turtle, and their adjacent internal node define a second clan that is highlighted with a gold oval. Although the bipartition helps define a clan, the bipartition (internal branch) itself is not part of either clan. In Fig. 7.4b, a more complex tree, in that more than four taxa are represented (arbitrarily named A to K), is presented with a few representative clans highlighted. This example demonstrates that the definition of a clan is very specific. Each internal node defines a clan, and clans may be nested, or embedded, inside of one another. For instance, the orange clan is nested inside the yellow clan in panel b of Fig. 7.4.

Figure 7.4 also reveals a serious limitation to phylogenies depicted as unrooted trees. One clan makes biological sense, in that Monkey and Badger are united by several morphological and genetic traits that are common to all mammals. The other clan implied by this tree is less intuitive, uniting Turtle with Banana. To better discern the evolutionary relationships among the four taxa depicted in this example, it is necessary to infer a root.

Rooting a tree

As has already been stated, rooting cannot be accomplished through sequence analysis alone because methods that are used to reconstruct phylogenies from sequence data produce unrooted trees. The root, which is an internal node representing the common

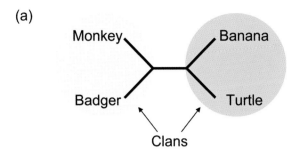

(a)

Monkey

Badger

Banana

Turtle

Clans

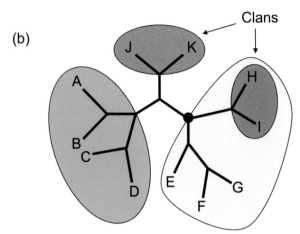

(b)

Clans

J   K

A

B

C

D

E   G

F

H

I

FIGURE 7.4 Clans describe lineages separated by a bipartition on a phylogenetic tree. (a) Monkeys and badgers form a clan on one side of a bipartition, while bananas and turtles form a clan on the other side. (b) Clans also may contain numerous taxa, with smaller clans nested within the larger clans. In this example, the orange clan, which is comprised of taxa H and I as well as their common ancestor, is nested within the yellow clan, which contains taxa E, F, G, H, and I as well as the node that defines the lineage leading to the yellow clan (exaggerated node). Illustration by Craig Herbold.

ancestor of all taxa in a tree, is considered a hypothetical taxon that is added to the tree in much the same way that taxa are added to existing branches of simpler trees to construct more complex trees (Fig. 7.1 to 7.3). Therefore, a branch leading to the root taxon must be attached to some preexisting branch in an unrooted tree. When this process is done, it is said that the tree is rooted on that particular branch or on a particular taxon. For instance, as shown in Fig. 7.5, the phylogeny of Monkey, Badger, Turtle, and Banana is rooted on the branch leading to Banana. There are generally two acceptable formats for displaying rooted phylogenetic trees, and both are presented in

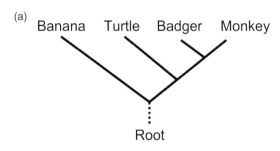

(a)

Banana   Turtle   Badger   Monkey

Root

FIGURE 7.5 Two formats are generally accepted to depict rooted phylogenies: (a) a fan shape, which grows upward from the root, and (b) a rectangular shape, in which branches extend from left to right. In the rectangular representation, nodes are represented as vertical lines while branches are represented as horizontal lines. Illustration by Craig Herbold.

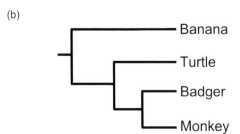

(b)

Banana

Turtle

Badger

Monkey

Fig. 7.5. In Fig. 7.5a, the root exists at the bottom of the tree and the tree "grows" upward, branching at appropriate divergence points. Overall, this style creates a fan-shaped tree with slanted branches. In Fig. 7.5b, the tree starts on the left and grows to the right, producing a dichotomous tree with rectangular branches. The two formats have identical tree topology and vary only in style, which is often a matter of personal choice. In addition, both trees are shown with unscaled branches, in that the length of the branches is not proportional to the number of nucleotide changes that have occurred over time. Sometimes it is useful to scale the branch length to reflect these molecular changes, but again it is a matter of preference.

Phylogenetic lineages created by rooted trees

Once the tree is rooted, another set of terms should be used to describe the phylogenetic relationships, including the order of descent, which now may be inferred. The concept of a clan, used in unrooted relationships, is replaced by the concept of a clade (also called a monophyletic group). A clade is defined as an internal node and all of the organisms that descended from that common node. In Fig. 7.6, this concept is reinforced using the same four-taxon example as initially explored in Fig. 7.5. It may be observed that the three animals, all vertebrates, are descended from a common internal node and thus form a clade highlighted with the green oval. Furthermore, a nested clade of mammals (yellow oval) exists within the larger clade of vertebrates. While it is stated that this tree is rooted on the branch leading to Banana, it by no means implies that vertebrates evolved from bananas! This phylogeny simply indicates that bananas and vertebrates have a common ancestor and that the vertebrates all are descended from a common ancestor that existed after the common ancestor of vertebrates and bananas.

A comparison of Fig. 7.4a and Fig. 7.6 shows that the Badger-Monkey clan (light blue oval in Fig. 7.4a) is the same as the Badger-Monkey clade (yellow oval in Fig. 7.6). The Banana-Turtle clan (gold oval in Fig. 7.4a), however, does not coincide with a Banana-Turtle clade (Fig. 7.6). As a general rule, it may be stated that a clan which contains the root cannot be a clade while a clan that does not contain the root must be a clade. In other words, the definition of a clan coincides with that of a clade only when the root is outside the clan. The arguments here come from the evolutionary directionality implied by rooting and the definition of clades and clans. In Fig. 7.6, the root is contained within the Banana-Turtle clan, and so these two taxa cannot form a clade.

FIGURE 7.6 A rooted representation of a four-taxon tree showing a clade of mammals that is nested within a clade of vertebrates. Illustration by Craig Herbold.

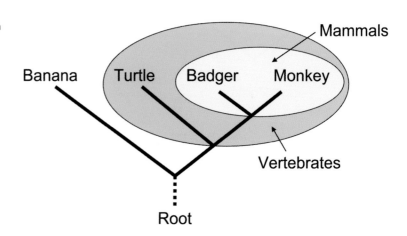

A group of organisms that form a clade are said to be monophyletic. Monophyletic groups consist of a single common ancestor (internal node) and all its descendents. Thus, if a set of taxa forms a clade, they may be considered monophyletic. Competing models of the universal tree of life provide excellent examples to illustrate the concept of monophyly. The well accepted three-domain model of the universal tree of life was calculated by Carl Woese using ribosomal RNA (rRNA) sequences for the small subunit (SSU rRNA). As shown in Fig. 7.7, the tree of life is represented showing three domains: *Bacteria, Archaea* and *Eucarya (Eukaryota)* (Woese et al., 1990). It readily can be seen that the *Eukaryota* all are descended from a single common ancestor. There is one node from which all eukaryotes, and no *Bacteria* or *Archaea*, emerge. Since the clade of eukaryotes perfectly matches the group known as *Eukaryota,* they constitute a monophyletic group. Likewise, under this model, the *Bacteria* may be interpreted as a group comprising a single node and all descendents of that node. Also under this model, the *Archaea* form a monophyletic group. This final observation provided the evidence necessary to elevate the *Archaea* to domain status. As a consequence, all life may be classified into one of three domains, the *Bacteria, Archaea,* or *Eukaryota.*

The grouping of *Archaea* and *Eukaryota* as sister taxa, in which the two domains share a common ancestor, has created a large problem for taxonomists who appreciate the word "prokaryote" to refer to both *Bacteria* and *Archaea.* As may be seen in Fig. 7.7, the root of the universal tree of life is placed on the branch leading to the *Bacteria.* This rooting places *Archaea* and *Eukaryota* as sister taxa. With this rooting, prokaryotes do not form a monophyletic group, and there is an active and spirited debate as to whether the word "prokaryote" should be abolished (Pace, 2006; see Martin and Koonin [2006] for a rebuttal). One argument against retaining the prokaryotic grouping arises because the clade defined by the node that represents the common ancestor of all prokaryotes (*Archaea* and *Bacteria*) also includes the *Eukaryota.* Thus, by excluding the branch leading to *Eukaryota,* the prokaryotes instead form a paraphyletic group. Like monophyletic groups, a paraphyletic group is also defined by a node that represents the common ancestor of all members of the group. However, this common ancestral node also is shared by descendents that are not considered members of the group.

FIGURE 7.7 Woese's three-domain model of life, showing the three monophyletic groups of life: *Bacteria, Archaea,* and *Eucarya* (Eukaryota). (Reprinted from Woese et al. [1990] with permission.)

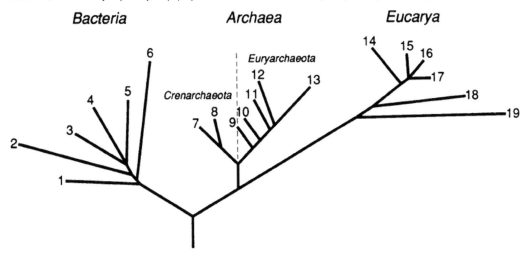

In Fig. 7.8, the rooted three-domain tree of life is reproduced to make the distinctions between monophyly and paraphyly more clear. The three monophyletic groups, *Bacteria, Archaea,* and *Eukaryota,* can be seen in Fig. 7.8a. Each is unambiguously defined by a common ancestor and all descendents of that common ancestor. In Fig. 7.8b, the prokaryotes are highlighted in turquoise. The common ancestor of both of these groups, the *Eukaryota* and the prokaryotes, is the common ancestor of all life. In other words, the common ancestor of all prokaryotes also is an ancestor of all eukaryotes. Thus, by definition, each domain constitutes a monophyletic group in which all descendents arise from a common ancestor unique to each domain. The prokaryotes, on the other hand, are a paraphyletic group due to the exclusion of the eukaryotes, which share the common ancestor for all three domains of life.

Implications for rooting the tree of life

While it may be clear from the rooted three-domain model of life that prokaryotes do not form a monophyletic group, it also should be noted that the three-domain model of life, and the rooting of the universal tree, is not the only model supported by the data. One competing theory to the three-domain model is known as the eocyte model. The eocyte model also was originally based upon analysis of SSU rRNA sequences (Lake, 1988) and later supported by an indel analysis of elongation factor EF-1α (Rivera and Lake, 1992). The eocyte model breaks the *Archaea* into a paraphyletic group, with the "eocytes" (also referred to as the *Crenarchaeota*) forming a clade with the eukaryotes. According to this model, *Bacteria, Eukaryota, Euryarchaea,* and eocytes (*Crenarchaeota*) all form well-supported monophyletic groups (Fig. 7.9a). Consequently, not only are

FIGURE 7.8 A simplified representation of Woese's three-domain model of the tree of life (Woese et al., 1990). (a) Each of the three domains as monophyletic groups; (b) the paraphyletic nature of the prokaryotes. Illustration by Craig Herbold.

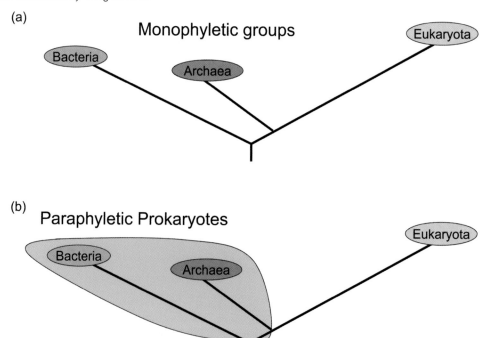

(a)

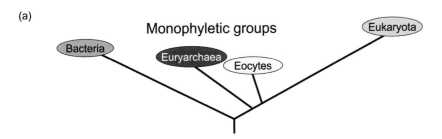

Monophyletic groups

(b)

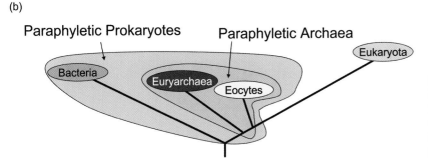

Paraphyletic Prokaryotes   Paraphyletic Archaea

FIGURE 7.9 A simplified representation of Lake's eocyte model of the tree of life (Lake, 1988). (a) Monophyletic groups within the tree; (b) the paraphyletic nature of both pro-karyotes and *Archaea*. Illustration by Craig Herbold.

prokaryotes paraphyletic (turquoise grouping in Fig. 7.9b), but so are *Archaea* (grey grouping in Fig. 7.9b).

While the factor which distinguishes these two models for the evolution of life (e.g., monophyly versus paraphyly of the *Archaea*) is interesting from an academic standpoint, they are presented here only as examples of how monophyly and paraphyly affect the interpretation of phylogeny. Nothing affects these arguments more than selecting the position for the root of the universal tree. For example, the root to the tree of life for both the three-domain model and the eocyte model is located on the branch leading to *Bacteria*. However, the root is not actually known with a great deal of confidence, and its position is sure to remain controversial for quite some time (Doolittle and Brown, 1994; Bapteste and Brochier, 2004). Some analyses, for instance, favor a root of the universal tree of life on a branch that exists within the *Bacteria*, not on the branch leading to the *Bacteria* (Cavalier-Smith, 2002; Skophammer et al., 2007). As depicted in Fig. 7.10, a root for the tree of life within the *Bacteria* not only makes this domain paraphyletic (turquoise grouping) but also would suggest that the earliest divergence in the history of life was between two bacterial lineages, whose modern-day descendents are represented as purple and pink ovals in Fig. 7.10. An ancestor of the pink group of bacteria gave rise to another lineage that eventually became the *Archaea* and the *Eukaryota*. Thus, if the root of the tree of life is contained within the *Bacteria*, the ancestral node common to all bacterial taxa also is common to all of life. When one compares the competing models of the rooting and branching order of the tree of life, the only group that enjoys undisputed monophyly is the *Eukaryota*. The *Bacteria* and the *Archaea*, which comprise the two other major groups of life, each may be paraphyletic. In summary, the eocyte hypothesis questions the monophyly of the *Archaea* and the alternative rooting schemes question the monophyly of the *Bacteria*.

Using outgroups to root a tree

The biggest challenge when it comes to rooting the universal tree of life is that there exists no outgroup, which is defined as a taxon that is not contained within a clade and is used to determine the branching order within the clade. One important assumption

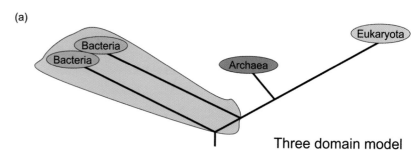

(a)

Three domain model

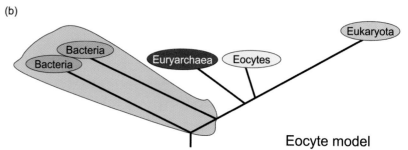

(b)

Eocyte model

FIGURE 7.10 Simplified representations of Woese's three-domain model (a) and Lake's eocyte model (b) for the tree of life, with emphasis on how alternative rooting schemes make the *Bacteria* paraphyletic. Illustration by Craig Herbold.

made when using outgroups to root a tree is that a well-defined clade of organisms exists. Clades such as this are called ingroups and comprise specific organisms known to reproducibly and robustly form monophyletic groups. Remember that clades consist of a common ancestor and all of its descendents. The outgroup taxon must have branched off the tree of life before the ancestor of the clade arose. This logic reduces rooting to a two-taxon rooting problem. One taxon is represented by the entire ingroup clade, and the other taxon is the outgroup. An example of this is shown in Fig. 7.11. Because the vertebrates form a proper clade, Banana can act as the outgroup. An important assumption here is that vertebrates share a common ancestor that existed sometime after the common ancestor of both vertebrates and bananas. Thus, wherever the outgroup connects to the ingroup will orient the ingroup clade and show where that clade can be rooted. For the taxa included in Fig. 7.6, the vertebrate clade is rooted on the branch leading to Turtle, with the mammals forming a nested clade within the vertebrates.

Phylogenetic taxonomy

A great deal of emphasis is placed on whether a set of taxa forms a clade and thus can be classified based on evolutionary ancestry. In other words, for the phylogeny of a set of taxa to depict a particular taxonomic group (e.g., phylum, class, order, family, genus, or species), there must be a single node that unites an ancestor with all of its descendents.

FIGURE 7.11 The banana may be used as an outgroup in determining branching order within the vertebrates. Illustration by Craig Herbold.

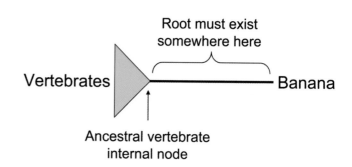

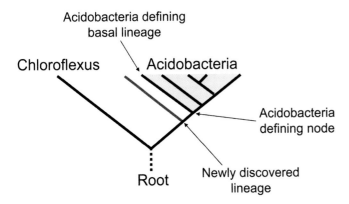

Chloroflexus Acidobacteria

Acidobacteria defining basal lineage

Acidobacteria defining node

Root

Newly discovered lineage

FIGURE 7.12 Classifying a newly discovered lineage as a member of an ingroup versus a sister taxon to an ingroup. Described *Acidobacteria* are all related through a defined node. A newly discovered bacterium that branches prior to this node may not be included within the *Acidobacteria* without redefining the ancestral node of the *Acidobacteria*. Illustration by Craig Herbold.

The application of phylogeny to the classification of organisms is referred to as cladistics. In Fig. 7.12, the *Acidobacteria* are defined by the inclusion of two or more ingroup taxa that are related through the clade-defining node, and the tree is rooted using *Chloroflexus* as the outgroup. This node-based interpretation has a few consequences that affect classification strategies. For instance, a scenario is presented in Fig. 7.12 in which a new lineage of bacteria is discovered (red line). This lineage groups very strongly with the *Acidobacteria*, and indeed shows the greatest similarity to *Acidobacteria* at the sequence level. With a node-based definition of the *Acidobacteria*, though, this newly discovered lineage cannot be classified as *Acidobacteria* without changing the *Acidobacteria*-defining node as well. This group may instead be a sister taxon to the *Acidobacteria*.

So how does one discern whether a newly discovered lineage should be considered a sister taxon or should be used to change the placement of the *Acidobacteria*-defining node? First, assume that the *Acidobacteria* are defined by a single node from which all *Acidobacteria* are descended. This node, in its simplest form, may be represented by two acidobacterial taxa, as in Fig. 7.13. The two acidobacterial representatives must be related through the *Acidobacteria*-defining node. Since the two *Acidobacteria* are related through the *Acidobacteria*-defining node, there exists a simple four-taxon test of whether a new lineage belongs within or outside the *Acidobacteria* clade. If the newly discovered sequence nests within the *Acidobacteria*, the new sequence must group with one of the acidobacterial lineages. This simple test is shown in Fig. 7.14. In this figure, the rooted three-taxon tree (from Fig. 7.13) with *Chloroflexus* as the outgroup is depicted again with the *Acidobacteria*-defining node shown as an enlarged node. Possible attachment points for a new lineage are shown as colored arrows. If a four-taxon relationship (rooted or unrooted) places the new lineage on the branch leading to the *Chloroflexus* lineage, as is

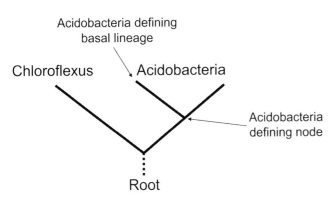

Chloroflexus Acidobacteria

Acidobacteria defining basal lineage

Acidobacteria defining node

Root

FIGURE 7.13 To construct a test for classification into a clade, one must first determine the clade-defining node with at least two ingroup taxa and one outgroup taxon. Illustration by Craig Herbold.

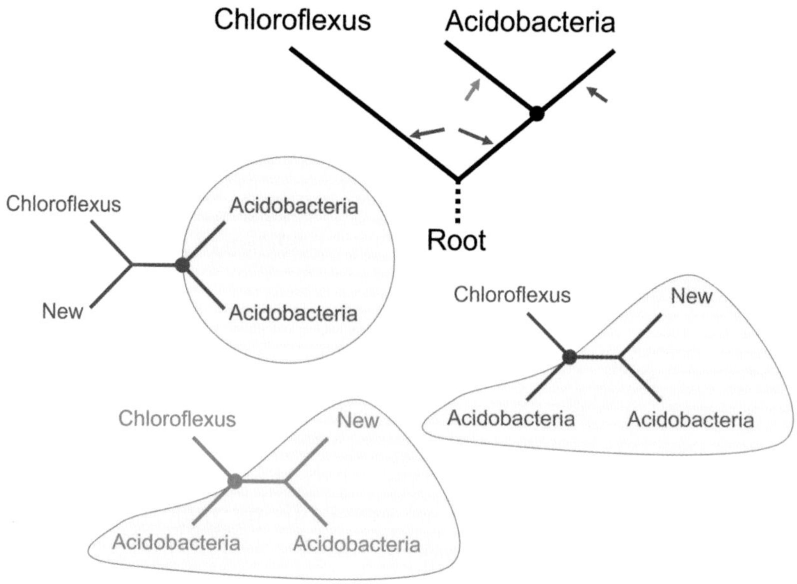

FIGURE 7.14 A taxonomic test for inclusion of a newly discovered sequence in the *Acidobacteria*. If the new sequence is grouped with either of the acidobacterial taxa (green or red arrows), it will be nested within the group and may be classified as *Acidobacteria*. If the new taxon is related through other locations (marked with blue arrows), the new sequence may not be classified as *Acidobacteria*. Illustration by Craig Herbold.

shown by the blue arrows and the resulting blue tree, then the new lineage exists outside of the *Acidobacteria*. It might be stated that the lineage is a sister taxon to the *Acidobacteria*; however, strong statements about including the new sequence within the *Acidobacteria* or strong claims about the discovery of a novel lineage should be avoided. Future phylogenetic experiments, preferably using cultivated isolates or type strains, may yield results capable of repositioning the node defining *Acidobacteria* clade, which consequently would cause this new lineage to be considered a member of that clade. Phylogeny alone should never be used to change a node defining a taxonomic group. On the other hand, two alternative topologies that could be produced by the four-taxon test, shown in Fig. 7.14 as red and green trees, define the new sequence as an acidobacterial sequence. The new sequence shares the *Acidobacteria*-defining node as an ancestor and is thus nested within the *Acidobacteria*.

This procedure, summarized in Fig. 7.15, may be extended to any group which is monophyletic and for which an ancestral node can be defined. To classify an organism into a taxonomic group, a few conditions must be met. First, two members of the

FIGURE 7.15 Summary of the steps and criteria in the generalized taxonomy test.

Generalized Taxonomy Test

STEPS:

(1) Define a clade-defining node with two ingroup taxa and one outgroup taxon

(2) Reduce the tree to a four-taxon relationship
(1 outgroup, 2 ingroup members, and 1 new sequence)

(3) Assess grouping of new sequence based on statistical significance
(Ex: bootstrapping)

RESULTS & INFERENCES:

New sequence on the branch leading to outgroup ⟶ Sister taxon

New sequence nested within ingroup taxa ⟶ Ingroup member

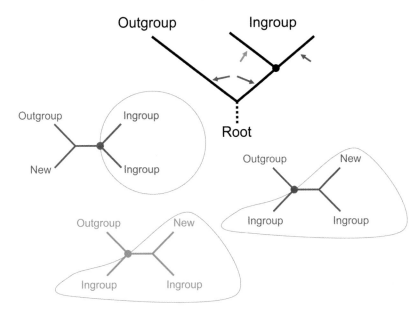

FIGURE 7.16 The generalized taxonomy test. The placement of a newly discovered taxon at two particular locations, shown by red and green arrows, meets the necessary criteria required for classification. The other two locations, shown in blue, do not meet the criteria to classify the newly discovered sequence as a member of the ingroup. Illustration by Craig Herbold.

ingroup that define an ancestral node must be included in the analysis. Second, there must be an outgroup. The sequence to be classified must group with one of the two members of the ingroup, and this grouping must pass statistical-significance tests (i.e., bootstrapping, to be discussed in the next section). The example shown in Fig. 7.14 provides an example of a bacterial phylum level test, but this procedure is just as applicable to any taxonomic level, including species. In Fig. 7.16, the method is generalized. Again, a major assumption to this analysis is that the members of an ingroup form a clade.

The cladistics approach, which can be applied to the classification of both cultivated and uncultivated environmental sequences, is rigorous and requires a great deal of attention to detail. When choosing taxa to define the ingroup node, great care should be taken to ensure that those sequences come from true members of that taxonomic group. It is not acceptable to use sequences obtained through environmental PCR to define an ancestral node unless the sequence can be traced to a published study and independently evaluated. Specifically, in a study that assigns a taxonomic level to environmental sequences, one must confirm that type strains were used to define the ingroup. Rather than using the uncharacterized sequences from such a study, it may then be better to use the same node-defining taxa as the publication does and give that publication the proper credit (i.e., cite their work).

KEY TERMS

Bifurcating event Creation of an internal node in which one ancestral branch divides into two new lineages.

Bipartition An internal branch connecting two internal nodes in an unrooted phylogenetic tree; fundamental unit of phylogenetically informative tree construction.

Branch A feature that defines the relationship between taxa. Only one branch connects any two adjacent nodes. Branches represent the intermediate sequence of the same gene from now-extinct organisms that existed during the course of evolution.

Branch length The length proportional to the number of changes that have occurred in the branch connecting two nodes as a function of time.

Clade A group of taxa in a rooted phylogenetic tree that includes the common ancestor and all of its descendents (ingroups); also called a monophyletic group or subtree.

Cladistics The hierarchical classification of organisms based on evolutionary ancestry; also called phylogenetics or phylogenetic taxonomy.

Clans Taxa on an unrooted phylogenetic tree that are separated by a bipartition, or internal branch.

Common ancestor An internal node defining the bifurcating event that gave rise to two immediate lineages (sister taxa).

Divergence A bifurcating event, or branch point, in which a single ancestral branch on a phylogenetic tree is split into two lineages (sister taxa), creating an internal node representing a common ancestor.

Evolutionary distance The sum of the physical distance, or number of changes that have occurred as a function of time, separating organisms on a phylogenetic tree; inversely proportional to evolutionary relatedness (e.g., the shorter the evolutionary distance, the more closely related the organisms are likely to be); the sum of all base pair differences between any two sequences.

Homologous Similar in nucleotide or amino acid sequences due to a biological relationship and not simply due to chance.

Ingroup A group of taxa that form a well-defined clade, displaying a nested hierarchy (ordered branch points) among taxa in that clade.

Internal node A node representing an ancestor from which terminal nodes descended; also referred to as a branch point arising from a bifurcating event.

Lineage A terminal node on a phylogenetic tree.

Monophyletic group *See* Clade.

Nested Embedded within larger lineages (i.e., clans and clades) in reference to phylogenetic trees.

Node A taxonomic unit, or lineage.

Orthologs Homologous sequences in two different organisms derived from a common ancestral gene; functionally similar and useful in phylogenetic analysis.

Outgroup A taxon that is less closely related to the taxa within a clade than the taxa are to each other. The sequence representing an outgroup must be sufficiently conserved to align with sequences from ingroup taxa and be functionally homologous. Used to determine the order of branching within a clade.

Paralogs Two independent copies of a gene in the genome of a single organism, created by a duplication event—one copy is no longer functionally constrained and thus may evolve a new function; homologous sequences in two different organisms that are descendents of independent copies of a gene derived by its duplication in a common ancestor.

Paraphyletic group A group of taxa in a rooted phylogenetic tree that contains the most recent common ancestor but does not contain all the descendents of that ancestor.

Phylogenetic tree A map, or graphical representation, depicting evolutionary relationships among members of a group of organisms; describes the pattern or timing of events that occurred as organisms diversified and new lineages appeared; documents which taxa are more closely related to one another based on sequence similarity.

Root An internal node representing the common ancestor of all taxa in a tree.

Sister taxa The two taxonomic lineages on either side of a bifurcating event in which the parental branch splits into two, independent daughter branches; two lineages that have a common ancestor.

Speciation Interpretation of divergence in eukaryotic phylogenies.

Terminal (external) nodes Nodes representing an extant, or existing, organism.

Topology The overall shape attributed to a phylogenetic tree; describes the branching pattern of the lineages depicted in a tree and reflects the order in which lineages diverged.

Type strains Microorganisms that have been acquired, authenticated, and preserved in a repository for distribution, serving as standard reference strains in research.

Unrooted tree A tree that shows evolutionary relationships among taxa but does not specify the order of evolutionary events that give rise to lineages in the tree or provide the position of the common ancestor to all taxa in the tree.

REFERENCES

Bapteste, E., and C. Brochier. 2004. On the conceptual difficulties in rooting the tree of life. *Trends Microbiol.* **12**:9–13.

Cavalier-Smith, T. 2002. The neomuran origin of archaebacteria, the negibacterial root of the universal tree and bacterial megaclassification. *Int. J. Syst. Evol. Microbiol.* **52**:7–76.

Doolittle, W. F., and J. R. Brown. 1994. Tempo, mode, the progenote, and the universal root. *Proc. Natl. Acad. Sci. USA* **91**:6721–6728.

Lake, J. A. 1988. Origin of the eukaryotic nucleus determined by rate-invariant analysis of rRNA sequences. *Nature* **331**:184–186.

Martin, W., and E. V. Koonin. 2006. A positive definition of prokaryotes. *Nature* **442**:868.

Pace, N. R. 2006. Time for a change. *Nature* **441**:289.

Rivera, M. C., and J. A. Lake. 1992. Evidence that eukaryotes and eocyte prokaryotes are immediate relatives. *Science* **257**:74–76.

Skophammer, R. G., J. A. Servin, C. W. Herbold, and J. A. Lake. 2007. Evidence for a Gram-positive, eubacterial root of the tree of life. *Mol. Biol. Evol.* **24**:1761–1768.

Wilkinson, M., J. O. McInerney, R. P. Hirt, P. G. Foster, and T. M. Embley. 2007. Of clades and clans: terms for phylogenetic relationships in unrooted trees. *Trends Ecol. Evol.* **3**:114–115.

Woese, C. R., O. Kandler, and M. L. Wheelis. 1990. Towards a natural system of organisms: proposal for the domains Archaea, Bacteria, and Eucarya. *Proc. Natl. Acad. Sci. USA* **87**:4576–4579.

Web Resource

NCBI, a Science Primer: Phylogenetics www.ncbi.nlm.nih.gov/About/primer/phylo.html

STATISTICAL TESTS: CONFIDENCE AND BIAS

Bipartition support and bootstraps

For clarification purposes in the previous section, it was assumed that the bipartitions shown in the examples of different phylogenetic trees were well supported, in that they passed statistical tests measuring significance. This is not always the case, and often a given bipartition cannot be considered reliable with any degree of confidence. There are several ways to test whether a bipartition is robust. An accepted method used to assess confidence in a given bipartition is called bootstrapping (Hall, 2008). In the most general sense, bootstrapping is a statistical procedure in which raw data are resampled numerous times to estimate model parameters such as the mean or variance of a set of data. This procedure emphasizes values that are common and deemphasizes rarely observed data and outliers. For phylogeny, the raw data being resampled are the alignment, and the model parameters may include topology and branch length. The probability that members of a given clade produced by an alignment and tree analysis are always members of that clade can be assessed by bootstrap analysis.

An example phylogenetic bootstrap analysis with three replicates is shown in Fig. 7.17. First, the original alignment is reinterpreted as a set of nucleotide patterns corresponding to individual columns of aligned nucleotides. Each pattern occurs with a frequency that can be estimated by counting the number of occurrences of that pattern. In this example, the very first column, highlighted in yellow, is represented by the pattern GGCCG. This pattern occurs only once in a set of 50 positions in the original alignment. In contrast, the pattern GGGGG occurs 11 times. After the pattern occurrences are counted, a frequency is assigned for each pattern. In this case, GGCCG occurs at 2% of the alignment positions (1 of 50) and GGGGG occurs at 22% of the alignment positions (11 of 50). Finally, new pseudoalignments, also called bootstrap replicates, are constructed by randomly selecting alignment patterns in accordance with the frequency at which the patterns were observed in the original alignment. These new pseudoalignments should be the same length as the original alignment. In this case, the first alignment consisted of 50 positions; therefore, each bootstrap replicate consists of 50 positions.

There is no guarantee that any particular pattern will be resampled for a single bootstrap replicate because the resampling is carried out using a random process. The probability of selecting a particular pattern for a bootstrap replicate is high if that pattern is common, and the probability of selecting a particular pattern in a bootstrap replicate is low if that pattern is rare. Figure 7.17 highlights two rare patterns, one in yellow and one in blue. Each pattern is observed only once in the original alignment. Due to the randomness associated with resampling data, the alignment pattern highlighted in yellow was not randomly selected for in any of the three bootstrap replicates shown, whereas the alignment pattern highlighted in blue is represented in bootstrap alignment 2 and bootstrap alignment 3.

Each bootstrap alignment is used to construct a phylogeny via one of the possible phylogenetic reconstruction methods (to be discussed in Section 7.3). The process of resampling data allows one to assess the reliability, or robustness, of any phylogeny or any feature of that phylogeny. In other words, one must make sure that the phylogeny constructed is not dependent on rare data but is reflective of the common data. Another way of thinking about this is to imagine that there is a phylogenetic signal contained

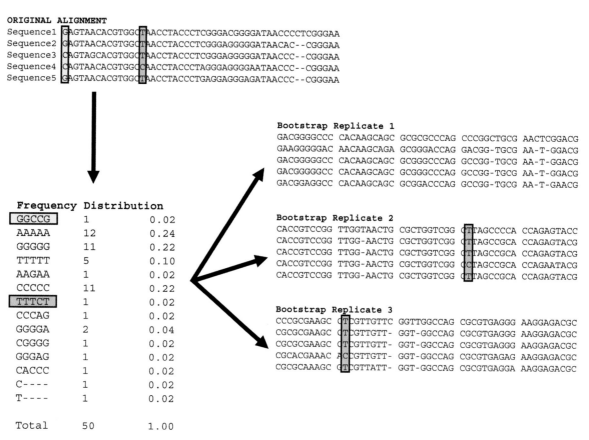

FIGURE 7.17 Example of resampling alignment data during bootstrap analysis. The nucleotide patterns, defined by the columns of an original alignment, are randomly sampled to produce new "pseudo-bootstrapped" alignments called bootstrap replicates. Each replicate is the same length as the original alignment, and each is used separately in the calculation of a phylogenetic tree. In the above example, one tree will be calculated using the original alignment and three bootstrapped trees will be calculated from the three bootstrap replicates for a total of four tree calculations. Illustration by Craig Herbold.

within the data, but also quite a bit of noise. Resampling the data helps to boost the signal-to-noise ratio by averaging out the noise. If there is a robust signal, it will be detected in the majority of the bootstrap replicates. By analogy, wet-lab scientists may test the reliability of a particular result by repeating the experiment with independent trials performed by multiple people—the more frequently a result is obtained, the more confidence one has in the data itself and any conclusions therefrom.

The bootstrapping process may be used to assess whether a given bipartition is robust. To be robust, a bipartition must exist in a majority of the phylogenies calculated. A general guideline is that at least 70 to 75% of the bootstrap replicates must produce a tree in which a particular bipartition is detected before bestowing any confidence in the existence of that bipartition. Figure 7.18 provides an example calculation of bootstrap values. In practice, at least 100 to 2,000 bootstrap replicates should be used to assess bipartition support (Hall, 2008). It is necessary to simplify, however, for the purposes of explaining the calculation in the example presented in Fig. 7.18. Four bootstrap replicates (pseudoalignments) were used to construct four different phylogenetic trees (left panel). Each phylogenetic tree may then be summarized by a number of specific bipartitions. For example, the tree at the top of the left-hand side of Fig. 7.18 contains five bipartitions. One bipartition separates taxa G and H from all other taxa. Another bipartition separates

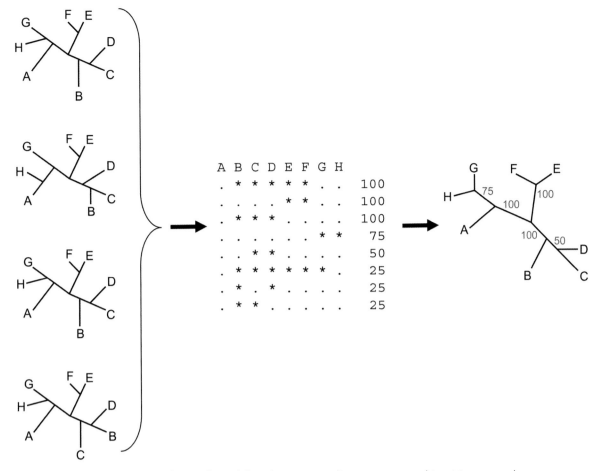

FIGURE 7.18 Using bootstrapped trees derived from bootstrap replicates to support bipartitions on a phylogenetic tree. All trees are examined for the presence of bipartitions, and all bipartitions observed (in all trees) are compiled into a list. In this list, all taxa on one side of the bipartition are marked with an asterisk and the taxa from the other side are marked with a period. The percentage of trees in which a particular bipartition is found is taken as the percentage bootstrap support. These bootstrap supports may then be reported on a tree. Illustration by Craig Herbold.

taxa G, H, and A from taxa B, C, D, E, and F. A third bipartition separates taxa E and F from all other taxa. A fourth bipartition separates taxa B, C, and D from taxa A, E, F, G, and H. Finally, a fifth bipartition separates taxa C and D from all other taxa.

Several bipartitions observed in the first tree also are found in the other trees shown in Fig. 7.18. Additional bipartitions also are detected. For example, the second tree from the top contains a bipartition that separates taxa B and C from all other taxa. This differs from the first tree, which contains a bipartition that places taxa C and D together as a group, separated from all other taxa. All bipartitions observed from all the bootstrap replicate phylogenies are tabulated to determine the percentage of bootstrap replicates that support a particular bipartition. A tabulated set of bipartition patterns for the bootstrap analysis is shown in the middle panel of Fig. 7.18. The first entry in the list places taxa A, G, and H together on one side of a bipartition and taxa B, C, D, E, and F on the other side. This particular bipartition is observed in 100% of the bootstrap replicates. That is, regardless of the ordering of A, G, and H with respect to one another and regardless of the ordering of taxa B, C, D, E, and F with respect to one another, the taxa are separated into these two sets of taxa in all the bootstrap replicates. Remember that for a bipartition to be considered robust, it must be observed in at least 70% of the bootstrap replicates. From this analysis, it can be seen that there are four robust bipartitions and four nonrobust bipartitions.

After the bootstrap analysis is complete, the calculated bootstrap support may then be mapped back onto a representative phylogenetic tree (right panel in Fig. 7.18). The bootstrap support for any given bipartition is placed as a label on that bipartition.

To interpret the bootstrap support, it is useful to think of a null hypothesis in which all taxa are equally unrelated (Fig. 7.19). After a phylogenetic calculation, a fully resolved tree is obtained, onto which bootstrap support may be shown. Bootstrap analyses may then be used to assess whether a given bipartition meets the criteria to reject the null hypothesis and group two or more taxa together into a clan. Any bipartition that passes this test may be used to interpret the relationships between organisms. Any bipartition that does not pass this test should be ignored, or at least should not be taken as evidence of a particular grouping. The fully resolved tree in Fig. 7.19 contains four bipartitions that are well supported. A fifth bipartition, which separates taxa C and D from taxon B and all other taxa, is not well supported (50% bootstrap support) and may be collapsed. There is simply not enough support for this bipartition to make the claim that taxa C and D are grouped to the exclusion of taxon B. While taxa B, C, and D are always separated from the other taxa with 100% support, the branching order cannot be determined with any degree of confidence. The null hypothesis has not been disproved for determining the relationship among these three taxa even though the separation of these three taxa from the other taxa is well supported.

Bipartition support is sometimes said to offer support for a node. This practice is somewhat useful for interpreting clades found within a rooted tree; however, it is very important to remember what exactly is being bootstrapped so that faulty conclusions may be avoided. In Fig. 7.20, one unrooted and two rooted representations of the phylogeny from Fig. 7.18 are presented with bootstrap labels. The phylogenies were rooted on the bipartition that separates taxa E and F from all other taxa. Taxa E and F are

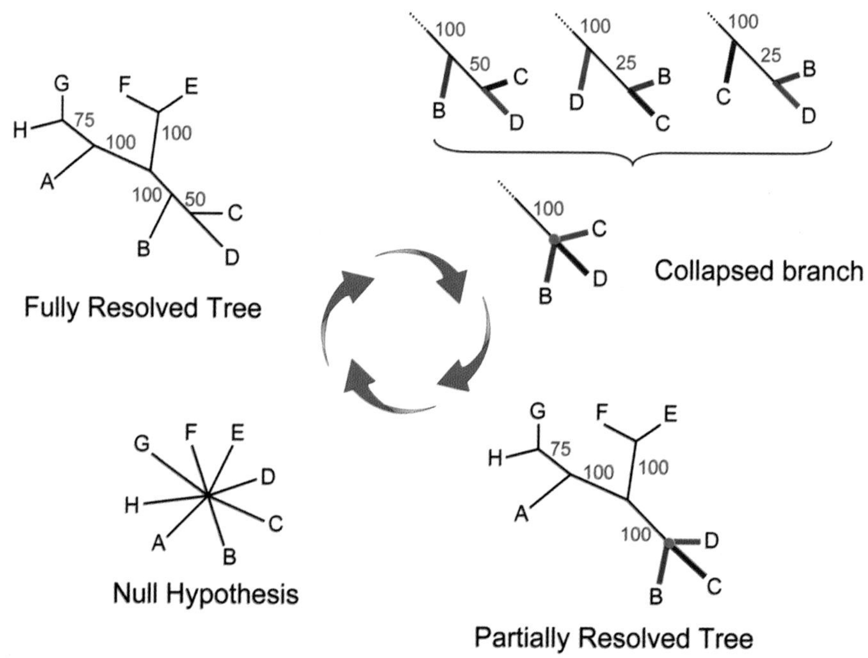

FIGURE 7.19 Each bipartition in a candidate phylogeny may be tested as an alternative hypothesis against an uninformative null hypothesis. If sufficient support exists for a particular bipartition, it may be accepted. Otherwise, that bipartition should not be accepted as evidence of a grouping. Illustration by Craig Herbold with modifications by Cori Sanders (iroc designs).

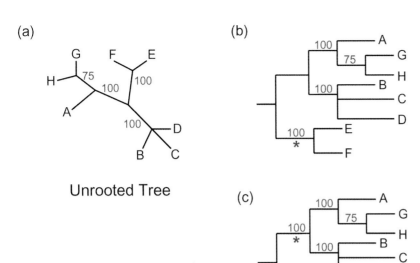

(a) Unrooted Tree

(b)

(c) Rooted Trees

FIGURE 7.20 Rooting a partially resolved tree on an internal branch. (a) Unrooted, partially resolved tree from Fig. 7.18. (b and c) Result of rooting the partially resolved tree on the internal branch leading to taxa E and F. Both rooted trees shown here are equivalent to the unrooted tree in panel a, but only one value was calculated for the internal branch on which the root was placed (red asterisks). Therefore, if support for the bipartition is used to assess node support, only one node may be supported. Illustration by Craig Herbold.

separated from all other taxa with 100% bootstrap support. With this rooting, it can be seen that a clade of taxa B, C, and D is supported with 100% of the bootstrap replicates. Likewise, a clade of taxa A, G, and H is supported with bootstrap support of 100%. While taxa E and F are separated from the other taxa with 100% support, the decision is arbitrary as to whether a clade of taxa E and F is supported with 100% bootstrap, as shown in the rooted phylogeny in Fig. 7.20b, or whether a clade of all taxa except taxa E and F is supported with 100% bootstrap, as shown in the rooted phylogeny in Fig. 7.20c. These difficulties arise from attempting to assign a direction to an unrooted test. Bootstraps measure bipartition support, not directed node support. As such, it is the bipartitions in the unrooted phylogeny that are tested, not specific nodes of the rooted phylogeny. The bipartitions support particular groupings which, assuming a root, may be interpreted to support clade assignments. In Fig. 7.20, taxa B, C, and D form an unresolved clade. Regardless of the exact branching order of the taxa within that clade, these taxa are all related through a common unresolved node, which also may be referred to as a multifurcating node.

The burden of proof that a rigorous bootstrap analysis imposes may create problems for overzealous taxonomic assignment. Recall from Section 7.1 (Fig. 7.16) that a clade must be defined by two ingroup taxa and a "new" taxon must group with one of the ingroup taxa for it to be considered a member of that taxonomic group. This grouping also must be robust. As shown in Fig. 7.21, a high bootstrap support for a phylogeny that places a newly discovered taxon as a sister taxon to one of the known ingroup taxa safely defines the new taxon as a member of the ingroup. A low bootstrap support does not reject the null hypothesis, and the relationship of the new taxon with the ingroup and outgroup remains obscure. The new taxon may very well branch prior to the ingroup-defining node, and therefore it cannot be safely categorized as a member of the ingroup.

When large data sets with representatives of several phyla are analyzed, as is usually the case with SSU rRNA phylogenies, taxonomic assignment is dependent upon two

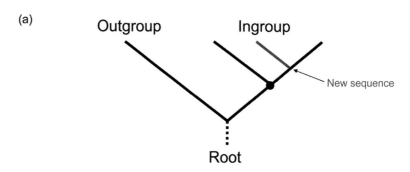

(a)

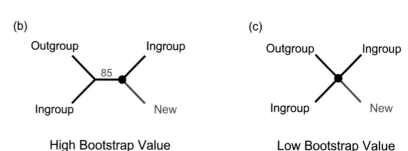

(b) High Bootstrap Value

(c) Low Bootstrap Value

FIGURE 7.21 Taxonomic assignment carries with it a burden of proof. (a) Rooted tree placing new sequence with one of the ingroup sequences. (b) Bootstraps must support the internal branch that positions a new sequence within the ingroup clade. (c) Without strong bootstrap support, it remains possible that the new sequence branches off prior to the clade-defining node and therefore cannot reliably be classified as a member of the ingroup. Illustration by Craig Herbold.

bipartition bootstrap values. First, a grouping must be well supported and contain at least two ingroup taxa, such as *Acidobacteria,* and the sequence to be assigned (Fig. 7.22). This bipartition establishes the existence of a clan containing the ingroup and the taxon to be assigned. All other taxa in the analysis are in a separate clan. These two clans must be separated by a robust bipartition; otherwise, there is no evidence to support the existence of that ingroup at all. The second bipartition must place the new taxon within the ingroup by robustly grouping it with one of the taxa within the ingroup. This bipartition establishes that the branching order of the ingroup is known and that the new taxon is not the first branch. Instead, the new taxon is nested within the ingroup. If the new taxon is the earliest branch, then it branched before the ingroup-defining node. In this case, the burden of evidence has not been met, and a taxonomic assignment would be irresponsible.

Long-branch attraction

A major problem with phylogeny is the phenomenon known as long-branch attraction (Felsenstein, 2004). Long branches have a tendency to group with other long branches in phylogenetic trees regardless of the true evolutionary history (Fig. 7.23). If an outgroup

FIGURE 7.22 Taxonomic assignment depends on two bootstrap values. In this figure, two outgroup taxa *(Firmicutes* and *Chloroflexi)* are used to calculate a bootstrap value that defines the *Acidobacteria* group. A second bootstrap value then defines the unknown sequence as nested within the group. Illustration by Craig Herbold.

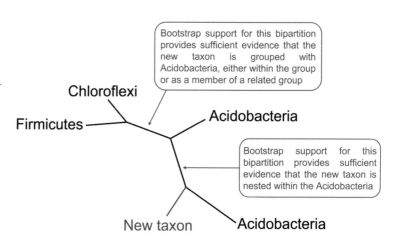

Bootstrap support for this bipartition provides sufficient evidence that the new taxon is grouped with Acidobacteria, either within the group or as a member of a related group

Bootstrap support for this bipartition provides sufficient evidence that the new taxon is nested within the Acidobacteria

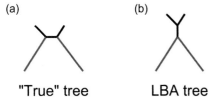

(a)　　　(b)

"True" tree　　　LBA tree

FIGURE 7.23 Effect of long branches on a phylogenetic tree. (a) Long branches (blue lines) in a phylogenetic tree, representing the "true" tree in this example. (b) When long branches are mistakenly grouped, a long-branch artifact (LBA) tree is observed. Illustration by Craig Herbold.

branch is too long, it arbitrarily pulls other long branches to the base of a group. This could happen, for instance, if one particular lineage of a group evolves very quickly while the others evolve very slowly. Therefore, the choice of an outgroup should be carefully evaluated to minimize any long-branch effects.

Excellent examples of long-branch artifacts exist within the literature. Figure 7.24 shows a phylogeny calculated using SSU rRNA (Pace, 1997). This tree places the microsporidia near the base of the eukaryotic domain *(Eucarya)*. It has been shown, however, that the microsporidia most probably comprise a rapidly evolving group within the fungi (Gill and Fast, 2006). Similar concerns exist for the placement of protozoan *Trichomonas* species, because these lineages have elevated evolutionary rates as well (Germot and Phillipe, 1999). With respect to the tree of life, prokaryotic sequences act as an extremely long-branch outgroup and exert an attractive force on the long branches of the fast-evolving eukaryotes.

A close examination of Fig. 7.24 reveals another potential long-branch artifact that might result in a monophyletic *Archaea* group. The branches leading to *Bacteria* and *Eucarya* are both relatively long, while the branches leading to the two groups of *Archaea* are comparatively short. Part of the controversy over the different tree-of-life models (three-domain versus eocyte) is that the three-domain model can be predicted if a long-branch artifact exists, while the eocyte model appears to correct for long-branch attraction (Fig. 7.25). Indeed, the eocyte hypothesis arose out of an attempt to minimize the rate effects that contribute to long-branch attraction (Lake, 1988). It is difficult to assess whether a long-branch artifact actually exists in the tree of life or whether the steps taken to minimize long-branch attraction went too far and biased the analysis against the three-domain model. It also is exceedingly difficult to completely rid an analysis of long-branch attraction. It is observed only after the analysis is complete and may be caused by many different things, including the choice of method used to construct an alignment or phylogeny.

Alignments may be the cause of some long-branch artifacts. Alignment methods that use clustering algorithms (such as CLUSTAL) tend to align closely related sequences with one another before aligning sequences that are more distantly related. In the tree of life (Fig. 7.24), this would include aligning all *Archaea* together before aligning the *Archaea* to the *Bacteria* or to eukaryotes. Aligning the *Archaea* to one another first tends to decrease branch length between the two groups of *Archaea* because their similarity scores are maximized. Remember from Unit 6 that the maximum-alignment score may "hide" specific evolutionary changes and make sequences appear more similar to one another, which decreases the pairwise distance between those short-branch sequences. This trend would be expected to create a long-branch artifact by making the short-branch groups cluster together. One can imagine that a progressive clustering scheme could easily group all short-branch sequences together before attempting to group the shorter-branch sequences with the long-branch sequences. By shortening the distances between short-

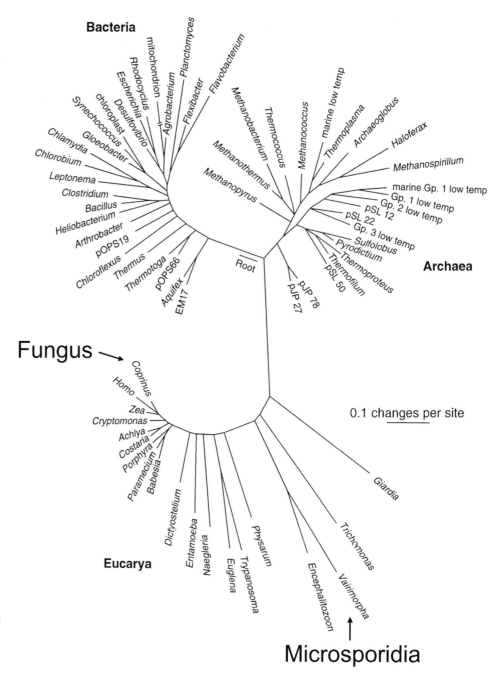

Bacteria

Archaea

Root

0.1 changes per site

Fungus →

Eucarya

Microsporidia

FIGURE 7.24 The three-domain model for the tree of life as calculated by Norman Pace. In this tree, the fast-evolving microsporidia (e.g., *Vairimorpha*, a genus of microsporidium parasites) are placed at the base of the eukaryotes, far away from the other fungi (e.g., *Coprinus*, a genus of mushrooms). (Reprinted from Pace [1997] with permission.)

FIGURE 7.25 A simplified comparison of the eocyte tree (a) with the three-domain tree (b), illustrating how the latter model can be predicted if a long-branch artifact exists. Illustration by Craig Herbold.

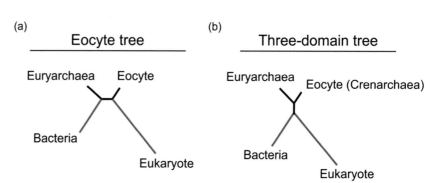

(a)

Eocyte tree

Euryarchaea   Eocyte

Bacteria

Eukaryote

(b)

Three-domain tree

Euryarchaea   Eocyte (Crenarchaea)

Bacteria

Eukaryote

branch taxa, the analysis will be more likely to group short branches together, causing the longer-branch taxa to be placed together. The final sequences or groups of sequences to be aligned will be the two sequences or groups of sequences that possess the longest branches in the phylogenetic tree.

Even in the best-case scenario, phylogenetic methods tend to group long branches together artificially. The worst offender in this regard is most likely maximum parsimony (for a critique, see Lake [1991]); however, no phylogenetic method is completely impervious to this bias. Pairwise distances used for constructing trees are estimated and therefore contain some level of uncertainty. The uncertainty in the distance calculation becomes higher as the distance increases, or, to put it another way, as the evolutionary relatedness decreases. High evolutionary rates exacerbate the problem, and the long branches would tend to be located near some "middle" part of a tree, away from sequences that reliably can be grouped together. This is likely to be part of the problem with the microsporidia example of Fig. 7.24.

KEY TERMS

Bootstrapping A statistical sampling method used by phylogeneticists to estimate the reliability of the internal branches (bipartitions) of a tree.

Mean The average value within a distribution of values.

Multifurcating node A node with more than two emergent branches, or immediate lineages; depicts unresolved relationships among taxa on a phylogenetic tree.

Outliers Observations that are so different from any others in a data set that they may cause misleading conclusions to be

made; values that are so far outside the normal distribution that they probably belong to a different sample. Outliers may arise from erroneous procedures during data sampling.

Variance Deviation from an expected value (e.g., the mean); captures the degree to which a distribution is spread out within a particular sample.

REFERENCES

Felsenstein, J. 2004. *Inferring Phylogenies*. Sinauer Associates, Inc., Sunderland, MA.

Germot, A., and H. Philippe. 1999. Critical analysis of eukaryotic phylogeny: a case study based on the HSP70 family. *J. Eukaryot. Microbiol.* **46:**116–124.

Gill, E. E., and N. M. Fast. 2006. Assessing the microsporidia-fungi relationship: combined phylogenetic analysis of eight genes. *Gene* **375:**103–109.

Hall, B. G. 2008. *Phylogenetic Trees Made Easy: a How-To Manual*, 3rd ed. Sinauer Associates, Inc., Sunderland, MA.

Lake, J. A. 1988. Origin of the eukaryotic nucleus determined by rate-invariant analysis of rRNA sequences. *Nature* **331:**184–186.

Lake, J. A. 1991. Tracing origins with molecular sequences: metazoan and eukaryotic beginnings. *Trends Biochem. Sci.* **16:**46–50.

Pace, N. R. 1997. A molecular view of microbial diversity and the biosphere. *Science* **276:**734–740.

SECTION 7.3
METHODS OF EVOLUTIONARY ANALYSIS

Phylogenetic estimation

As discussed in Unit 6, obtaining a reliable multiple sequence alignment is the critical first step in constructing a phylogenetic tree. Choosing a method to calculate a phylogeny is the second step. Regardless of the method selected, an underlying model of evolution is always assumed. For instance, *parsimony methods* are akin to Occam's razor, where the simplest explanation is the most likely explanation. In parsimony, the topology of a tree is determined by counting the minimum number of changes for every possible topology and taking the one topology that would require the fewest changes as the one most likely to represent the true phylogeny. *Distance-based methods*, on the other hand, rely on pairwise distances to build the most likely tree relating a group of taxa. The distances are used to determine the topology of the tree. These distances are calculated using a specified model of evolution, and depending on which model is used, very different distances may be calculated. Even within a given model, the uncertainty on the distance estimate may make it difficult to assess whether a given bipartition exists. *Likelihood methods* combine distances and correlated nucleotide changes to calculate the likelihood of a given model of evolution. The model consists of distances (which are defined by probabilistic evolutionary models) and topology. *Maximum-likelihood methods* find the single model that exhibits the highest likelihood of any of the models. Each model parameter (substitution rate, branch length, topology, etc.) is assigned the single value that maximizes the overall likelihood of the phylogeny. *Bayesian likelihood methods* differ from maximum-likelihood methods by examining a whole range of the model parameter values, not just the maximum-likelihood value, to assess the statistical significance of phylogenetic features such as bipartitions. While likelihood methods are not discussed in detail in this text, students with a specific interest in phylogeny should examine appropriate textbooks (Felsenstein, 2004; Li, 1997) and the primary literature (see Holder and Lewis [2003] for a review and Suchard et al. [2005] for an example of a Bayesian method).

Parsimony. Parsimony is a tree-searching method in which many trees are constructed and then a criterion is applied to the resulting trees that allows for the selection of the "best" tree, or the one that meets the criterion. The best, or most parsimonious, tree is the one in which the fewest nucleotide changes are necessary for two lineages to diverge from a common ancestor. Specifically, penalties are assessed for nucleotide substitutions that occur at positions within the alignment. The topology that is most parsimonious is the one that requires the fewest nucleotide substitutions; thus, an analysis of parsimony requires enumeration of all the possible topologies. The idea is that this criterion minimizes the total amount of evolutionary change that has occurred among the taxa in the analysis. In other words, simpler explanations are favored over more complex explanations.

In Fig. 7.26, all possible five-taxon topologies for a set of five sequences (A, B, C, D, and E) are enumerated. An example of a parsimony calculation is shown using the Fitch algorithm (Fitch, 1971). This topology was rooted on taxon A to make calculations straightforward; however, the placement of the root does not change the parsimony counts. The 14 other topologies are shown in their unrooted form along with the par-

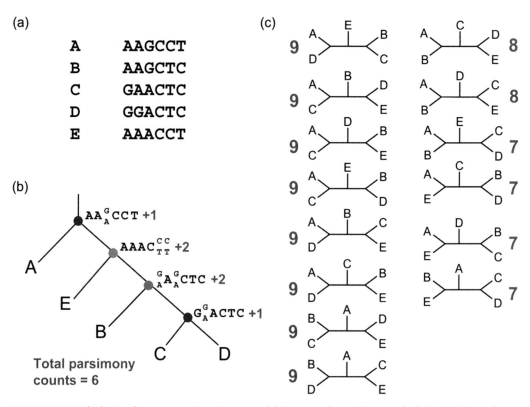

(a)

A	AAGCCT
B	AAGCTC
C	GAACTC
D	GGACTC
E	AAACCT

(b)

AA^G_ACCT +1

AAAC$^{CC}_{TT}$ +2

^G_AA^G_ACTC +2

G^G_AACTC +1

Total parsimony counts = 6

FIGURE 7.26 Calculating the most parsimonious tree. (a) An example parsimony calculation is shown for the most parsimonious tree relating sequences from taxa A, B, C, D, and E. (b) This topology provides a parsimony score of 6. (c) Parsimony scores for all other possible topologies are also calculated and are shown next to their respective topologies. Illustration by Craig Herbold.

simony counts for each topology. The most parsimonious topology has a parsimony score of 6, meaning that if the given sequences are related to one another according to this branching pattern, a minimum of six nucleotide substitutions would have been required. All other topologies had higher parsimony scores. The Fitch algorithm works by inferring the possible sequences at each internal node. For instance, the internal node that is adjacent to taxa C and D, colored blue in Fig. 7.26, would be predicted to have a sequence very similar to both taxa C and D. Since taxa C and D both contain a G at position 1, the blue node is assigned a G at position 1. Since taxa C and D differ at position 2, the internal node is assigned both characters A and G. The consequence of allowing an additional character as a possible character state for a position is that one change is required along one of the branches. Therefore, a penalty of +1 is assessed. Because taxa C and D possess the same character states at positions 3 through 6, the total penalty for joining taxa C and D is 1.

For inferring the sequence found at the green internal node, the sequence inferred for the blue node is examined as well as the sequence for taxon B. At position 1, the blue node has a G character and taxon B has an A character. Therefore, the green node may be a G or an A at position 1. Remember that allowing an additional character at that position requires the addition of a penalty of +1. A similar situation occurs at position 3, except that taxon B has a G character while the blue node has an A character. Just like position 1, the inclusion of both A and G at position 3 requires a penalty. At position 2, taxon B has an A character. Since position 2 at the blue node may have been an A or a G, position 2 at the green node is assigned an A character with no further penalty.

The penalty was assessed when the sequence at the blue node was inferred. Positions 4 though 6 are the same in taxon B and at the blue node. Inference of the sequence found at the green node required two penalties, one for position 1 and one for position 3. This procedure was repeated for the orange node (two additional penalties for positions 5 and 6) and for the purple node (one additional penalty for position 3). Therefore, the total parsimony score for this topology is 6. To ensure that this topology is the most parsimonious (requires the fewest changes), the Fitch algorithm must be repeated with each possible topology. For five taxa, this is relatively straightforward, since there are only 15 possible unrooted topologies. Additional taxa, however, make the analysis much more computationally demanding. The number of possible unrooted topologies is given by the equation:

$$\text{Number of possible trees} = -\frac{(2t - 5)!}{2^{t-3}(t - 3)!}$$

In the case of 20 taxa, there are more than 10^{20} possible unrooted topologies! The large number of possible topologies makes enumerating all trees a computational challenge, which may be relieved by the use of heuristic algorithms that limit the number of possible topologies examined.

Neighbor joining. Distance-based phylogenetic reconstruction, which provides a measurement of the amount of evolutionary change between any two sequences since divergence from a common ancestor, is much less computationally intensive than either parsimony or likelihood-based methods. Distance-based methodologies are typified by the neighbor-joining (NJ) algorithm (Saitou and Nei [1987] with modifications by Studier and Keppler [1988]), which is one of the most widely used methods for building phylogenetic trees. Rather than enumerating all possible trees and evaluating each one by some metric, neighbor joining is used to construct a single tree that best summarizes the relationships among taxa.

The first step in building a neighbor-joining tree is to construct a distance matrix. A distance matrix is a way of tabulating all pairwise distances between all possible taxon pairs. Using the sequences from Fig. 7.26 and a simplified distance metric (e.g., the number of differences between two sequences), the distance matrix in Fig. 7.27 was constructed by using MEGA4 (Tamura et al., 2007). This distance matrix summarizes all pairwise relationships between the sequences. For example, sequences A and B differ at positions 5 and 6. Since the distance metric used in this example is simply the number of differences and all base changes are considered equal (i.e., no particular substitutions are given larger weight than others), the distance between taxa A and B is 2 whereas the distance between taxa A and D is 5. The matrix can be modified to assess gap penalties in an alignment, giving more weight to insertions or deletions than to nucleotide substitutions. It also is possible to apply corrections that account for multiple changes at a single site resulting in homoplasy, or the possession of identical character states acquired through either convergent evolution or reversal to the ancestral character state. In an effort to keep the explanation simple, these types of substitutions will be ignored for now. However, it also is important to keep in mind that these sorts of complications are common when assessing relationships among distantly related taxa or for sites that evolve relatively rapidly.

(a)

A	AAGCCT
B	AAGCTC
C	GAACTC
D	GGACTC
E	AAACCT

(b)

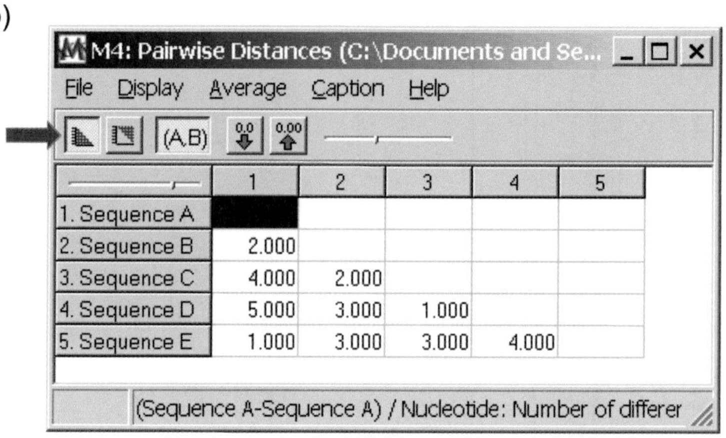

FIGURE 7.27 Pairwise distance matrix using edit distance. (a) Five short DNA sequences (A to E) are shown. (b) Using MEGA4, a pairwise distance matrix can be generated showing the edit distance between all five sequences (Tamura et al., 2007). For clarity, the upper half of the distance matrix has been omitted; only the lower half is shown. In MEGA4, one can toggle between the two halves of the matrix by pressing the buttons on the tool bar below the File menu (red arrow).

M4: Pairwise Distances (C:\Documents and Se...

File   Display   Average   Caption   Help

	1	2	3	4	5
1. Sequence A					
2. Sequence B	2.000				
3. Sequence C	4.000	2.000			
4. Sequence D	5.000	3.000	1.000		
5. Sequence E	1.000	3.000	3.000	4.000	

(Sequence A-Sequence A) / Nucleotide: Number of differer

The distance matrix is then used to build a phylogeny. The calculation is not very complex but does require several steps. The first step is to calculate a summarized distance from each sequence to all other sequences. This value is designated d_i, which is the summarized distance for taxon i and can be calculated using the following equation:

$$\text{Summarized distance } (d_i) = \sum_{j:j\neq I}^{n} D_{ij}/(n-2)$$

where n is the number of sequences (taxa) and D_{ij} is derived from the pairwise edit distance between taxa i and j. For example, in Fig. 7.27 the summarized distance for sequence A (d_A) is calculated as $(2 + 4 + 5 + 1)/(n-2)$, where $n = 5$. Therefore, $d_A = 12/3 = 4$. Taking sequence C as another example, we can calculate $d_C = (4 + 2 + 1 + 3)/3 = 3.33$. The summarized distance for each sequence (d_i) in the matrix is shown in Fig. 7.28.

The summarized distances are used to calculate a pairwise measurement that facilitates clustering among taxa. These clustering values usually, but not always, cluster taxa that contain the fewest differences between them. Once clustered, these taxa are considered neighbors. For each taxon pair, a cluster value is calculated using the following equation:

$$\text{Cluster value} = D_{ij} - d_i - d_j$$

where D_{ij} is the pairwise distance between taxa i and j, d_i is the summarized distance for taxon i, and d_j is the summarized distance for taxon j. The pair that has the smallest clustering value will be the first neighbors joined, or consolidated, by the neighbor-joining method. Clustering methods such as neighbor joining usually link the least distant pairs of taxa, followed by successively more distant pairs of taxa, until one final unrooted tree is produced. As discussed in Section 7.2, long-branch artifacts would be caused by consistently clustering the most closely related taxa. This problem is avoided, in principle,

$$\dfrac{D_{ij} - d_i - d_j}{}$$

AB	-5.33
AC	-3.33
AD	-3.33
AE	-6.67
BC	-4.67
BD	-4.67
BE	-4.00
CD	-6.67
CE	-4.00
DE	-4.00

	d_i
Sequence A	4.00
Sequence B	3.33
Sequence C	3.33
Sequence D	4.33
Sequence E	3.67

FIGURE 7.28 Using summarized distance (d_i) to calculate the cluster value from pairwise distance information. Summarized distances (d_i), which are derived from the pairwise edit distance values in the matrix in Fig. 7.27, are given for each sequence (i). These numbers are used to calculate a value that will be used to cluster two sequences. This clustering value is calculated as the pairwise distance (D_{ij}) between two sequences (i and j) minus the two summarized distances for each of the sequences (d_i and d_j). For example, the clustering value between sequences B and D is $D_{BD} - d_B - d_D = 3 - 3.33 - 4.33 = -4.66$. Note, however, that some of the summarized distances have a repeating decimal. Rather than display the repeated numerals (e.g., $10/3 = 3.3333333...$), the numbers were rounded to three significant figures. The cluster values in the figure (e.g., -4.67 instead of -4.66 for B and D) were calculated with rounded summarized distance values. This particular distance matrix establishes a node that connects taxon A and taxon E based on a minimum cluster value ($D_{AE} = -6.67$). Note that D_{CD} has an identical cluster value to D_{AE}, but to simplify the explanation, the discussion focuses on D_{AE}.

by using the calculated clustering values, rather than pairwise distances, to determine neighbors. In practice, no method consistently avoids long-branch artifacts.

The clustering values for the sequences in the example are listed in Fig. 7.28. The pairwise distance between taxa A and B was 2. Since the summarized distance (d_i) for taxon A is 4.00 and the summarized distance for taxon B is 3.33, the clustering value is $(2 - 4.00 - 3.33) = -5.33$. As seen in Fig. 7.28, two pairs produce -6.67 as a clustering value, taxon pair A and E and taxon pair C and D. Either choice is acceptable as the first neighbors, so for this example, taxa A and E will be joined first.

Once a neighbor pair is identified, the terminal branch lengths leading to each member of the pair must be calculated. The terminal branch length (L_i) for a particular taxon (i) is calculated using the following equation:

$$L_i = [0.5 \times (D_{ij} + d_i - d_j)]$$

For taxon A, the terminal branch length (L_A) is calculated as $[0.5 \times (D_{AE} + d_A - d_E)]$ $= [0.5 \times (1 + 4.00 - 3.67)] = 0.67$, while the terminal branch length for taxon E (L_E) is calculated as $[0.5 \times (D_{AE} + d_E - d_A)] = [0.5 \times (1 + 3.67 - 4)] = 0.33$. Notice that the pairwise distance is maintained. That is, the distance between sequences A and E is still 1 ($0.67 + 0.33 = 1$). Now that taxa A and E have been joined, there exists an internal node that connects these taxa to the rest of the tree.

The next step is to find the distances from each of the remaining taxa to the internal node adjacent to taxa A and E. These distances are listed in Fig. 7.29. The distance from taxon B to the AE internal node (AE_{int}) is 2. It was calculated by the following formula:

$$D_{(AEnode)B} = [0.5 \times (D_{AB} + D_{EB} - D_{AE})]$$

(a)

	A	B	C	D	E	AE$_{int}$
Sequence A		2	4	5	1	
Sequence B	1		2	3	3	2
Sequence C	4	2		1	3	3
Sequence D	5	3	1		4	4
Sequence E	1	3	3	4		
AE internal node		2	3	4		

(b)

	B	C	D	AE$_{int}$
Sequence B		2	3	2
Sequence C	2		3	3
Sequence D	3	3		4
AE internal node	2	3	4	

FIGURE 7.29 Forming internal nodes for pairwise distance calculations. (a) Because the clustering value between taxa A and E is the lowest (-6.67), they are joined as neighbors first (denoted as AE internal node, or AE$_{int}$). The distance matrix is adjusted for the taxon cluster, treating AE$_{int}$ as a taxon. The pairwise distance between taxa A and E, as well as the summarized distances for both taxa A and E, is used to determine the distance from each of the remaining taxa (B, C, and D) to the internal node that connects taxa A and E (AE$_{int}$). (b) These values are used to construct a new distance matrix, where a new row and column corresponding to the AE internal node replaces the rows corresponding to sequences A and E. Note that both the upper and lower halves of the matrices are shown in both panels. Recall that the upper half was omitted for clarity in Fig. 7.27, but it is shown here in grey to distinguish it from the values calculated for the lower half of the matrix.

As shown in Fig. 7.29a, a new column and row are added to the original distance matrix, which contains the distances from each taxon to the AE internal node (AE$_{int}$). Then the columns and rows containing taxa A and E can be stripped from the distance matrix, resulting in the new matrix shown in Fig. 7.29b. The entire procedure, which involves calculating the summarized distance and cluster values for the sequences in the new matrix, is repeated. Once all taxa have been grouped using this clustering method, an unrooted phylogenetic tree is produced. For this example, the resulting tree is shown in Fig. 7.30. Notice that the pairwise distances from Fig. 7.27 (e.g., $D_{AD} = 5$) are not necessarily recovered but are similar [$(0.67 + 1.5 + 1.5 + 1) = 4.67$ in Fig. 7.30].

Neighbor joining uses distance metrics to determine the topology of the tree; thus, it is very important to use a realistic distance. In this example, edit distance was used as a distance metric; however, this is usually not the most realistic measure of distance. Instead, distances that are calculated using probabilistic models of evolution are much better for estimating true phylogenetic distance. Two such models for nucleotide substitutions, the Jukes-Cantor model and the Kimura two-parameter model, are discussed later in this section.

Distance calculation

In the previous section, edit distance was used as a distance metric. Edit distance (Jones and Pevzner, 2004) is simply the number of changes that would be required to transform one sequence into the other. The number of differences between two aligned sequences represents the total amount of change that the sequences have undergone in their evolution. Many phylogenetic methods use a distance measure which asserts that the evo-

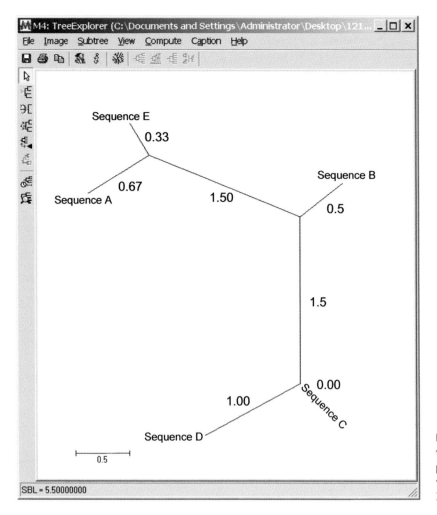

FIGURE 7.30 The unrooted neighbor-joining tree calculated using edit distances as the pairwise metric. This tree was calculated and visualized by using MEGA4 (Tamura et al., 2007).

lutionary distance (D) is the substitution rate (γ) multiplied by time (t), or $D = \gamma t$. If the substitution rate is expressed in units of substitutions per position per unit time, then the effect of multiplying by time is to calculate the substitutions per position. Thus, the units of evolutionary distance for nucleotides are expressed as substitutions/position.

Phylogenetic distance shares many properties with spatial distance. Although the exact path from one point to another, or the exact length of time spent following the path, may not be known, the overall distance "the way a crow flies" can be measured. Despite the intuitive feeling, what is being measured with phylogenetic distance is not really time. A high substitution rate and a long duration both would have the effect of increasing phylogenetic distances. Since distance determines branch length, long branches may indicate an ancient lineage or a more recently diverged but very quickly evolving lineage.

Edit distance is ultimately the simplest metric, but it is generally undesirable for calculating distance. First, it assumes that any nucleotide in the sequence is subject to change with equal probability, without regard to the history of that nucleotide. This means that the history of nucleotide changes does not matter. Each nucleotide may change to another nucleotide with the same probability as its neighbor regardless of its history. In other words, the model of evolution has no memory. The current sequence state is all that affects the probability of change.

Imagine, for a moment, a sequence that has 100 nucleotides. Now imagine that, over time, positions change according to a random, probabilistic process in which any nucleotide has the same probability of change as any other. If five positions change every million years, how many observable nucleotide changes would be expected after 20 million years? The answer is not 100 (i.e., 20 × 5). It is actually around 64. Because the changes are governed by a probabilistic process, the fact that a nucleotide changes in the first million years does not affect its probability of subsequent change in the next million years. In Fig. 7.31, this principle is expressed graphically. The expected number of original nucleotides is shown to decrease in a nonlinear manner. The equation that predicts such behavior is:

$$N_t = N_0 e^{-\gamma t/N_0}$$

where N_t is the expected number of original nucleotides remaining after a duration of time (t), N_0 is the number of original nucleotides, t is time, and γ is the mutation rate expressed in number of changes expected per unit time.

Figure 7.31 shows that the number of nucleotide positions in an alignment retaining their original nucleotide identity decreases nonlinearly. It also implies that several nucleotide positions have changed more than once. For this example, we would expect 18 nucleotide positions to have changed twice in 20 million years. Furthermore, six have changed three times and two have even changed four times. Twenty-six nucleotide positions, a quarter of the total of 100, have changed multiple times.

When a nucleotide substitution occurs, the original nucleotide can be replaced by any one of the other three nucleotides. For now, assume that there is no preference for any given nucleotide to change to one or another nucleotide. Making these assumptions, we expect about 9 of the 26 nucleotides that have changed multiple times to have changed back to their original identity (homoplasy). This means that although 64 positions may have changed at least once, only 55 differences are observable (since 9 changed back to the original identity).

Figure 7.32 shows how phylogenetic distance can be underestimated if these probabilistic effects are not properly taken into account. When the timescale is very short, the number of observed differences and actual differences is very close. Over time, the probabilistic nucleotide substitution process has the effect of hiding some of the changes. Therefore, two nucleotide sequences that are separated by 20 million years of evolution at a rate of 5 nucleotide changes per million years would be expected to have approximately 55 observable nucleotide substitutions. This is a very different number from the 100 nucleotide substitutions that have actually occurred (true distance = 1 substitution/

FIGURE 7.31 The number of original nucleotides, which have not changed to any other nucleotide, decreases with time in a nonlinear fashion. Although after 20 million years there have been, on average, 100 nucleotide changes, some sites have changed more than once while others have not changed at all.

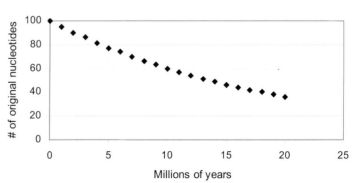

The number of original nucleotides remaining when $\gamma = 5$ per million years

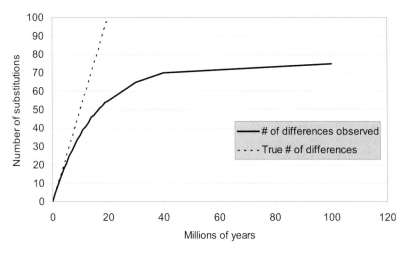

Comparison of the true number of substitutions
to the number of observable substitutions

FIGURE 7.32 The number of observable nucleotide substitutions is approximately the same as the number of true substitutions between sequences that are nearly identical. As the number of substitutions accumulates, the number of observable substitutions becomes quite different from the actual number of substitutions. As the number of substitutions continues to accumulate, the number of observable substitutions reaches a maximum of about 75% of the total number of nucleotides. At this point no recoverable information is contained within the sequence data. It was assumed for this graph that a nucleotide will change to any other nucleotide with the same frequency.

position [sub/pos]). Using the number of nucleotide differences with no correction would result in an estimation of the phylogenetic distance to be only about half of the true value (estimated distance = 0.55 sub/pos; true distance = 1.0 sub/pos). After 40 million years, it is expected that only about 70 out of 200 nucleotide substitutions will be observable. Consequently, estimated phylogenetic distance would be only about a third of the true value (estimated distance = 0.70 sub/pos; true distance = 2.0 sub/pos). Luckily, the same probabilistic equations which predict that phylogenetic distance is not linearly proportional to the number of differences observed between two sequences can be used to more accurately estimate phylogenetic distance. This is accomplished by specifying an evolutionary model.

Models of molecular evolution

Several evolutionary models exist that can be used to estimate phylogenetic distance. In an evolutionary model, a set of probabilities is assigned for the change of each nucleotide into each of the other nucleotides. This concept is best illustrated with a transition matrix. The transition matrix is a 4 × 4 matrix where each row and each column represent a nucleotide type (A, G, C, and T). It may be interpreted that the nucleotide representing row X changes to the nucleotide representing column Y with the probability found at the coordinates (X, Y). In Fig. 7.33, the transition matrix for the Jukes-Cantor model of nucleotide substitution is shown (Jukes and Cantor, 1969). For the Jukes-Cantor model, the probability of any nucleotide change is represented by γ and the probability that there is no change is $(1 - 3\gamma)$. Therefore, the probability of the A-to-G transition is the same as the probability of the A-to-C, A-to-T, or even C-to-T tran-

$$
\begin{array}{c@{\quad}c@{\quad}c@{\quad}c@{\quad}c}
 & \mathbf{A} & \mathbf{G} & \mathbf{C} & \mathbf{T} \\
\mathbf{A} & 1-3\gamma & \gamma & \gamma & \gamma \\
\mathbf{G} & \gamma & 1-3\gamma & \gamma & \gamma \\
\mathbf{C} & \gamma & \gamma & 1-3\gamma & \gamma \\
\mathbf{T} & \gamma & \gamma & \gamma & 1-3\gamma
\end{array}
$$

FIGURE 7.33 Transition matrix for the Jukes-Cantor model of nucleotide substitution.

sition. Since each row must add up to 1, the probability that there is no change at all is simply $1 - 3\gamma$. That is, the probability of changing plus the probability of not changing equals 1.

The simplicity of the Jukes-Cantor model allows one to estimate phylogenetic distance. Details on deriving this equation are omitted here (but may be found in Felsenstein [2004]). The Jukes-Cantor calculation is remarkably straightforward:

$$D = \gamma t = -\tfrac{3}{4}\ln\left[1 - (4 \times p)/3\right]$$

where p is the proportion of nucleotide differences between two sequences. Thus, if $p = 0$, then $D = 0$, indicating that two sequences are identical. On the other hand, if 20% of the nucleotides differ between two sequences, the Jukes-Cantor distance (D) is 0.23. The true model parameter (γ) cannot be calculated since γ and t (time) are intrinsically linked and one cannot be calculated without assuming a value for the other. This should not be worrisome, though, since phylogenetic distance calculations require the product of the two, which is readily calculated by this equation.

A few interesting properties emerge from the Jukes-Cantor model. Imagine a hypothetical sequence made completely of A characters, and allow it to evolve under the Jukes-Cantor model. After an infinite amount of time, there would be an equal number of A, G, C, and T characters, and all would have frequencies of approximately 0.25. This is known as a stationary distribution in molecular evolution terms. The stationary distribution is usually denoted by π, representing a vector that contains a π value for each nucleotide. Thus for the Jukes-Cantor model, the stationary distribution is

$$\pi = \{\pi_A, \pi_G, \pi_C, \pi_T\} = \{0.25, 0.25, 0.25, 0.25\}$$

Any sequence following the Jukes-Cantor model of nucleotide substitution would be expected to approach this stationary distribution. This stationary distribution also implies that any two sequences will be identical at 25% of the alignment positions even if they are completely unrelated to one another.

The Jukes-Cantor model is the simplest model that may be used in phylogenetic analysis, but several biological realities prevent it from being universally applicable. For example, for some sequences the probability of a transition may be higher than the probability of a transversion. To allow for a difference in probability between transitions and transversions, the model must be relaxed so that each nucleotide does not change to all others with equal probability, but instead is more likely to undergo transitions than transversions.

In the Jukes-Cantor model, there is a single parameter, γ, which is used to denote the probability of any nucleotide substitution. This parameter refers to the probability of all substitutions, regardless of whether the substitution can be classified as a transition or a transversion. To use a model in which transitions and transversions occur with differing probabilities, two parameters are needed. If the probability of an A character undergoing a transition to a G character is denoted α, the probability of a transversion to a C character is denoted by β, and the probability of a transversion to a T character also is denoted by β, then the relationship of γ to α and β is

$$3\gamma = \alpha + 2\beta$$

Because two parameters are required to describe this model, it is known as the Kimura two-parameter model (Kimura, 1980). The transition matrix for the Kimura two-parameter model is shown in Fig. 7.34. Any transition is just as likely as another (α), and any transversion is as likely as another (γ). The self-to-self probabilities have been modified accordingly so that each row adds up to 1.

Under the Kimura two-parameter model, observable nucleotide differences are classified as either transitions or transversions and the associated probabilities of a transition (α) or a single transversion (β) are taken as the two parameters. Sometimes these two parameters are expressed as mutation rate (μ) and transition-to-transversion ratio (κ) as follows:

$$\mu = \alpha + 2\beta$$
$$\kappa = \alpha/\beta$$

Any evolutionary model makes several assumptions; the simpler the model, the more assumptions are made. For instance, the Jukes-Cantor model assumes several aspects of the evolutionary model, only one of which is that transitions are as likely as transversions. The Kimura two-parameter model relaxes this assumption but, like the Jukes-Cantor model, assumes that, given enough time, the stationary distribution (π) will be achieved. Models become more complex as fewer assumptions are made. Therefore, a simpler model is always a special case of a slightly more complex model.

A thorough discussion of more extensive models is beyond the scope of this text, but students are encouraged to think about what assumptions they are making when choosing an evolutionary model and to be able to justify the model selected for their analysis. Ask the question, "Are the assumptions inherent to the model being used appropriate for the data being examined?" The decision about which model should be used is, ultimately, the choice of the phylogeneticist, but it is sound advice to use the simplest model in which the assumptions are justified. For further information on this topic, students are encouraged, once again, to consult appropriate textbooks (Felsenstein [2004], for example).

There exist assumptions that are inherent in all models, from the most simple to the most complex. For instance, all models assume that alignment positions evolve independent of one another. This assumption is necessary to make a solution mathematically tractable. There are methodological frameworks that allow one to relax this assumption somewhat. One way to relax the assumption of independence, for instance, is to analyze data in a site-specific fashion, where the evolutionary model required to analyze a particular alignment position depends on that alignment position. This allows a model to be constructed where the presence of a nucleotide at one particular position may influence the probability of a nucleotide at another position. These methods require a great deal of data to develop and are not regularly used for phylogenetic inference of rRNA

$$
\begin{array}{c c c c c}
 & \mathbf{A} & \mathbf{G} & \mathbf{C} & \mathbf{T} \\
\mathbf{A} & 1-\alpha-2\beta & \alpha & \beta & \beta \\
\mathbf{G} & \alpha & 1-\alpha-2\beta & \beta & \beta \\
\mathbf{C} & \beta & \beta & 1-\alpha-2\beta & \alpha \\
\mathbf{T} & \beta & \beta & \alpha & 1-\alpha-2\beta
\end{array}
$$

FIGURE 7.34 Transition matrix for the Kimura two-parameter model of nucleotide substitution.

genes because, ultimately, they require far more assumptions than are required for the assumption of independence.

Another universal assumption is that of reversibility. The reversibility assumption says that the probability of changing from an A to a G is the same as the probability of changing from a G to an A. While the specific parameters that produce this probability may be adjusted for compositional bias in complex evolutionary models, the probabilities contained within the transition matrix will match one another. This is a fundamental assumption that is required for the mathematics behind the evolutionary model to work out properly. It also is the reason that phylogenetic modeling cannot be used to root a given phylogenetic tree.

A third major assumption made in using evolutionary models for phylogenetic inference is the assumption that all alignment positions evolve according to a single, identical model, which includes the evolutionary rate. When this assumption is violated, long-branch artifacts may arise. Luckily a strategy has been developed to address this assumption. The nucleotide patterns found in the alignment may be classified into rate categories according to how much change is observed between sequences. For instance, if a protein-coding nucleotide sequence was examined, the third codon position (the "wobble" position) would appear to evolve at a very high rate compared to the first and second positions. Each alignment position may be classified into a rate category according to how different or similar all sequences are at that position. Then the patterns observed at each alignment position may be analyzed separately, and positions with lower rates of nucleotide substitution may be weighted as being more likely to be true than those with higher rates of nucleotide substitution. It also is possible to incorporate a rate category for invariant sites, referring to a set of nucleotides that do not ever change. By allowing a class of invariant sites, the rate variation among the remainder of the alignment positions changes as well.

None of the aforementioned models deal with gaps in the alignment. There is really no consensus on how frequent insertions or deletions are or whether these nucleotide changes can be modeled reliably as an evolutionary process. To address this issue, most researchers ignore any column that contains gaps. When pairwise distances are calculated using MEGA4 (Tamura et al., 2007), there is a choice of whether to include positions containing gaps. Two steps are required before a distance may be calculated: (i) choose two taxa and (ii) strip columns containing gaps. MEGA4 allows the choice of which order to perform these steps. Complete deletion means that all gap-containing columns are stripped from the alignment before any taxa are chosen. This choice ensures that only sites containing nucleotides in all sequences are analyzed and all are treated equally. There is an additional option, referred to as pairwise deletion, in which two sequences are selected first and then gap-containing columns are removed. With this procedure, only the columns that contain gaps in those two sequences are ignored for the pairwise distance calculation. The choice of the method to be used depends on the specific analysis and whether too many data would be lost in the complete-deletion option. To be thorough, phylogenies should be calculated by both methods. If the calculated phylogenies are drastically different, the alignment should be reexamined, the cause behind the discrepancy should be determined, and the alignment should be adjusted as necessary.

KEY TERMS

Convergent evolution The acquisition of the same biological trait or molecular character (e.g., DNA or protein sequence) though independent lineages, which are those without a common ancestral origin.

Distance matrix A method by which all pairwise distances between all sequence pairs can be tabulated for use in the construction of a neighbor-joining phylogenetic tree.

Edit distance The number of nucleotide changes that distinguish any two sequences being compared in a pairwise alignment (e.g., the number of mismatches and gaps).

Heuristic A strategy applied to rapidly obtain a solution to a problem by using informal rules or principles (e.g., an educated guess).

Homoplasy Similarity between two sequences that has evolved independently and is not indicative of a common ancestral origin; as an example, two independent mutations in a DNA sequence at a site resulting in the reversal of a substitution to the original state:

Sequence 1 TCG**C**TA T A G C (3 nucleotide changes)

Sequence 2 CTC**C**TG T ⟶ C (1 nucleotide change)

Invariant sites Nucleotide positions in an alignment that never change.

Transition A nucleotide change (point mutation) in which a purine (A or G) is replaced by a different purine (G or A) or a pyrimidine (C or T) is replaced by a different pyrimidine (T or C).

Transition matrix A matrix in which, for nucleotide sequences, there are four rows and four columns, each representing one of the four nucleotides (A, G, C, or T), with each coordinate depicting a probability equation based on a particular model of nucleotide substitution. The sum of the equations in each row and each column must be equal to one.

Transversion A nucleotide change (point mutation) in which a purine (A or G) is replaced by a pyrimidine (C or T) or a pyrimidine (C or T) is replaced by a purine (A or G).

REFERENCES

Felsenstein, J. 2004. *Inferring Phylogenies*. Sinauer Associates, Inc., Sunderland, MA.

Fitch, W. 1971. Toward defining the course of evolution: minimum change of specified tree topology. *Syst. Zool.* **20**:406–416.

Hall, B. G. 2008. *Phylogenetic Trees Made Easy: a How-To Manual*, 3rd ed. Sinauer Associates, Inc., Sunderland, MA.

Holder, M., and P. O. Lewis. 2003. Phylogeny estimation: traditional and Bayesian approaches. *Nat. Rev. Genet.* **4**:275–284.

Jones, N. C., and P. A. Pevzner. 2004. *An Introduction to Bioinformatics Algorithms*. MIT Press, Cambridge, MA.

Jukes, T. H., and C. R. Cantor. 1969. Evolution of protein molecules, p. 21–132. *In* H. N. Munro (ed.), *Mammalian Protein Metabolism*. Academic Press, Inc., New York, NY.

Kimura, M. 1980. A simple method for estimating evolutionary rate in a finite population due to mutational production of neutral and nearly neutral substitution through comparative studies of nucleotide sequences. *J. Mol. Biol.* **16**:111–120.

Li, W. 1997. *Molecular Evolution*. Sinauer Associates, Inc., Sunderland, MA.

Saitou, N., and M. Nei. 1987. The neighbor-joining method: a new method for reconstructing phylogenetic trees. *Mol. Biol. Evol.* **4**:406–425.

Studier, J. A., and K. J. Keppler. 1988. A note on the neighbor-joining algorithm of Saitou and Nei. *Mol. Biol. Evol.* **5**:729–731.

Suchard, M. A., R. E. Weiss, and J. S. Sinsheimer. 2005. Models for estimating Bayes factors with applications to phylogeny and tests of monophyly. *Biometrics* **61**:665–673.

Tamura, K., J. Dudley, M. Nei, and S. Kumar. 2007. MEGA4: Molecular Evolutionary Genetics Analysis (MEGA) software version 4.0. *Mol. Biol. Evol.* **24**:1596–1599.

READING ASSESSMENT (SECTIONS 7.1 TO 7.3)

1. Which is the optimal DNA multiple alignment and why? What alignment parameter(s) could be changed to convert alignment i, below, into alignment ii?

   Choice i: `AT-C-GG`
   `AT-C-GG`
   `ATTCTGG`

   Choice ii: `ATC--GG`
   `ATC--GG`
   `ATTCTGG`

2. In the text, a four-taxon test was provided for determining whether a sequence belongs to a taxonomic group (Fig. 7.14). What is the effect on the "ancestral node" if the two acidobacterial sequences were from two recently divergent species (related through a recent node)? How will this manifest itself in the interpretation of whether to classify new sequences as *Acidobacteria*?

3. If you have a sequence that is 1,500 nucleotides long and it is evolving at a rate of 1 change per 1 million years, how many original nucleotides are expected to remain after 1 billion years (i.e., how many positions never changed?) See the equation under "Distance calculation" above.

4. Two rooted phylogenies are presented here that differ only in the placement of the root. For each phylogenetic tree, how many clades are there? Specify which taxa are in each clade. Use a bootstrap cutoff value of 75. Are there any nested clades?

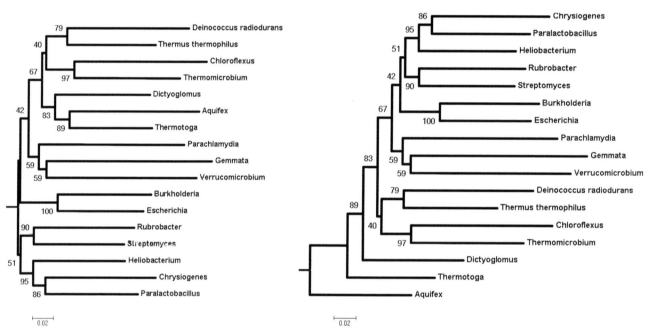

5. The following tree appears to exhibit an artifact of long-branch attraction (LBA). Which taxa may be artificially placed together as a result of LBA? What are two possible explanations for the appearance of this artifact?

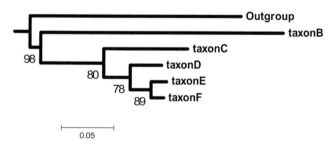

6. The figure below shows the phylogeny estimated for a sample of flowering plants (angiosperms) from *PHYTOCHROME A* and *PHYTOCHROME C,* a pair of genes that duplicated prior to the origin of the angiosperms. The figure is reprinted from S. Mathews and M. J. Donoghue, *Science* 286:947–950, 1999, with permission.

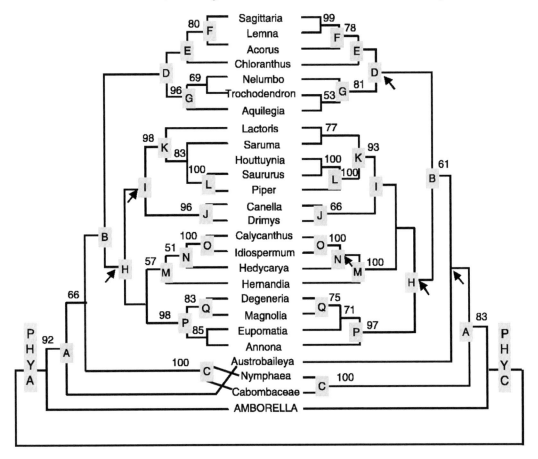

Which of the following sets of taxa constitute a clade on one gene tree, but not on the other tree, using a bootstrap value of 70 or higher?

a. *Degeneria-Magnolia-Eupomatia*

b. All angiosperms except *Amborella*

c. *Austrobaileya-Nymphaea-Cabombaceae*

d. *Nelumbo-Trochodendron-Aquilegia*

7. Part I: Given the following six nucleotide sequences, calculate a distance matrix using Edit Distance, and then calculate a second distance matrix using Jukes-Cantor distances. Fill in the appropriate table below accordingly.

Sequence 1 AGTCAAGGTTTCCTGAGCCG

Sequence 2 GGTGAAGGCATCACAAGCGG

Sequence 3 ACCCAAGATATCCTAAGCCG

Sequence 4 AGCCAAGCAATCCTGAGCCG

Sequence 5 CGACAAGCAATCCTAAGAGG

Sequence 6 GGACAGGCCATCCTAAGCCG

Edit Distance	1	2	3	4	5	6
Sequence 1						
Sequence 2						
Sequence 3						
Sequence 4						
Sequence 5						
Sequence 6						

Jukes-Cantor	1	2	3	4	5	6
Sequence 1						
Sequence 2						
Sequence 3						
Sequence 4						
Sequence 5						
Sequence 6						

Part II: Below are two neighbor-joining trees, one calculated from the correct Edit Distance matrix for the six sequences given in Part I and the other for the correct Jukes-Cantor distance matrix calculated for the same six sequences. Are the two trees identical? Describe similarities and differences between the two tree topologies.

Neighbor-joining tree for "Edit Distance" matrix:

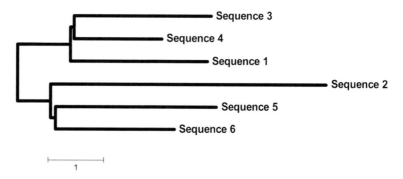

Neighbor-joining tree for "Jukes-Cantor" distance matrix:

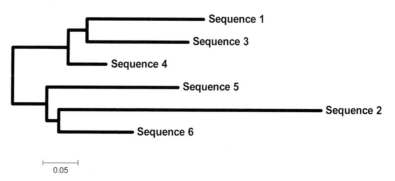

SECTION 7.4
RECONCILING BACTERIAL TAXONOMY: AN EVOLVING LEXICON FOR THE 21ST-CENTURY MICROBIOLOGIST

The astounding diversity of microorganisms that abound in the biosphere can be attributed to their metabolic flexibility and genetic promiscuity (e.g., horizontal gene transfer [HGT]). The former creates a prokaryotic population with dynamic phenotypic properties, which enable them to colonize environments as commonplace as the kitchen counter or as extreme as the icy surface of a glacier. The latter is indifferent to the cohesive barriers that traditionally impede genetic exchange among crown eukaryotic species (i.e., plants and animals), thus masking the ancestral origin of many bacterial or archaeal lineages. Attempts to bring systematic order to the immense diversity of the prokaryotic world by using classification systems that rely on phenotypic and genetic properties have left taxonomists with a system for categorizing *Bacteria* and *Archaea* that is functional at broader taxonomic levels (phylum, order, class, and family) but inadequate beyond the genus level. In this section, the classical taxonomic approaches to organizing the realm of prokaryotes are discussed within a historical framework, noting the impact of more recent technological advances in microbial ecology and genomics.

History of the classification of organisms

The use of a methodical approach to determine the relationships among organisms is the core of any classification system. The basis of taxonomy is simply to assign scientific names to organisms on the basis of established conventions or rules (i.e., nomenclature). Carolus Linnaeus, a Swedish botanist and physician in the 1700s (Fig. 7.35), is credited with development of a hierarchical system to classify organisms based on observable physical characteristics, starting with kingdoms, which were divided into classes, which

FIGURE 7.35 Carolus Linnaeus (1701–1778). Etching based on a portrait by Martin Hoffman (1737). Source: http://en.wikipedia.org/wiki/File:Carl_Linnaeus_dressed_as_a_Laplander.jpg.

in turn were divided into orders and then into genera and species. In this system, organisms were grouped according to similarities that do not necessarily mirror evolutionary history, thus representing a practical yet unnatural classification scheme. His concept of the living world, in which organisms can be categorized into a ranking system, has resonated throughout the biological sciences, with progression to a revised structure in which modern taxonomic ranks reflect the evolutionary principles of Charles Darwin (Fig. 7.36). The idea now is that a natural classification scheme should be viewed as a hypothesis of evolutionary relationships. Although phylogenetic taxonomy provides a framework upon which life can be classified (refer to the Tree of Life Web Project [Maddison and Schulz, 2007] under Web Resources), by definition the relationships among organisms are not bound by static rules and an absolute hierarchy. Instead, evolutionary relationships, and consequently perhaps taxonomic rank, are subject to change if, with the advent of new data, technology, or methods of analysis, such relationships are not supported (i.e., the hypothesis is refuted).

Phylogenetic taxa are defined as monophyletic groups (Hennig, 1965). For the ideologically motivated systematist, taxonomic assignments not supported by monophyly are not valid under the modern classification system. Nevertheless, paraphyletic groups such as prokaryotes (as well as fish and dicots, among others), although not recognized as phylogenetic taxa, still may be acknowledged as a useful classification scheme within this system (Freeman and Herron, 2004).

Within the context of a phylogenetic tree, taxonomic ranks often are nested, and some comprise sister groups with genotypes and phenotypes indicative of convergent evolution. For example, if one separates the *Bacteria* into their respective phyla (there are thought to be at least 100 phyla, 30 of which have at least one species that is cultivatable [Schloss and Handelsman, 2004]) and creates a phylogenetic arrangement of the phyla based on analysis of SSU rRNA, one might obtain a tree similar to that depicted in Fig. 7.37, which contains a number of nested clades along multiple branches in the tree. Interestingly, if one compiles a list describing the defining characteristics of each bacterial phylum, one finds that phylogeny and physiology are not necessarily linked. For instance, several phylogenetic groups are known to fix nitrogen, and these tend to be scattered randomly throughout the *Bacteria* (note the asterisks in Fig. 7.37) (Raymond et al., 2004).

FIGURE 7.36 Charles Darwin (1809–1882). Photograph by Julia Margaret Cameron (1869). Source: http://en.wikipedia.org/wiki/File:Charles_Darwin_01.jpg.

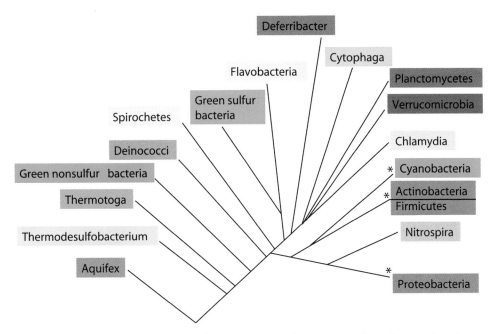

FIGURE 7.37 Major phylogenetic groups in domain *Bacteria*. Those nitrogen-fixing lineages found among the major bacterial phyla are denoted by an asterisk (*). Illustration by Erin Sanders based on a figure from Madigan and Martinko (2006).

The most widely received view of the basic organization of life was the five kingdoms: Plantae, Animalia, Fungi, Protista, and Monera (Whittaker, 1969). This division system recognized two fundamentally different kinds of cells, prokaryotic and eukaryotic, a dichotomy proposed in 1937 but not popularized until the early 1960s (Chatton, 1937; Stanier and van Niel, 1962). Whittaker simply placed the unicellular prokaryotes into their own kingdom, Monera. Although this allows prokaryotes to share equal taxonomic rank with the organisms representing the other four kingdoms, it represents a cytologically based division of life. Eukaryotes (Plantae, Animalia, Fungi, and Protista) are joined by the possession of organelles. Whittaker described three kingdoms of multicellular eukaryotes distinguished primarily by nutritional modes. Plantae were autotrophs, Animalia were heterotrophs, and Fungi were saprotrophs. The remaining kingdom, Protista, was a term concocted to describe the eukaryotic "leftovers" that did not conform to the definitions ascribed to members of the other four kingdoms. This definition of Protista lacked any unifying positive character traits. Instead, Protista was unified by containing organisms lacking the traits of the other eukaryotic groups. Prokaryotes (Monera) were classified according to a similar convention, in that they also were unified by a lack of the characteristics that define eukaryotic cells, a definition that also is devoid of shared positive character traits. Despite this, the prokaryotes were assumed to be a monophyletic group.

The prokaryotic-eukaryotic dichotomy and the five-kingdom system prevailed for more than 20 years. It was not until the pioneering work of Carl Woese (Fig. 7.38) in the 1970s that the monophyly of the prokaryotes (Monera) was finally disputed (Woese et al., 1977). By sequencing SSU rRNA, Woese was able to demonstrate that the prokaryotes (Monera) could be divided into two distinct lineages, the Eubacteria (now domain *Bacteria*) and the Archaebacteria (now domain *Archaea*). Furthermore, these two

FIGURE 7.38 Carl Woese. (Reprinted from J. Whitfield, *Nature* **427**:674–676, 2004, with permission of Nature Publishing Group, © 2004.)

groups of unicellular organisms did not comprise a single clade. Given advances regarding the placement of the root of the universal tree of life, it now appears that the term "prokaryote" and the kingdom designation Monera describe a paraphyletic lineage (see Fig. 7.7 in the first section of Unit 7). Although an oversimplification of the unicellular world of microbes, the word "prokaryote" has persisted as a common descriptive term used to convey information about certain cytological (e.g., unicellular, absence of nucleus), physiological (e.g., coupled transcription and translation), and genetic (e.g., 70S compared to the eukaryotic 80S ribosome, operon structure) features of cells in a consistent way (Martin and Koonin, 2006; Sapp, 2006; Madigan and Martinko, 2006).

The evolution of genes: variations on a theme leading to the universal phylogeny

While the use of molecular phylogeny has brought some consistency to taxonomic categories, it has caused the dissolution of others. The current view of life is based almost exclusively on genetic data, particularly sequences for the SSU rRNA. This gene was selected for use in the construction of the universal tree because it was thought to be a reliable and appropriate evolutionary chronometer, or a macromolecular "time clock" for measuring evolutionary change (Madigan and Martinko, 2006). For a gene to be considered a reliable evolutionary chronometer, it must meet the following six criteria:

1. The gene must be universally distributed across all members of the group under study. Thus, phylogenies relating all members of the group may be reconstructed.
2. The gene must be functionally orthologous among organisms within the group under study. In other words, functional and sequence similarity must be due to descent from a common ancestor and not to gene duplication or convergent evolution. This criterion establishes that the different parts of the sequence are under similar selection pressure and are related through a common ancestor.
3. The gene must have regions of sequence conservation to facilitate alignment, or matching up, of nucleotides or amino acid residues for the analysis.
4. The sequence changes that have occurred in the gene must have done so at a rate appropriate for evolutionary distance to be measured. In other words, the rate of change must be suitable for the group under study. For example, the rate should

be lower for a comparison of different phyla (deep phylogenetic branches on a tree) and higher for a comparison of members of the same genus (branches near the tips of a tree).

5. The sequence being studied must not have been horizontally transferred during its evolution. This criterion establishes that a calculated gene phylogeny reflects organismal phylogeny. If the gene being studied has been horizontally transferred from one taxonomic group to another, the gene phylogeny calculated with that gene would no longer accurately represent the organismal phylogeny.

6. The sequence changes that have occurred in the gene must be independent and identically distributed. In other words, a nucleotide change in one location in a gene cannot induce other nucleotide changes within the same gene or in another gene, as is the case of overlapping genes; all changes are independent of one another. Furthermore, when changes do occur, transitions and transversions to other nucleotides occur with the same probabilities in all parts of the gene. We say that they are identically distributed because the probability distributions for nucleotide changes are the same throughout the sequence.

The assumption is that the SSU rRNA gene meets all of these criteria. As discussed in more detail below, this assumption generally is true except for criterion 6. Generally speaking, the mechanism by which genetic information is stored and processed appears to be universally consistent among prokaryotes and eukaryotes, in that the information is maintained in the form of DNA and is converted to RNA and protein by functionally homologous transcription and translation systems. Prokaryotic and eukaryotic ribosomes, central components of the translation machinery, both are comprised of a large and a small subunit, and each subunit is made up of multiple proteins and RNAs (Table 7.1). This structural uniformity is considered to be a reflection of a common ancestral origin for the essential elements of all living cells.

The 16S rRNA and the 18S rRNA are key constituents of the SSU in prokaryotes *(Bacteria and Archaea)* and eukaryotes, respectively, and are thus referred to as SSU rRNAs, providing an acronym to denote their universal distribution across all forms of life. Similarly, RNAs associated with the large subunit, including the 23S rRNA in prokaryotes, are referred to as large-subunit (LSU) rRNAs. In *Bacteria*, the 16S rRNA mediates a specific interaction between the ribosome and the messenger RNAs (mRNAs),

Table 7.1 Structure of ribosomes

Property	*Bacteria* and *Archaea*	Eukaryotes
Overall size[a]	70S	80S
SSU	30S	40S
No. of proteins	ca. 21	ca. 30
RNA size (length)	16S (1,500 nt)	18S (2,300 nt)
LSU	50S	60S
No. of proteins	ca. 34	ca. 50
RNA size (length)	23S (2,900 nt)	28S (4,200 nt)
	5S (120 nt)	5.8S (160 nt)
		5S (120 nt)

[a]In Svedberg units (S).

wherein, via complementary base pairing, the 3′ end of the 16S rRNA binds to a sequence in the 5′ end of the mRNA called the Shine-Dalgarno sequence, facilitating recognition of the appropriate start codon so that protein synthesis may be initiated in the proper reading frame (Madigan and Martinko, 2006). rRNAs also play several other roles in the translation process, wherein both the 16S and 23S rRNAs interact with the transfer RNAs (tRNAs) during elongation steps. The formation of peptide bonds in the growing polypeptide chain occurs on the 50S subunit and is catalyzed by the 23S rRNA. The 23S rRNA also plays a role in the translocation process, interacting with certain elongation factors. Finally, the 16S rRNA is involved in some way in the termination phase of the translation process.

The regions of SSU rRNA that play essential functional roles during the translation process experience selective pressures to maintain a certain degree of sequence conservation. Specifically, residues required for catalytic activity, structural integrity, or mediation of interactions with other RNA or protein constituents during the translation process exhibit correlated levels of sequence variation. As shown in Fig. 7.39, which portrays the secondary structure predicted for the SSU rRNA, there is a common core of helices (stem-loop structures) shared by thousands of members of the *Bacteria, Archaea,* and eukaryotes (black helices). There also are regions common only to *Bacteria* (blue helices), common only to eukaryotes (red and pink helices), or shared by eukaryotes and *Archaea* (orange helices) or by *Bacteria* and *Archaea* (green helices). In Fig. 7.40, the relative substitution rate of each nucleotide site has been superimposed by means of a color code on the secondary-structure model for the SSU rRNA shown in Fig. 7.39. What is immediately apparent is a distribution of substitution rates that correspond loosely to positions, and thus potential function, of the various helices. Specifically, one can distinguish a color consistency among the core of helices common to all three domains, wherein the nucleotides are assigned colors corresponding to low site variability, in comparison to domain-specific helices, which exhibit much higher nucleotide substitution rates. For example, examine helices numbered 1 to 3, 5, 7, 12 to 16, 20, 22, 27, 31 to 35, and 38 to 42, which are common to all three domains, and compare the nucleotide color assignments to helices 8, 10, 23, 43, and 45, which are specific to the *Eucarya* or *Eucarya* plus *Archaea,* or helices 37b1 and b2, which are specific to *Bacteria.* In Fig. 7.41, the color-coded substitution rates are superimposed on a stereo view of the tertiary structure of the SSU rRNA. It is obvious that substitution rates are generally low near the center of the ribosomal subunit whereas solvent-exposed sites at the surface are largely more variable. The central core of the SSU rRNA contains functional sites such as those required for tRNA binding, and thus reflects necessary sequence conservation. Overall, the analysis by Wuyts et al. (2001) indicates that regions of the SSU rRNA should permit alignment of sequences derived from organisms across all domains of life, and there also are sufficiently variable regions exhibiting rates appropriate for measuring evolutionary distance at deep taxonomic levels (i.e., at the level of genus and above), but not at the level of species.

The final criterion that SSU rRNA sequences are challenged to meet is that nucleotide substitutions must be independent of one another. On the one hand, it is improbable that a nucleotide change in the SSU rRNA gene will induce a change in another gene; however, it is likely that a substitution in the SSU rRNA gene will influence the composition of bases within the SSU rRNA gene itself, especially if a secondary mutation preserves structure and consequently function of the ribosome. For example, the stem-loop secondary structures depicted in Fig. 7.39 and 7.40 form based on complementary

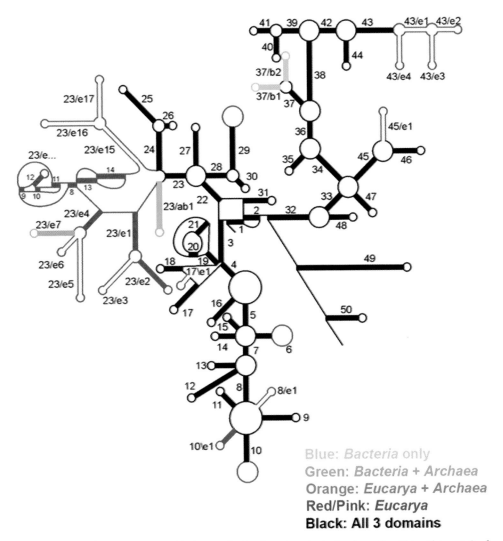

FIGURE 7.39 Secondary structure and helix numbering (running clockwise from the 5′ to 3′ termini) of SSU rRNA. The presence of helices (stem-loop structures) in the three domains is indicated according to the color scheme in the figure. Solid-colored stems occur in all species of the domain(s), whereas those shown in outline occur only in a subset of species of a domain. Large loops may indicate that the sequence forms an additional, as yet unknown, structure in some species. (Reprinted from Wuyts et al. [2001] with permission.)

base pairing between nucleotide partners that reside distal to one another in the primary sequence (see Fig. 6.25 for an example of nucleotide numbering in *E. coli*). It is thus plausible that nucleotide changes in one position may persist if an additional substitution takes place in its base-pairing partner. It therefore becomes apparent that the SSU rRNA gene does not conform to the last criterion of a molecular chronometer. However, recall from Unit 6 that the Ribosomal RNA Database Project (RDP-II) as an alignment tool places appreciable weight on secondary-structure contributions to the evolution of the molecule. Because this program considers the secondary-structure nuances associated with RNA molecules, the concerns for major evolutionary consequences can at least be minimized for SSU rRNA.

The work by Woese and others using the SSU rRNA sequences led to a proposal for the organization of life that formally recognized the three domains, *Bacteria*, *Archaea*, and *Eucarya* (Woese et al., 1990; Pace, 1997). The three-domain model, which became

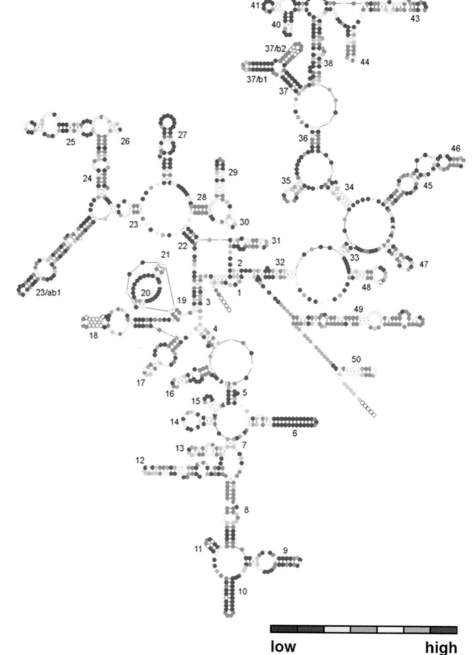

FIGURE 7.40 Variability map of bacterial SSU rRNA superimposed on the secondary-structure model of the molecule from *T. thermophilus*. Provided that the sites were occupied in ≥25% of the sequences in the alignment, the sites are subdivided into seven groups according to their relative substitution rate as shown on the scale bar, with purple indicating the lowest rate and red indicating the highest rate. (Reprinted from Wuyts et al. [2001] with permission.)

low　　　　　　　　**high**

widely accepted in the scientific community by the mid-1990s, was subsequently examined by Brown and Doolittle, who demonstrated that there are several other purported molecular chronometers that can be used to test the universal phylogeny depicted from analysis of the SSU rRNA gene (refer to Fig. 7.42a; also see Table 5.1 [Brown and Doolittle, 1997]). Their efforts established that recovery of the three-domain model can be influenced by gene choice, in that various genes produce different relationships among the domains (Fig. 7.42b and c). The challenge was to rationalize the discordance among phylogeny estimates obtained from different genes. One reasonable explanation was that a number of the housekeeping genes that produced alternative phylogenies were actually acquired by HGT, rather than strict vertical evolution (Boucher et al., 2003). With the

(a)

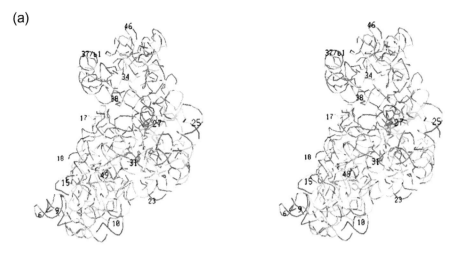

(b)

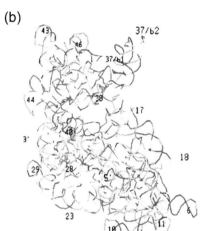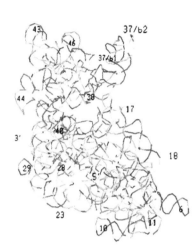

FIGURE 7.41 Variability map from Fig. 7.40 superimposed on the tertiary structure of the rRNA in the *T. thermophilus* SSU. A stereo drawing is shown of the SSU from the side of the interface with the large subunit (a) and from the solvent side (b). Each nucleotide is represented by a colored bar connecting the coordinates of the two adjoining phosphate atoms. Colors for substitution rates are the same as in Fig. 7.40. The most easily recognized helices are numbered, and the 5′ and 3′ ends are indicated. (Reprinted from Wuyts et al. [2001] with permission.)

advent of whole-genome sequencing, one can now recognize such genes, called xenologs, to distinguish them from orthologs and paralogs. Xenologs and orthologs may meet the criterion of a molecular chronometer in that they are functionally homologous; however, each has a very different evolutionary origin that ultimately affects the phylogeny it produces. Compared to other genes in the genome of a single microorganism, xenologs may have a G+C content (percentage of guanine plus cytosine) that differs markedly from that of the majority of other genes (presumably orthologs). Unusual codon usage also turns out to be a clue that a gene may be a xenolog. Although the genetic code is degenerate, whereby more than one codon can specify a single amino acid, not all codons are used at equal frequencies, and codon usage varies among microorganisms. Genes that use a different assortment of codons to specify an amino acid are likely xenolog candidates. For instance, in *E. coli*, 1 of every 20 isoleucines is encoded by AUA; the remaining 19 are encoded by either AUU or AUC. If one finds a gene in *E. coli* in which 10 of every 20 isoleucines is encoded by AUA, then this finding may suggest that this gene was acquired by HGT.

It is important to emphasize here that the work by Brown and Doolittle did not disprove the three-domain model, nor did it provoke fervent skepticism in the utility of the SSU rRNA gene as an appropriate tool for evolutionary analysis. Instead, their analysis brought the notion of HGT as a major contributor to the organization of genomes

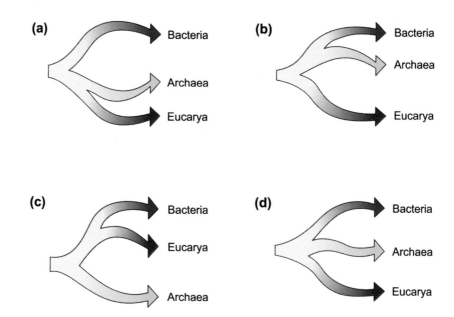

FIGURE 7.42 Different genes give different estimates of the universal phylogeny. (a) Phylogeny based on SSU rRNA and other housekeeping genes including ATPase subunits α and β, RNA polymerase subunits α and β, and elongation factors G and Tu. The root positions the *Archaea* and eukaryotes as sister groups. (b to d) Alternative rootings of the same three-taxon tree presented in panel a, based on discrepancies in gene distribution across the three domains, paraphyly or polyphyly in the domains, and in some cases, distance arguments and midpoint rooting of the gene phylogenies. (b) Phylogeny based on glutamate dehydrogenase, glutamine synthetase, gyrase B, and heat shock protein 70 positions *Archaea* as a paraphyletic group within the *Bacteria*. (c) Phylogeny based on enolase, glyceraldehyde 3-phosphate dehydrogenase, and 3-phosphoglycerate kinase positions the *Bacteria* and eukaryotes as sister groups. (d) Phylogeny based on acetyl-coenzyme A synthetase, dihydrofolate reductase, and photolyase produces a tree in which the relationships among the three domains are unresolved. See Brown and Doolittle (1997) and references therein for full list of genes used to construct the phylogenies presented in this figure. Illustration by Cori Sanders (iroc designs).

to the forefront of scientific investigation. Clearly, HGT was prevalent among the community of organisms in the early history of life (Fig. 7.43) (Doolittle, 1999). The evolutionary path depicted by the SSU rRNA gene certainly reflects a successful mechanism by which cells can store and process genetic information, which is why all cells now share homologous genes for this purpose. However, there are indications that even the phylogeny of the 16S rRNA gene may be influenced by HGT (Yap et al., 1999; Gogarten et al., 2002; Boucher et al., 2004).

In general, it is thought that the origins for genes such as the SSU rRNA gene reside among some of the deepest branches of the universal tree and have been propagated to progeny along a vertical evolutionary path despite other changes in the genetic repertoire leading to divergence. It follows, then, that a single genome is the product of multiple evolutionary paths, in which an amalgam of genes ultimately generates the organism that harbors them. Individual genes have distinct phylogenies, making reconstruction of the evolutionary history of organisms less simple and, consequently, phylogenetic taxonomy (in its purest sense) more difficult. However, it also is important to bear in mind that classification schemes exist to bring order to the immense diversity of organisms in the biosphere for pragmatic purposes. The users of taxonomic designations, including researchers, forensic scientists, and medical and clinical professionals, need to understand

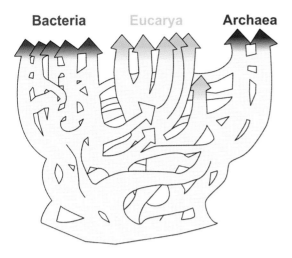

Bacteria Eucarya **Archaea**

FIGURE 7.43 Contributions of horizontal gene transfer to the early history of life. Rather than viewing the last common ancestor of all extant organisms as a single organism, the three domains emerged from a tangled base representing a community of organisms that readily traded genes (Doolittle, 1999). Illustration by Cori Sanders (iroc designs).

and manage ecosystems, cases of bioterrorism, and human infections, and thus they require a common dialect for information about microorganisms to be conveyed in a consistent, organized and meaningful way (Buckley and Roberts, 2007).

Currently, it is practical and scientifically sound to employ gene phylogenies to reconstruct the evolutionary history of organisms for classification purposes. The phylogenetic placement of an organism can be surmised by employing a "majority rules" criterion, in which categorical ambiguities can be resolved by examining phylogenies produced by the majority of genes under analysis. Gene choice ultimately depends upon the taxonomic group under study, provided that an ortholog can be selected with a suitable evolutionary rate and an ancestral origin of appropriate phylogenetic distance. If one understands the limitations of the analysis and employs the proper lexicon, the results can be compelling but necessarily contextualized.

A polyphasic approach to microbial taxonomy provides an operational definition for the bacterial species

In the late 1800s, a German biologist, Ferdinand Cohn (Fig. 7.44), applied the archetypal taxonomic framework proposed by Linnaeus for the classification of plants and animals to the description of microorganisms. Specifically, he used a binomial nomenclature, in which microbes could be categorized according to genus- and species-level assignments. New bacterial species initially were classified into broader taxonomic groups according to morphology and reaction to the Gram stain, as well as other properties such as substrate requirements for growth and pathogenic potential. Together, these properties were used to make finer and more detailed taxonomic-level assignments. Denoting a unification of the nomenclature and criteria used to identify bacteria among microbiologists, the first edition of *Bergey's Manual* was published in 1923 as a resource for information on all recognized bacterial species. Now in its second edition, it has been expanded to include archaeal species as well (Garrity et al., 2001; Garrity, 2005). Table 7.2 lists the taxonomic ranks and numbers of recognized divisions within each category according to the most recent edition of *Bergey's Manual*.

It was not until the mid-1950s that chemotaxonomy was incorporated into the designation of a bacterial species. Specifically, assays were developed that revealed chemical differences in the structure of cellular components such as proteins, sugars, or fatty acids.

FIGURE 7.44 Ferdinand Cohn (1828–1898). Photograph origin unknown. Source: http://en.wikipedia.org/wiki/File:Ferdinand_Julius_Cohn_1828-1898.jpg.

In the early 1960s, technological advances allowed at least a superficial classification of bacteria according to overall base composition (i.e. G+C base pair content) of their genomes (reviewed by Rosselló-Mora and Amann [2001]). Later, a quantitative molecular method was described that compared the relatedness of two microbial species based on differences in their genetic material (McCarthy and Bolton, 1963). Using a hybridization procedure (discussed below), this technique measures the degree to which pools of DNA sequences extracted from two organisms can form complementary base pairs with one another. By the mid-1960s, it was projected that a particular category of biological molecules, which included DNA, RNA, and polypeptides, could be used as the basis for deducing the molecular phylogeny of organisms (Zuckerkandl and Pauling, 1965). In 1973, John Johnson became the first to use DNA-DNA hybridization (DDH) as a means to classify bacterial species on the basis of homology data (Johnson, 1973). Then, of course, Carl Woese tested this hypothesis by sequencing genes for SSU rRNA, consequently transforming perceptions for the organization of life from a five-kingdom system to a three-domain model (Woese et al., 1977).

Table 7.2 Taxonomic ranks and numbers of recognized divisions for *Bacteria* and *Archaea*[a]

	No. of divisions		
Rank	*Bacteria*	*Archaea*	Total
Domain	1	1	2
Phylum	25	4	29
Class	34	9	43
Order	78	13	91
Family	230	23	243
Genus	1,227	79	1,306
Species	6,740	289	7,029

[a]Sources: Garrity et al. (2001), Garrity (2005), and Madigan and Martinko (2006).

The combined advancements in molecular, chemotaxonomic, and physiological methods have facilitated the classification of microbes at finer taxonomic levels, leading to a consensus-based definition of a bacterial species. Today, bacterial species are delineated after analysis and comparison of phenotypic, genotypic, and phylogenetic parameters— an approach called polyphasic taxonomy (Colwell, 1970; Vandamme et al., 1996; Stackebrandt, 2007). As a formal classification system, its aim is to integrate different kinds of data and arrive at a consensus that has the fewest contradictions regarding the identity of a bacterial species (Gevers et al., 2005).

Using a polyphasic approach, a collection of bacterial isolates may be screened initially by rapid methods that allow clustering of closely related isolates, thereby making an immediate distinction between unrelated isolates. Routinely used methods include whole-cell fatty acid methyl ester (FAME) analysis (Eder, 1995; Cavigelli et al., 1995), DNA-based typing assays such as PCR ribotyping (restriction fragment length polymorphism [RFLP]) (Bouchet et al., 2008) or amplified fragment length polymorphism (AFLP) fingerprinting (Vos et al., 1995), and matrix-assisted laser desorption ionization–time-of-flight (MALDI-TOF) mass spectrometry (MS) (Keys et al., 2004; Ryzhov and Fenselau, 2001). FAME allows characterization of the classes of fatty acids present in the cytoplasmic membrane (as well as lipopolysaccharides in the outer membrane of gram-negative bacteria). The fatty acid composition can be highly variable due to differences in chain lengths, degree of branching, and the extent of saturation (e.g., double bonds). As outlined in Fig. 7.45, the procedure involves replacement of a proton on the carbonyl of a fatty acid with a methyl group, causing volatilization of the derivatives, which can then be detected by gas chromatography. The pattern of peaks on the chromatogram can be compared to a database containing profiles for reference organisms, with a match permitting identification of the bacterial isolate.

Like FAME, molecular ribotyping and AFLP methodologies also rely on pattern recognition, which can be used to catalog organisms. When genomic DNA from a bacterial isolate is subjected to a restriction digest, the fragments corresponding to regions in the ribosomal operons can be visualized in a number of ways. With the more traditional RFLP system, restriction fragments can be revealed in a Southern blot with labeled DNA probes (Fig. 7.46) (McCartney et al., 1996). More recently, the technology has progressed to enable PCR-based detection of DNA fragments as in AFLP fingerprinting. Other variations of DNA-based typing assays include T-RFLP (discussed in Unit 1), which involves PCR amplification of the 16S rRNA gene first, followed by restriction digest of the fluorescently labeled PCR products. The ribotyping methods are similar in that the diversity of restriction fragments is dependent upon the rate at which point mutations occur in the genome, causing a single base pair change in an endonuclease recognition site that results in the loss of a site and a consequential change in the RFLP pattern (Bouchet et al., 2008).

Rapid analysis of the surface components of intact bacterial cells can be achieved using MALDI-TOF MS (Keys et al., 2004; Ryzhov and Fenselau, 2001). As a high-throughput system, this technique stands to reveal information about low-abundance biomolecules such as those associated with virulence and pathogenicity as well as microbial physiology (i.e., electron transport, ATP synthesis, and signal transduction), which thus far have eluded systematic study for chemotaxonomic applications. Intact bacterial cells are embedded within a matrix material on a slide. As shown in Fig. 7.47, a laser is used to irradiate the cells, causing vaporization and ionization of the surface components, which form a plume of gaseous ions that desorb from the matrix. A TOF analyzer is used to

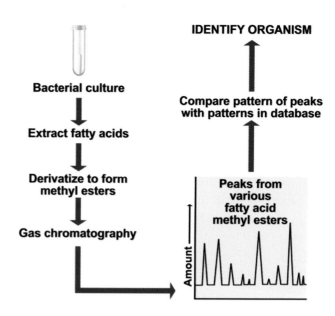

FIGURE 7.45 FAME analysis. Illustration by Cori Sanders (iroc designs) based on a figure from Madigan and Martinko (2006).

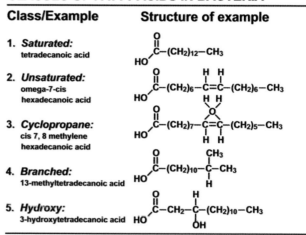

detect the ion plume and separate the ions according to their mass *(m)* and charge *(z)*. The computational output of the analysis is a mass spectrum (Fig. 7.48) in which the normalized peak intensities are plotted versus the mass-to-charge ratio of ions *(m/z)* in the plume. The spectral profile generated by an isolate is compared to a database of reference organisms and identified based on the conformity of peak patterns.

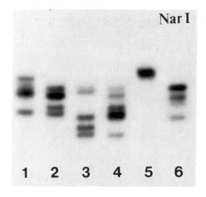

FIGURE 7.46 Autoradiograph showing ribotypes of six *Bifidobacterium* isolates. DNA digested by NarI. (Reprinted from McCartney et al. [1996] with permission.)

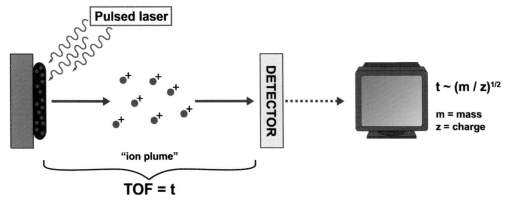

FIGURE 7.47 Overview of MALDI-TOF MS. Illustration by Cori Sanders (iroc designs).

Once a large number of bacterial isolates have been categorized into relatedness groups (clusters) by automated, standardized procedures like those described above, a 16S rDNA sequence analysis may be conducted on a subset of isolates representative of each cluster (Gevers et al., 2005). The classification of microorganisms based on sequence similarity of a universally distributed trait initially involves use of a sequence database (e.g., GenBank and RDP-II), wherein the identity of the isolate ostensibly can be resolved based on a comparison to other rDNA sequences in the database. If there is 99% identity or lower between two sequences, there are enough differences to characterize the isolate as a distinct and novel species (Stackebrandt and Ebers, 2006). This cutoff is based on a correlation with genome hybridization data (discussed next). However, one should not rely solely on a database match to assign a taxonomic classification because the genes

FIGURE 7.48 Example of a spectral profile from MALDI-TOF MS analysis. (Reprinted from Keys et al. [2004] with permission.)

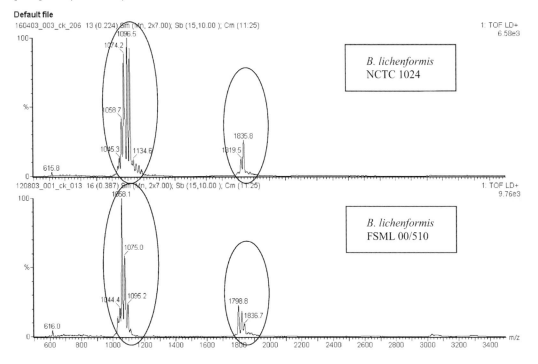

present in the databases are not necessarily annotated reliably (for a discussion, see Unit 6). Although a more rigorous phylogenetic analysis also should be applied to ascertain the identity of the isolate, using only 16S rRNA gene sequences does not permit species-level resolution if the isolate demonstrates at least 99% identity to the best match in the databases. Such isolates also tend to lack well-defined morphological or phenotypic characteristics. Alternative approaches, including DDH experiments and multilocus sequence analysis (MLSA), should be employed as well.

Compared to animals, which display a narrow range of GC ratios (35 to 43%), microbes span the gamut, with *Bacteria* demonstrating the greatest range in DNA base composition from as low as 18 to 19% to as high as 78% (Madigan and Martinko, 2006). However, it is more than the relative percentage of nucleotides that distinguishes the genomes of two different organisms. It is also the order in which the nucleotides are organized within each genome because therein lies the genetic information that engenders a unique species. Genetic relatedness between two organisms is dictated by the degree to which their gene sequences are similar or even identical. DDH experiments measure the degree of sequence similarity between two entire genomes; the method is based on the premise that only homologs will hybridize. DDH is time-consuming, requires every new prospective species to be tested against all closely related species, and is not robustly comparable between laboratories (Gevers et al., 2005). However, it offers sensitivity to the subtle variations in genes from two closely related organisms and thus is useful for differentiating species-level taxonomic relationships where 16S rDNA sequences fail to reveal definitive differences.

As outlined in Fig. 7.49a, in a DDH experiment, genomic DNA from one organism (a type strain) is isolated and sheared mechanically into smaller fragments (<1 kb) and radioactively labeled. The genomic DNA from a second organism (isolate) is prepared the same way, except that it is left unlabeled. The two DNA pools are heated, allowing the duplex DNA fragments to separate completely. The denatured DNA fragments from the two organisms are mixed and cooled slowly to allow annealing of matching DNA strands (Fig. 7.49b). Excess unlabeled DNA from the isolate is added to prevent labeled DNA from the type strain from hybridizing to itself. For similar regions in the two genomes, the radioactive strand from the type strain anneals to a nonradioactive strand of the isolate, forming radiolabeled heteroduplexes that can be separated from any remaining unhybridized DNA fragments. To determine what percentage of DNA fragments from the isolate hybridized with the type strain, the level of radioactivity must be compared to what would be obtained in a control reaction conducted identically to that described above except with the DNA from the type strain being allowed to hybridize to itself (Fig. 7.49c). The control is considered the 100% hybridization value.

For isolates exhibiting 99% sequence identity to the type strain in the 16S rRNA gene, a DDH value of at least 70% is considered sufficient evidence to assign an isolate to the same species as the type strain (Stackebrandt and Ebers, 2006). The species boundary was established based on a trend observed in a series of experiments that compared 16S rRNA sequence similarities and DDH values (Fig. 7.50) (Hagström et al., 2000; Rosselló-Mora and Amann, 2001; Stackebrandt and Ebers, 2006). There are no examples of organisms with less than 98.5% 16S rDNA sequence identity and a DDH value of greater than 70%. Further inspection of these data shows that although an isolate may demonstrate 99% 16S rDNA sequence identity, it may or may not meet the 70% DDH criterion for inclusion in the same species. For those that do meet the DDH cutoff, the

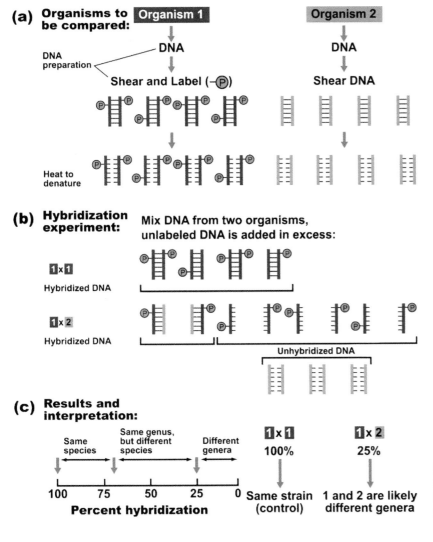

(a) Organisms to be compared:

Organism 1   Organism 2

DNA preparation

DNA → Shear and Label (–Ⓟ)

DNA → Shear DNA

Heat to denature

(b) Hybridization experiment:

Mix DNA from two organisms, unlabeled DNA is added in excess:

1×1 Hybridized DNA

1×2 Hybridized DNA

Unhybridized DNA

(c) Results and interpretation:

| Same species | Same genus, but different species | Different genera |

100   75   50   25   0

Percent hybridization

1×1 100%   1×2 25%

Same strain (control)   1 and 2 are likely different genera

FIGURE 7.49 Overview of the DDH technique used as a taxonomic tool. Illustration by Cori Sanders (iroc designs) based on a figure from Madigan and Martinko (2006).

result potentially could be misleading due to one major shortcoming of the technique, in that it may not be accurate for comparison of closely related species because differences between orthologous sequences from two genomes may be overwhelmed by hybridization of paralogous sequences from within a single genome. For those that do not meet the DDH cutoff, a different method must be applied.

The limited resolving power of 16S rRNA gene phylogenies can be overcome by using a sequence-based approach that involves genetic loci that evolve faster than 16S rDNA. However, the phylogeny of a single gene may be influenced by the effects of horizontal gene transfer (Gogarten et al., 2002), so techniques have been developed that utilize multiple genes for construction of a single, consensus tree. MLSA provides a classification system for placement of isolates into species-specific lineages, resolving phylogenetic relationships at the interspecies and even the intraspecies level (Gevers et al., 2005). The MLSA procedure is outlined in Fig. 7.51. Genomic DNA is purified from a microbial isolate, and PCR is used to amplify, in this case, six or seven target genes. These genes tend to have housekeeping functions, are ubiquitously expressed in the taxon under study, and are present in single copy within the genome. Once the individual genes are sequenced, they can be concatenated, or lined up end-to-end to form a contiguous

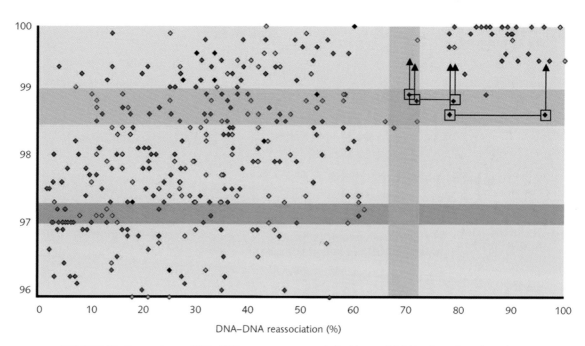

FIGURE 7.50 Comparison of 16S rRNA gene sequence similarities and DDH values. Data have been compiled from all publications containing species descriptions in volume 55 of the *International Journal of Systematic and Evolutionary Microbiology* (2005). The color of a particular data point refers to the DDH method employed to obtain the value: microtiter plate technique (red), spectrophotometric technique (dark blue), membrane filter method (light blue), and other methods such as dot hybridization or not defined (black). Horizontal lines between open squares indicate data for the same organism obtained by two different reassociation methods. Arrows point to the position of in silico-recalculated binary 16S rRNA gene sequence similarity values (i.e., predicted DDH values for sequences in which erroneous nucleotides were corrected). The horizontal blue bar indicates the threshold range above which it is now recommended to perform DDH experiments. The horizontal red bar indicates the threshold values published previously. The vertical green bar denotes the threshold DDH value. (Reprinted from Stackebrandt and Ebers [2006] with permission.)

sequence, and used to generate a phylogenetic tree. The methodology assumes that the isolate being characterized has previously been assigned to a family or genus, such that the strains selected for constructing the MLSA tree are confined to this lineage.

To exemplify the resolving power of MLSA, Hanage et al. (2005) used this approach to sort *Neisseria* isolates into genotypic clusters. Using a large sample (770 strains) of several named *Neisseria* species, some pathogenic and others commensal, these authors compared the phylogenetic trees produced by single genes to that generated by MLSA. As shown in the left panel of Fig. 7.52, the single-gene trees result in anomalous clustering of the *Neisseria* isolates, with no obvious or consistent pattern emerging from any of the depicted phylogenies, whereas in remarkable contrast, the MLSA concatamer tree shown in the right panel clearly resolves the three most numerous named *Neisseria* species in the sample: *N. lactamica* (blue), *N. meningitidis* (red), and *N. gonorrhoeae* (green). Interestingly, the clustering pattern also makes sense in an ecological context, with two groups known to colonize the human nasopharynx, one pathogenic *(N. meningitidis)* and the other commensal *(N. lactamica)*, and a third group recognized for causing genital disease *(N. gonorrhoeae)* (Gevers et al., 2005). One other notable observation comes upon inspection of the boundaries between the three *Neisseria* species, which the authors describe as "fuzzy" because the isolates exhibit intermediate genotypes or mosaic genomes (Hanage et al., 2005). The fuzzy species may be resolved by eliminating recombinant

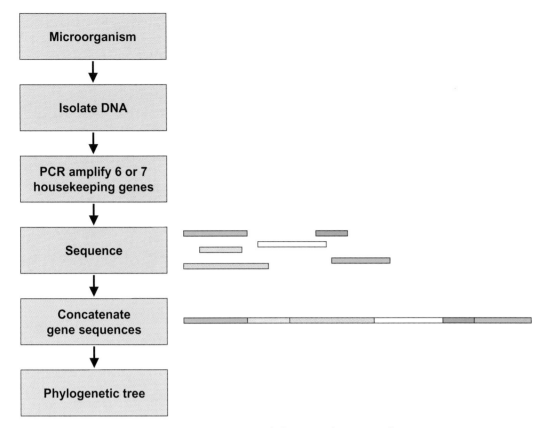

FIGURE 7.51 Overview of MLSA as a taxonomic tool. Illustration by Erin Sanders.

segments of the genome from the concatemer (Koeppel et al., 2008; Didelot and Falush, 2007). Using a recombination-free phylogeny may not get rid of the fuzziness of the species, but it does result in an accurate classification. The results may suggest that the remaining eight *Neisseria* species are not genetically distinct populations and thus should not be given species status. Whole-genome sequencing of the *Neisseria* isolates ultimately may reveal differences that cannot be detected by MLSA, and these differences may provide necessary justification for their species-level designation. Clearly, a more comprehensive analysis is needed to determine whether bacteria can be consistently partitioned into unambiguous, biologically meaningful clusters at all.

Differences in phenotype are required for the description and naming of a new species. Although the polyphasic approach relies on a set of criteria and identity thresholds to establish clusters with phylogenetic and phenotypical similarities, there is no consensus among members of the microbiology community about which tests and results are requisite for the identification of an isolate as a novel species. For instance, genomic techniques that capture the sequence diversity within a population of microorganimsms (i.e., metagenomics), that characterize the metabolic potential of a single organism (i.e., whole-genome sequencing and annotation), or that compare differences in the metabolic potential among multiple species and strains (i.e., comparative genomics) are not widely used in microbial taxonomy (Buckley and Roberts, 2007; Coenye et al., 2005). Although powerful in discovering whether a gene is present in the genome of a particular microbe or within a community of microorganisms, genomic methods do not reveal whether the gene is actually expressed or under what conditions the activity takes place in the natural

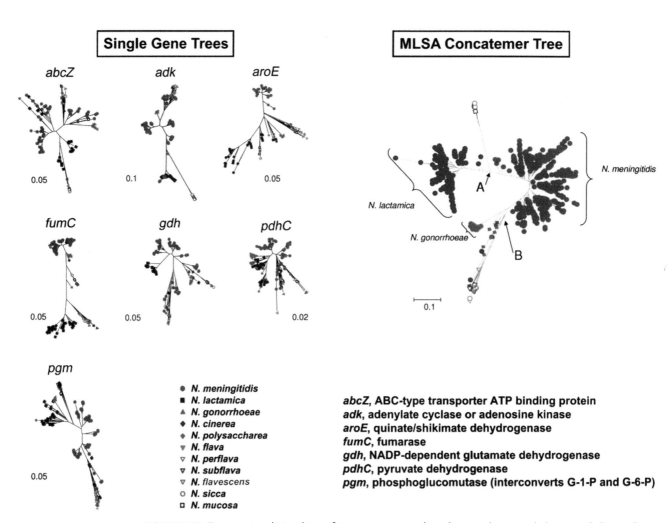

Single Gene Trees

abcZ
0.05

adk
0.1

aroE
0.05

fumC
0.05

gdh
0.05

pdhC
0.02

pgm
0.05

MLSA Concatemer Tree

N. lactamica

N. gonorrhoeae

A

B

N. meningitidis

0.1

- ● *N. meningitidis*
- ■ *N. lactamica*
- ▲ *N. gonorrhoeae*
- ◆ *N. cinerea*
- ◆ *N. polysaccharea*
- ▽ *N. flava*
- ▽ *N. perflava*
- ▽ *N. subflava*
- ▽ *N. flavescens*
- ○ *N. sicca*
- ◻ *N. mucosa*

abcZ, ABC-type transporter ATP binding protein
adk, adenylate cyclase or adenosine kinase
aroE, quinate/shikimate dehydrogenase
fumC, fumarase
gdh, NADP-dependent glutamate dehydrogenase
pdhC, pyruvate dehydrogenase
pgm, phosphoglucomutase (interconverts G-1-P and G-6-P)

FIGURE 7.52 Taxonomic relationships of *Neisseria* species based on single-gene phylogenies (left panel) compared to MLSA phylogeny (right panel). The names of the 11 *Neisseria* species and functions of the 7 housekeeping genes in the study are indicated. (Reprinted from Hanage et al. [2005] with permission.)

environment or even laboratory conditions. Therefore, detection of a gene does not demonstrate performance of a biological process, which provides organisms with phenotypic properties that can be observed and measured as part of a conventional taxonomic characterization procedure. The merger of functional genomics, wherein the complexities associated with gene and protein expression are explored, with sequence-based approaches will undoubtedly provide a more compelling and meaningful link between phenotype and genotype. However, the information needs a phylogenetic framework and ecological context to achieve a rigorous taxonomic classification.

With the exception of PCR-based 16S rRNA gene sequence analyses, these techniques share a very significant limitation in that only cultivated strains can be characterized by polyphasic taxonomy. Strict application of the polyphasic approach to the delineation of microbial species has caused the view of all microorganisms to be shaped by data only from cultivated lineages, which represent less than 1% of the total diversity present in the environment (Amann et al., 1995; Torsvik et al., 1990; Torsvik and Øvreås, 2002; Buckley and Roberts, 2007). With the infusion of genomic and metagenomic sequences into the public databases, a call is in order for a new system devised to classify all microbial life.

The bacterial ecotype: a modern, theory-based species concept

To define a microbial species, one might investigate the creation of a more conceptual model instead of or in addition to an operational model such as polyphasic taxonomy. A conceptual model implies that there exists a predictive mechanism of speciation that takes specific genetic and environmental features of an organism into account (Buckley and Roberts, 2007). In introductory biology courses, students are taught the classical species concept as zoologists apply it to animals and as botanists relate it to plants (Mayr, 1942; Stebbins, 1950). In this context, a biological species is defined as a population that can naturally interbreed and produce fertile offspring. Furthermore, it is delineated as a population that is reproductively isolated, in that there is a lack of gene flow between it and other organismal populations. Circumscribed as such, the biological species serves as a critical measure for quantifying biodiversity, providing a definition for conservation-based initiatives such as the Endangered Species Act designed to preserve organisms and ecosystems at risk of becoming extinct. One major problem, even with eukaryotes, is that this sort of definition applies only to organisms that reproduce sexually, thereby excluding anything (including *Bacteria* and *Archaea*) that reproduces asexually. Another issue is that the definition cannot be tested with fossils, leaving paleontologists with only morphology and evolutionary history, albeit limited to a correlation of rock strata and their age according to a geologic time scale (Jurassic, Cambrian, etc.), as parameters with which to create a definition of species.

An alternative to the biological species concept which considers the evolutionary history of organisms already has been introduced (Rosselló-Mora and Amann, 2001). The criterion for identifying a taxon by using a phylogenetic species concept is monophyly, in which a taxonomic group is defined as all known descendents of a single, common ancestor (Hennig, 1965). A species would thus be defined as the smallest monophyletic group on a phylogenetic tree containing closely related members of a given population. Such a classification is constrained by two parameters: (i) the populations must be evolutionarily independent, in that there is a lack of gene flow (i.e., HGT) between populations, and (ii) the populations must have existed long enough for synapomorphies to emerge (Freeman and Herron, 2004). Recall from Unit 6 that sequence changes arise and can be used to construct alignments, which in turn become the building blocks for assembly of a phylogenetic tree. Some of those sequence changes potentially denote evolutionary branch points in a phylogenetic tree (e.g., a bifurcation event in which two populations evolve into two independent lineages). Tracing a tree from its root to the tips is a walk through ancient time, with a nested hierarchy of extant phylogenetic taxa created by successive branching events. Once split, two lineages evolve independently, with homologous sequences accumulating mutations or other changes in the gene pool as a result of natural selection, genetic drift, or founder effects. The resulting genetic variations, referred to as synapomorphies, are shared by all members of a population and constitute unique traits derived from a common ancestor but long since modified as distinguishing characteristics of a population. Consider the example in Fig. 7.53. Here, the ancestral DNA sequence is CTGGCTATCT. At the first branch point in the phylogenetic tree, one of the two lineages acquired a mutation that converted an A to a T at position 7 of the sequence while the other lineage retained the A at the same position. The A-to-T transversion is a synapomorphy that defines the first branch point. At a second branch point in the tree, one lineage sustained a mutation that changed a G to an A at position 4 while the other lineage did not change. The G-to-A transition is a synapomorphy that defines the second branch point. So on it goes as the tree is built,

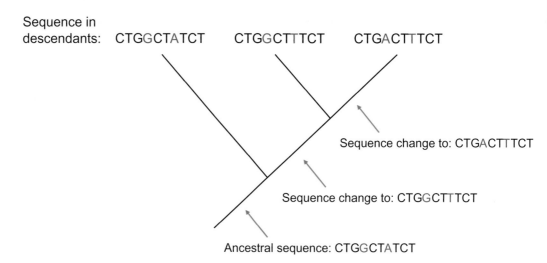

FIGURE 7.53 Phylogenetic tree with examples of synapomorphies defining each monophyletic group. An A-to-T transversion denotes the first branch point, while a G-to-A transition identifies the second branch point in the nested hierarchy. Illustration by Erin Sanders-Lorenz based on a figure from Freeman and Herron (2004).

with shared, derived traits identifying branch points and with branch points generating monophyletic groups that can be assigned taxonomic levels in a hierarchical but nested fashion.

The benefit to using phylogeny to define a species, a method also referred to as cladistics or phylogenetic systematics, is that the approach can be applied to any living organism, irrespective of reproductive mechanisms. However, it is limited in that it requires access to the genetic features of an organism, which still excludes fossils. In addition, the influence of gene duplications, HGT, gene loss, and other chromosomal rearrangements continues to pose challenges to the elucidation of a consistent phylogenetic taxonomy (Coenye et al., 2005).

Rather than subscribe to the notion that the species concept for microorganisms is fundamentally different from that for eukaryotes, some microbial ecologists have proposed a more modern and broad interpretation of the species concept, one that can be applied to all organisms and that considers genetic, evolutionary, and ecological attributes (Cohan, 2006). Central to the modern species concept are four elements, summarized by Cohan as follows:

1. Species show genetic cohesion, in that some force acts to constrain divergence within a population. In the case of the highly sexual animals and plants, genetic exchange is a powerful cohesive force and occurs only within a population defined as a species. However, in bacteria recombination is extremely rare, occurring at about the rate of mutation, and thus is a much less cohesive force. In the case of asexual and rarely sexual organisms, such as bacteria, the most powerful force constraining diversity is periodic selection and in some cases genetic drift (Cohan and Koeppel, 2008). It is interesting to recognize that even bacteria recombine sexually by way of conjugation, demonstrating a spectrum of population structures from strictly clonal to highly recombinogenic (Smith et al., 1993; Feil et al., 2001). Bacteriophages also act as agents of genetic exchange, whereby generalized transduction facilitates homologous recombination (Gogarten et al., 2002). As discussed

above, microorganisms also exchange genetic material via HGT; however, this phenomenon is not considered a cohesive force because it can occur between distantly related organisms.

2. Species are irreversibly separate, in that there are no cohesion forces (i.e., genetic exchange, periodic selection, or genetic drift) between individuals from distinct species. In the case of the crown eukaryotes, a species cannot become irreversibly separate until it breaks free of frequent recombination with its closest relatives. In the case of the bacteria, recombination is so rare that a decrease in its frequency is not required for populations to adaptively diverge. Bacteria need only be ecologically distinct to inhabit different evolutionary tracks.

3. Species are monophyletic, in that all individuals within a group comprising a species have a common ancestor on a phylogenetic tree.

4. Species are ecologically distinct, in that multiple species can coexist within a community of organisms because they use dissimilar resources, are able to thrive under distinctive environmental conditions, or respond differently to predators and pathogens.

The bacterial counterpart to the modern species concept as it applies to plants and animals seems to correspond to an ecotype, or a group of bacteria that are ecologically similar to one another in that they occupy a particular ecological niche (Cohan, 2006). For instance, envision a population of cells, which originated from a single cell, that now share a specific environmental resource unique to the microbial habitat. This population of cells is referred to as an ecotype; as shown in Fig. 7.54, multiple ecotypes can coexist in the same habitat, but each occupies a distinct ecological niche (i.e., each uses a different set of environmental resources within the same habitat or in different habitats). Should an adaptive mutation occur within a cell of one ecotype, offering a selective

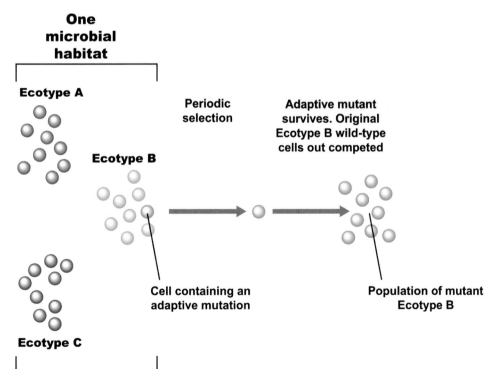

FIGURE 7.54 Ecotype model as a mechanism for microbial evolution. An adaptive mutant within ecotype B extinguishes genetic diversity within this ecotype, while ecotypes A and C are unaffected. Illustration by Cori Sanders (iroc designs) based on a figure from Madigan and Martinko (2006).

growth advantage to this adaptive mutant over its wild-type counterparts, the mutant and its progeny will survive and eventually replace the original ecotype completely, purging diversity at all genetic loci within the mutant's ecotype. This process, referred to as periodic selection, repeats itself many times, constraining diversity within the ecotype. New ecotypes are formed with adaptive mutations (or recombination events) that allow the mutant to occupy a novel ecological niche and to coexist with the original ecotype (compare Fig. 7.55a and b). Consequently, a bifurcation event, or branch point, on a phylogenetic tree may delineate the formation of a new ecotype or a speciation event (Fig. 7.55c).

Since different ecotypes do not compete for the same ecological niche, an adaptive mutant within one ecotype does not cause the extinction of members of other ecotypes. Thus, ecotypes, which are defined as ecologically distinct, cohesive groups, are irreversibly separate since there is no periodic selection between ecotypes. Moreover, because each is founded by a single individual, an ecotype can be identified as a monophyletic sequence cluster. Taken together, the properties of microbial ecotypes, not species, are consistent with the four elements of a modern species concept. There are several variations of this model, but such a discussion is beyond the scope of this text. Interested students are encouraged to consult the primary literature (Cohan and Perry, 2007; Cohan and Koeppel, 2008).

Efforts are under way to develop operational criteria for performing ecotype-based classification of bacterial diversity (F. Cohan, personal communication). Specifically, Cohan and colleagues have proposed that an ecotype should be recognized when it constitutes a well-supported clade (e.g., >70% bootstrap support in a recombination-free phy-

FIGURE 7.55 Effects of adaptive mutations on diversity within and between ecotypes. Summary of a periodic selection event (a) in comparison to a niche-invasion event (b). (a) An adaptive mutant within ecotype 1, indicated by an asterisk, extinguishes the genetic diversity within the same ecotype while ecotype 2 is unaffected. Diversity accumulates later within ecotype 1. (b) An adaptive mutant within ecotype 1, indicated by a plus sign, obtains an ability to utilize a different set of resources and thus founds a new ecotype. (c) Phylogenetic history of two closely related ecotypes based on the ecotype model discussed in the text. After each periodic selection event, indicated by an asterisk, only one ecotype variant survives. Later, genetic diversity accumulates (indicated by dashed lines), but with the next periodic selection event, again only one variant survives. Permanent divergence occurs when a niche-invasion mutation (or recombination event) creates a new ecotype. (Reprinted from Cohan [2006] with permission.)

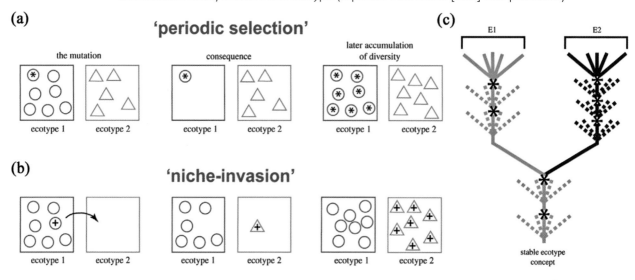

logeny) that is ecologically distinct from other closely related clades. Furthermore, the clade has to have shown a history of coexistence with other closely related ecotypes as revealed by algorithms that identify DNA sequence clusters that most likely correspond to ecotypes (Koeppel et al., 2008; Hunt et al., 2008). An ecotype must exhibit a phenotypic trait that specifies its ecological niche, distinguishing it from its closest relatives. And finally, recognition of the ecotype as a species must not require the reclassification of previously identified species. So, when a recognized species is found to consist of multiple ecotypes, the taxonomy of the species should be split into various ecospecies with the same species name (e.g., *Bacillus simplex* ecospecies Graminiphilus for a grassland-associated ecotype). Overall, this approach transitions the ecotype from a conceptual model to an operational definition that can be employed in the taxonomic organization of all prokaryotes.

A final word on bacterial taxonomy?

In practice, clustering methods such as those employed in polyphasic taxonomy can be used to differentiate bacterial species, provided that they yield the same demarcations as a theory-based approach in which the "species" label enables scientists to formulate predictions about all members of the population (Cohan, 2006). Thus, the end goal for defining a prokaryotic species, whether by an operational or theory-based model, should be to organize the vast diversity of *Bacteria* and *Archaea* in a practical but biologically meaningful way. Such a framework cannot be so rigid that it is unable to accommodate newly discovered microbes that appear at "fuzzy" borders between well-defined clusters, nor can it exclude a description of the uncultured microbial majority. When it comes to studies in microbial ecology and molecular evolution, tolerance for a flexible lexicon is certain to be an invaluable skill for the 21st-century microbiologist.

KEY TERMS

Annotation The interpretation of raw sequence data (reads) to open reading frames with predicted functions according to the conventions of gene structure and homology to previously characterized proteins.

Chemotaxonomy Use of the chemical structures of cell components as a measure of relatedness to separate bacteria into taxonomic groups; also called chemosystematics.

Cohesion The unification of individuals within a population based on the restriction of genetic exchange with individuals from outside the population, or the limitation of genetic diversity within a population by periodic selection or genetic drift.

Commensal Part of the normal human bacterial flora, living within the human body without hurting or helping it.

Comparative genomics Side-by-side comparison of the gene content and organization of two or more entire genomes.

Evolutionary chronometer A molecule whose nucleotide or amino acid sequence can be used as a comparative measure of evolutionary divergence.

Founder effects The loss of genetic variation that occurs when a new population is established by a very small number of individuals originating from a larger population.

GC ratio The combined percentage of G and C nucleotides present in the genomic DNA of a single organism; also called G+C content.

Genetic drift A process by which an allele frequency in a population changes entirely by chance from one generation to the next.

Housekeeping genes Genes that are generally constitutively expressed and are thought to be involved in routine cellular metabolism functions; they tend to encode essential metabolic processes.

Hybridization Reassociation of complementary nucleic acid strands once denatured.

Interspecies Occurring between species; involving members of multiple species.

Intraspecies Occurring within a species; involving members of a single species.

Metabolic potential The ability to carry out cellular processes.

Midpoint rooting A method to root a phylogenetic tree by placing the root somewhere in the center of the tree, equidistant from all organisms in the tree.

Natural selection The process by which advantageous traits become more common in successive generations of a population of reproducing organisms due to the increased frequency of the genotype responsible for the phenotype associated with that trait.

Niche A way of making a living, which may coincide with consumption of a particular set of resources and/or habitation in an environment with a particular set of physical conditions or resources.

Overlapping genes A single nucleotide sequence coding for more than one polypeptide; an arrangement of genes typically found in small genomes, where it provides a strategy for optimizing the amount of genetic information contained within a small region.

Paraphyly A group of taxa in a rooted phylogenetic tree that contains the most recent common ancestor but does not contain all the descendents of that ancestor.

Pathogenic Disease causing; capable of bringing about human illness or death.

Polyphasic taxonomy Assimilating all available phenotypic and genotypic data into the classification of a bacterial species.

Polyphyly A group of taxa in a rooted phylogenetic tree that excludes the most recent common ancestor; a group created by similarity, not descent.

Saprotrophs Organisms that obtain nutrients from nonliving organic matter; as decomposers, they absorb organic compounds from dead or decaying plant and animal matter.

Synapomorphies Shared, derived traits that characterize a lineage on a phylogenetic tree.

Xenologs Genes that are functionally similar to those found in other organisms but are present in a particular organism as a result of HGT.

REFERENCES

Amann, R. I., W. Ludwig, and K.-H. Schleifer. 1995. Phylogenetic identification and in situ detection of individual cells without cultivation. *Microbiol. Rev.* **59**:143–169.

Boucher, Y., C. J. Douady, R. T. Papke, D. A. Walsh, M. E. R. Boudreau, C. L. Nesbø, R. J. Case, and W. F. Doolittle. 2003. Lateral gene transfer and the origins of prokaryotic groups. *Annu. Rev. Genet.* **37**:283–328.

Boucher, Y., C. J. Douady, A. K. Sharma, M. Kamekura, and W. F. Doolittle. 2004. Intragenomic heterogeneity and intergenomic recombination among haloarchaeal rRNA genes. *J. Bacteriol.* **186**:3980–3990.

Bouchet, V., H. Huot, and R. Goldstein. 2008. Molecular genetic basis of ribotyping. *Clin. Microbiol. Rev.* **21**:262–273.

Brown, J. R., and W. F. Doolittle. 1997. Archaea and the prokaryote-to-eukaryote transition. *Microbiol. Mol. Biol. Rev.* **61**:456–502.

Buckley, M., and R. J. Roberts. 2007. *Reconciling Microbial Systematics and Genomics.* American Academy of Microbiology, Washington, D.C.

Cavigelli, M. A., G. P. Robertson, and M. J. Klug. 1995. Fatty acid methyl ester (FAME) profiles as measures of soil microbial community structure. *Plant Soil* **170**:99–113.

Chatton, E. 1937. *Titres et travaux scientifiques.* Sette, Sottano, Italy.

Coenye, T., D. Gevers, Y. Van de Peer, P. Vandamme, and J. Swings. 2005. Towards a prokaryotic genomic taxonomy. *FEMS Microbiol. Rev.* **29**:147–167.

Cohan, F. M. 2006. Towards a conceptual and operational union of bacterial systematics, ecology, and evolution. *Philos. Trans. R. Soc. B* **361**:1985–1996.

Cohan, F. M., and A. F. Koeppel. 2008. The origins of ecological diversity in prokaryotes. *Curr. Biol.* **18**:R1024–R1034.

Cohan, F. M., and E. B. Perry. 2007. A systematics for discovering the fundamental units of bacterial diversity. *Curr. Biol.* **17**:R373–R386.

Colwell, R. R. 1970. Polyphasic taxonomy of the genus *Vibrio*: numerical taxonomy of *Vibrio cholerae*, *Vibrio parahaemolyticus*, and related *Vibrio* species. *J. Bacteriol.* **104**:410–433.

Didelot, X., and D. Falush. 2007. Inference of bacterial microevolution using multilocus sequence data. *Genetics* **175**:1251–1266.

Doolittle, W. F. 1999. The nature of the universal ancestor and the evolution of the proteome. *Curr. Opin. Struct. Biol.* **10**:355–358

Eder, K. 1995. Gas chromatographic analysis of fatty acid methyl esters. *J. Chromatogr. B* **671**:113–131.

Feil, E. J., E. C. Holmes, D. E. Bessen, et al. 2001. Recombination within natural populations of pathogenic bacteria: short-term empirical estimates and long-term phylogenetic consequences. *Proc. Natl. Acad. Sci. USA* **98**:182–187.

Freeman, S., and J. C. Herron. 2004. *Evolutionary Analysis,* 3rd ed. Pearson Prentice Hall, Pearson Education, Inc., Upper Saddle River, NJ.

Garrity, G. M., D. R. Boone, and R. W. Castenholz. 2001. *Bergey's Manual of Systematic Bacteriology,* vol. 1. *The Archaea and the Deeply Branching and Phototrophic Bacteria,* 2nd ed. Springer-Verlag, New York, NY.

Garrity, G. M. 2005. *Bergey's Manual of Systematic Bacteriology,* vol. 2. *The Proteobacteria,* 2nd ed. Springer-Verlag, New York, NY.

Gevers, D., F. M. Cohan, J. G. Lawrence, B. G. Spratt, T. Coenye, E. J. Feil, E. Stackebrandt, Y. Van de Peer, P. Vandamme, F. L. Thompson, and J. Swings. 2005. Re-evaluating prokaryotic species. *Nat. Rev.* **3**:733–739.

Gogarten, J. P., W. F. Doolittle, and J. G. Lawrence. 2002. Prokaryotic evolution in light of gene transfer. *Mol. Biol. Evol.* **19**: 2226–2238.

Hagström, Å, J. Pinhassi, and U. L. Zweifel. 2000. Biogeographical diversity among marine bacterioplankton. *Aquat. Microb. Ecol.* **21**: 231–244.

Hanage, W. P., C. Fraser, and B. G. Spratt. 2005. Fuzzy species among recombinogenic bacteria. *BMC Biol.* **3:**6.

Hennig, W. 1965. Phylogenetic systematics. *Annu. Rev. Entomol.* **10:**97–116.

Hunt, D. E., L. A. David, D. Gevers, S. P. Preheim, E. J. Alm, and M. F. Polz. 2008. Resource partitioning and sympatric differentiation among closely related bacterioplankton. *Science* **320:** 1081–1085.

Johnson, J. 1973. Use of nucleic-acid homologies in the taxonomy of anaerobic bacteria. *Int. J. Syst. Bacteriol.* **23:**308–315.

Keys, C. J., D. J. Dare, H. Sutton, G. Wells, M. Lunt, T. McKenna, M. McDowall, and H. N. Shah. 2004. Compilation of MALDI-TOF mass spectral database for the rapid screening and characterization of bacteria implicated in human infectious diseases. *Infect. Genet. Evol.* **4:**221–242.

Koeppel, A., E. B. Perry, J. Sikorski, D. Krizanc, A. Warner, D. M. Ward, A. P. Rooney, E. Brambilla, N. Conner, R. M. Ratcliff, E. Nevo, and F. M. Cohan. 2008. Identifying the fundamental units of bacterial diversity: a paradigm shift to incorporate ecology into bacterial systematics. *Proc. Natl. Acad. Sci. USA* **105:**2504–2509.

Madigan, M. T., and J. M. Martinko. 2006. *Brock Biology of Microorganisms,* 11th ed. Pearson Prentice Hall, Pearson Education, Inc., Upper Saddle River, NJ.

Martin, W., and E. V. Koonin. 2006. A positive definition of prokaryotes. *Nature* **442:**868.

Mayr, E. 1942. *Systematics and the Origin of Species.* Columbia University Press, New York, NY.

McCarthy, B. J., and E. G. Bolton. 1963. An approach to the measurement of genetic relatedness among organisms. *Proc. Natl. Acad. Sci. USA* **50:**156–164.

McCartney, A. L., W. Wenzhi, and G. W. Tannock. 1996. Molecular analysis of the composition of the bifidobacterial and *Lactobacillus* microflora of humans. *Appl. Environ. Microbiol.* **62:**4608–4613.

Pace, N. R. 1997. A molecular view of microbial diversity and the biosphere. *Science* **276:**734–740.

Raymond, J., J. L. Siefert, C. R. Staples, and R. E. Blankenship. 2004. The natural history of nitrogen fixation. *Mol. Biol. Evol.* **21:** 541–554.

Rosselló-Mora, R., and R. Amann. 2001. The species concept for prokaryotes. *FEMS Microbiol. Rev.* **25:**39–67.

Ryzhov, V., and C. Fenselau. 2001. Characterization of the protein subset desorbed by MALDI from whole bacterial cells. *Anal. Chem.* **73:**746–750.

Sapp, J. 2006. Two faces of the prokaryote concept. *Int. Microbiol.* **9:**163–172.

Schloss, P. D., and J. Handelsman. 2004. Status of the microbial census. *Microbiol. Mol. Biol. Rev.* **68:**686–691.

Smith, J. M., N. Smith, M. O'Rourke, and B. G. Spratt. 1993. How clonal are bacteria? *Proc. Natl. Acad. Sci. USA* **90:**4384–4388.

Stackebrandt, E., and J. Ebers. 2006. Taxonomic parameters revisited: tarnished gold standards. *Microbiol. Today* **33:**152–155.

Stackebrandt, E. 2007. Forces shaping bacterial systematics. *Microbe* **2:**283–287.

Stanier, R. Y., and C. B. van Niel. 1962. The concept of bacterium. *Arch. Microbiol.* **42:**17–35.

Stebbins, G. L. 1950. *Variation and Evolution in Plants.* Columbia University Press, New York, NY.

Torsvik, V., J. Goksøyr, and F. L. Daae. 1990. High diversity in DNA of soil bacteria. *Appl. Environ. Microbiol.* **56:**782–787.

Torsvik, V., and L. Øvreås. 2002. Microbial diversity and function in soil: from genes to ecosystems. *Curr. Opin. Microbiol.* **5:**240–245.

Vandamme, P., B. Pot, M. Gillis, P. De Vos, and J. Swings. 1996. Polyphasic taxonomy, a consensus approach to bacterial systematics. *Microbiol. Rev.* **60:**407–438.

Vos, P., R. Hogers, M. Bleeker, M. Reijans, T. van de Lee, M. Hornes, A. Frijiters, J. Pot, J. Peleman, and M. Kuiper. 1995. AFLP: a new technique for DNA fingerprinting. *Nucleic Acids Res.* **23:**4407–4414.

Whittaker, R. H. 1969. New concepts of kingdoms of organisms. *Science* **163:**150–160.

Woese, C. R., W. E. Balch, L. J. Magrum, G. E. Fox, and R. S. Wolfe. 1977. An ancient divergence among the bacteria. *J. Mol. Evol.* **9:**305–311.

Woese, C. R., O. Kandler, and M. L. Wheelis. 1990. Towards a natural system of organisms: proposal for the domains Archaea, Bacteria, and Eucarya. *Proc. Natl. Acad. Sci. USA* **87:**4576–4579.

Wuyts, J., Y. Van de Peer, and R. D. Wachter. 2001. Distribution of substitution rates and location of insertion sites in the tertiary structure of ribosomal RNA. *Nucleic Acids Res.* **29:**5017–5028.

Yap, W. H., Z. Zhang, and Y. Wang. 1999. Distinct types of rRNA operons exist in the genome of the actinomycete *Thermomonospora chromogena* and evidence for horizontal transfer of an entire rRNA operon. *J. Bacteriol.* **181:**5201–5209.

Zuckerkandl, E., and L. Pauling. 1965. Molecules as documents of evolutionary history. *J. Theor. Biol.* **8:**357–366.

Web Resources

Bacterial (Prokaryotic) Phylogeny Webpage http://www.bacterialphylogeny.com/index.html

Bergey's Manual Trust (Taxonomic Outlines) http://www.bergeys.org/index.html

List of Prokaryotic Names with Standing in Nomenclature (LPSN) http://www.bacterio.cict.fr/

Maddison, D. R., and K.-S. Schulz (ed.), 2007. The Tree of Life Web Project http://tolweb.org

MicrobeWiki http://microbewiki.kenyon.edu/index.php/Microbial_Biorealm (student-edited microbiology resource at Kenyon College)

Todar's Online Textbook of Bacteriology http://textbookofbacteriology.net/index.html

1. Briefly explain the major problems plaguing the current bacterial taxonomic system, in which each microorganism is granted a genus and species designation.

2. Why is the prokaryotic species concept a controversial topic?

3. Do you think genomic techniques that reveal the metabolic potential of an organism should be considered part of a polyphasic approach to bacterial taxonomy? Why or why not? (*Hint:* Consider its strengths and limitations.)

4. Is the 16S rRNA gene appropriate for identifying microbes at the species level? Why or why not?

5. Assuming microorganisms can be clustered in groups with certain likenesses, how would you personally define a bacterial species? What approaches would you use to construct this definition?

UNIT 7

EXPERIMENTAL OVERVIEW

In Experiments 6.2 and 6.3, students performed BLAST and RDP-II searches using the DNA sequences for the 16S rRNA gene derived from bacterial isolates and clones. A taxonomic assignment can be proposed based on these results; however, conclusive evidence in support of these hypotheses awaits phylogenetic analysis using the methods described in Unit 7.

(Exps. 6.2 & 6.3)

>F08UCLA121ACH30R01
CGGATCGGCTATCTGTGGTACGTCAAACAGCAAG
GTATTAACTTACTGCCCTTCCTCCCAACTTAAAG
TGCTTTACAATCCGAAGACCTTCTTCACACACGC
GGCATGGCTGGATCAGGCTTTCGCCCATTGTCCA
ATATTCCCCACTGCTGCCTCCCGTAGGAGTCTGG
ACCGTGTCTCAGTTCCAGTGTGACTGATCATCCT
CTCAGACCAGTTACGGATCGTCGCCTTGGTAGGC

DNA sequences
from blastn and RDP-II

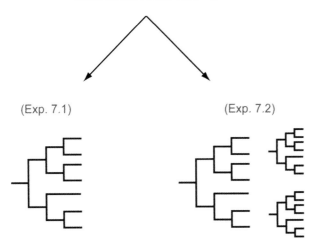

(Exp. 7.1)

Preliminary NJ trees
using RDP-II

1. Use Jukes-Cantor model
   for the alignment
2. Include OUTGROUP to
   root the tree
3. Display Bootstrap values

(Exp. 7.2)

Final NJ tree
using MEGA4

1. Use CLUSTALW to
   create the alignment
2. Include OUTGROUP to
   root the tree or exclude it
   for unrooted tree
3. Display Bootstrap values
   (2000 replicates)
4. Collapse branches and
   create subtrees

In Experiment 7.1, students will return to RDP-II and make a preliminary phylogenetic tree. Sequences aligned to the RDP-II structure-based model will be used to construct a rooted NJ tree using Jukes-Cantor distances (see Section 7.3 for review). In Experiment 7.2, students will use MEGA4 to build a phylogenetic tree. Recall that MEGA4 utilizes the CLUSTAL algorithm to create a multiple sequence alignment (refer to Section 6.4 for review). Students may decide whether to use an outgroup to root the MEGA4 tree. Furthermore, students should consider collapsing branches and creating subtrees to improve visualization of relationships among lineages in the tree. Finally, bootstrapping will be used to obtain statistical support for the bipartitions displayed in the phylogenetic trees produced by RDP-II and MEGA4 (for review, refer to Section 7.2).

EXPERIMENT 7.1 Tree Construction Using the Ribosomal Database Project

Students will be using the Ribosomal Database Project (RDP-II) Tree Builder to generate preliminary phylogenetic trees, acquiring FASTA sequence files for type strains which will be appended to a data set comprised of isolate or clone 16S rRNA gene sequences (Cole et al., 2005). The analysis program in Tree Builder takes both primary nucleotide sequence and secondary RNA structure into account when constructing the alignment used to make the phylogenetic trees. Tree Builder uses a weighted version of the NJ distance method, and allows you to construct phylogenetic trees with bootstrap confidence estimates. Tree Builder is available free, but an account must be created to perform data analysis. Instructions for creating an account are provided in Experiment 6.3.

METHODS

Part I Choose the "Nearest Neighbors" from SeqMatch

1. Access the RDP-II home page: http://rdp.cme.msu.edu/.

2. Log in to your myRDP account. You should be directed to the overview page. Click the box next to the group name to select all sequences. (*Hint:* Make sure the box is red, not grey, and that the number of aligned sequences in that group changes from 0 to a number greater than 1.)

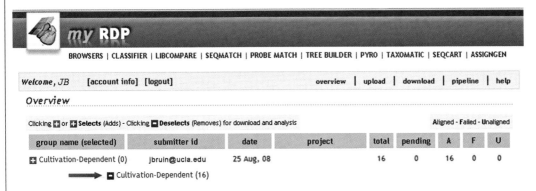

3. Select the SeqMatch link at the top of the page. Recall that the results obtained with SeqMatch can be filtered according to the specifications of the user. Select the same parameters as for your search in Experiment 6.3 or change the parameters if you wish to generate different results. Then press the button Do SeqMatch with Selected Sequences.

4. The results once again will be presented as a hierarchical view of the taxonomic categories, which can be expanded by clicking on View Selectable Matches. For each query sequence, choose two or three species with the highest similarity scores (highlighted in pink); these are considered the nearest neighbors for your query sequences. Make sure that each selection is a different species, and try to select the ones with a full genus and species name. If SeqMatch finds a possible match in more than one taxonomic group, select one sequence from each group. For example, if more than one genus is listed for a particular family, select one sequence from each genus. Check the box next to the sequence identifier, and then click the button Save Selection and Return to Summary.

Experiment continues

genus Massilia (0/17/77)

☐	S000253909	0.979	0.767	1374	Massilia sp. VA23069_03; AY445912
☐	S000322811	0.981	0.805	1328	beta proteobacterium Ellin152; AF408994
☑	S000426606	0.976	0.778	1374	Janthinobacterium sp. WSH04-01; AY753304
☑	S000437784	0.963	0.770	1418	Massilia timonae (T); timone; U54470
☐	S000472485	0.968	0.818	1370	Massilia timonae; CP183-9; AJ871463
☑	S000497324	0.976	0.841	1411	Janthinobacterium sp. IC161; AB196254

Save selection and return to summary

5. To save your progress, click on the SEQCART link at the top of the page. On the left side, it will show how many sequences are selected, separately identifying the isolate or clone sequences uploaded and aligned in the current group as well as the public sequences chosen from the list of SeqMatch results.

   To save the contents of the current sequence cart, click on Save Cart. You will be prompted to save the file to the hard drive of your computer. The file will have an .ids extension and can be retrieved later.

   Note: Anytime you want to completely start over with your public sequence selections, simply click Reset Cart.

   To download the contents of the sequence cart, click the Download link in the top right-hand corner.

Sequence Cart

[video tutorial | download]

Your sequence cart contains 59 sequences.

16 myRDP sequences, 16 aligned.
43 public sequences:

   3 type strain(s)
   0 environmental sequence(s)
   43 sequence(s) with length >= 1200 bases
   0 sequence(s) with low quality

RESET CART

Save your sequence cart (aligned only) for future use:

RDPX-Bacteria (59 seqs)   ▼   SAVE CART

Retrieve an existing file to current session:

File contains: ⦿ rdp id   ○ genbank accno

                         *choose upload: [   ] [Browse...]   RETRIEVE

6. There are numerous format options for downloading sequences. For the current application, use the following settings:

   a. "display Genbank" to display sequence descriptions using GenBank identifiers

   b. "fasta" to download files in FASTA format

   c. "aligned" and "show mask/structure" so the alignment reflects the secondary-structure model

   d. "jukes cantor" so model for distance matrix incorporates the Jukes-Cantor correction (for a review, see section 7.3)

○ fasta   ● aligned

○ phylip   ☑ show mask/structure

                           ○ unaligned

● display Genbank
    ACCNO / seq name

               ○ navigation tree
                  (ARB compatible) *more ARB help*

○ display RDP id

               ○ distance matrix   ○ uncorrected
                  (DNADist format)   ● jukes cantor

DOWNLOAD 59 SEQS FOR MODEL: RDPX-BACTERIA

Important: If the sequences are to be downloaded for use with MEGA in Experiment 7.2, the settings must be changed. Change setting 3 to "unaligned" and setting 4 to "uncorrected."

● fasta   ○ aligned

○ phylip   ☐ show mask/structure

                           ● unaligned

● display Genbank
    ACCNO / seq name

               ○ navigation tree
                  (ARB compatible) *more ARB help*

○ display RDP id

               ○ distance matrix   ● uncorrected
                  (DNADist format)   ○ jukes cantor

DOWNLOAD 59 SEQS

7. You will be prompted to save the file to the hard drive of your computer. The file will have a .fa extension and can be retrieved later. Make sure you modify the file name to reflect the format options used to download the sequences in step 6.

Part II Make a Preliminary Phylogenetic Tree

1. Before making a phylogenetic tree with the sequences downloaded from SeqMatch, you need to pick an outgroup from the Hierarchy Browser. Return to the results of the Classifier search from Experiment 6.3 (the information should have been printed or downloaded as a text file). Inspect the assignment detail information. For the outgroup, choose a 16S rDNA sequence from an organism that does not belong to the same phylum as any of your isolate or clone sequences previously uploaded to myRDP.

Experiment continues

To compile a list of potential outgroup sequences, click on the Browsers link at the top of the page. The Hierarchy View reveals a list of all phyla in domain *Archaea* and domain *Bacteria*. Note that there are a series of three numbers in parentheses to the right of each phylum name. Bacterial phyla with sequences available in RDP-II have a number greater than zero in the middle position. The number of sequences from a particular taxonomic group that have been selected for download is reflected by the number displayed in the first position.

⊞   domain Bacteria (43/240078/0)
⊞   phylum Aquificae (0/763/0)
⊞   phylum Thermotogae (0/312/0)
⊞   phylum Thermodesulfobacteria (0/77/0)
⊞   phylum Deinococcus-Thermus (0/505/0)

2. To choose an outgroup sequence, click on the name of an appropriate phylum, and then sequentially click on the class, order, family, and genus names, allowing you to view all possible species for that genus. Click on the box next to the species name and RDP identifier, and then click on the Root hyperlink under Lineage to include this sequence as your outgroup in the phylogenetic tree.

Note: You also may use Hierarchy Browser to select sequences more closely related to your query sequences than the outgroup but not as similar as the "nearest neighbors" obtained in Part I above. The addition of sequences with different levels of similarity to the query sequences will facilitate the creation of nested lineages in the phylogenetic tree. This effort is critical if a taxonomic assignment is to be given for the isolates and clones based on phylogenetic placement of their 16S rRNA gene sequences in a tree.

Lineage *(click node to return it to hierarchy view):*

<u>Root</u> (43/245557/0) ; Bacteria (43/240078/0) ; Aquificae (0/763/0) ; (0/729/0)

                    Click here to include the
                    sequence for the tree

Hierarchy View:

⊞ ▼ genus Aquifex (0/6/0) (selected/total/search matches)

☐ S000104174 Aquifex sp. Gri14L3B; AJ320223
☑ S000263494 Aquifex aeolicus; VF5; AJ309733
☐ S000392916 Aquificales str. CIR3017HO90; AF393377
☐ S000436807 Aquifex pyrophilus (T); Kol5a; M83548
☐ S000628091 Aquifex aeolicus VF5; AE000657
☐ S000628092 Aquifex aeolicus VF5; AE000657

3. Once all desired sequences have been selected in Hierarchy Browser, click on Tree Builder at the top of the page. From the Select Outgroup drop-down menu, choose the name of the species you just selected from Hierarchy Browser as your outgroup sequence. Then click the button Create Tree.

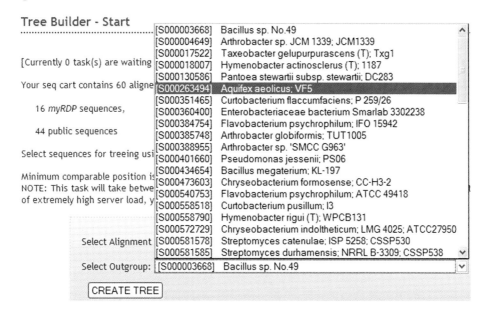

Tree Builder - Start

[Currently 0 task(s) are waiting

Your seq cart contains 60 aligne

   16 *myRDP* sequences,

   44 public sequences

Select sequences for treeing usi

Minimum comparable position i
NOTE: This task will take betwe
of extremely high server load, y

[S000003668]	Bacillus sp. No.49
[S000004649]	Arthrobacter sp. JCM 1339; JCM1339
[S000017522]	Taxeobacter gelupurpurascens (T); Txg1
[S000018007]	Hymenobacter actinosclerus (T); 1187
[S000130586]	Pantoea stewartii subsp. stewartii; DC283
[S000263494]	Aquifex aeolicus; VF5
[S000351465]	Curtobacterium flaccumfaciens; P 259/26
[S000360400]	Enterobacteriaceae bacterium Smarlab 3302238
[S000384754]	Flavobacterium psychrophilum; IFO 15942
[S000385748]	Arthrobacter globiformis; TUT1005
[S000388955]	Arthrobacter sp. 'SMCC G963'
[S000401660]	Pseudomonas jessenii; PS06
[S000434654]	Bacillus megaterium; KL-197
[S000473603]	Chryseobacterium formosense; CC-H3-2
[S000540753]	Flavobacterium psychrophilum; ATCC 49418
[S000558518]	Curtobacterium pusillum; I3
[S000558790]	Hymenobacter rigui (T); WPCB131
[S000572729]	Chryseobacterium indoltheticum; LMG 4025; ATCC27950

Select Alignment [S000581578] Streptomyces catenulae; ISP 5258; CSSP530
[S000581585] Streptomyces durhamensis; NRRL B-3309; CSSP538

Select Outgroup: [S000003668] Bacillus sp. No.49

CREATE TREE

4. Tree Builder generates a phylogenetic tree according to a distance matrix that uses the Jukes-Cantor corrected distance model, which considers that as two sequences diverge, the probability of a second substitution at any nucleotide site increases. Because multiple nucleotide changes may occur at a single alignment position, evolutionary distance is calculated using the Jukes-Cantor model (see Section 7.3 for a review). The phylogenetic tree created by Tree Builder is a weighted version of an NJ tree, or Weighbor tree, which gives significantly less weight to longer distances in the distance matrix. A sample tree is shown below.

Experiment continues

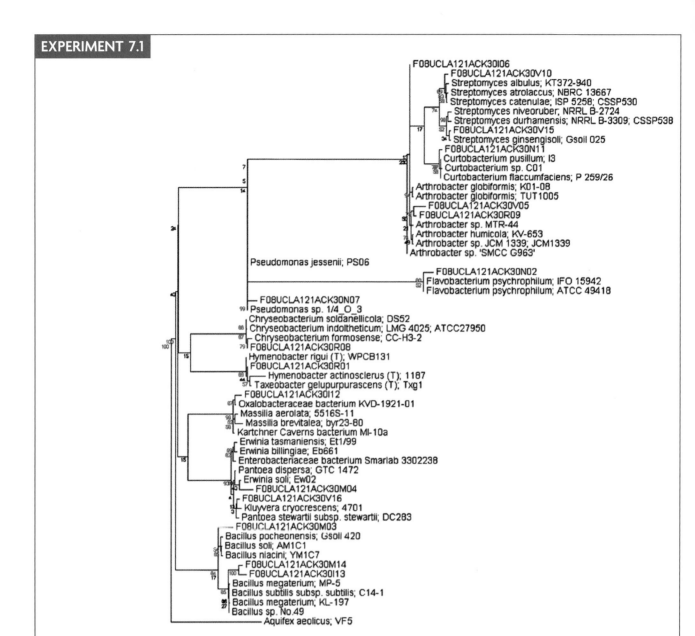

Notice that the results of the analysis are presented in an interactive java applet that allows you to rearrange the nodes of the tree and make other cosmetic changes by using the following commands:

a. Ctrl + N = Shows display name for sequences

b. Ctrl + I = Shows ID for sequences

c. Ctrl + D = Shows description for sequences

d. To zoom in, press minus (−) key

e. To zoom out, press equal (=) key

f. Alt + Click on node = Swap branches

g. To show or hide distance, press the d key

h. To show bootstrap values, press the b key

i. The Spacebar toggles Edit/Print mode

Bootstrapping is a statistical method for evaluating which branches are well-supported in a phylogenetic tree (refer to Section 7.2 for a review). The reliability of the sampling distribution is estimated by creating a pseudoalignment, in which random positions in the original alignment are sampled, replacing the original alignment. This sampling process is repeated 100 to 2,000 times, and a majority consensus tree is displayed showing the number (or percentage) of times a particular group appeared in the bootstrap replicates. A value above 70% provides strong support for, or confidence in, a particular node or clade.

The bootstrap value for each node in the tree produced by Tree Builder should be carefully inspected, and obvious problems (e.g., bootstrap values below 70%) should be fixed. For example, if the outgroup is nested within a clade that contains one of your isolate or clone sequences, choose a different outgroup and repeat the analysis (see Section 7.1 to review outgroup and nested hierarchy concepts). If an isolate or clone sequence exists on an isolated branch in the tree and does not cluster with its nearest neighbors, return to your SeqMatch results and pick sequences from alternative species in the list. You also may add more sequences from lineages intermediate to the outgroup and nearest neighbors from Hierarchy Browser.

5. Once you have finished with the tree manipulations, save your tree in Postscript format by pressing Ctrl + P. Although Mac users may use this format directly, PC users must convert this file to PDF format, using the free PS2PDR converter, by selecting the link (right click to open in new window or tab) within RDP to the website www.ps2pdf.com.

If the Adobe Acrobat suite of programs is available, you may open your Postscript file with Distiller and save your tree as a PDF or as a JPG (.jpg) or TIFF (.tif) file. The latter two options create images that can be imported into graphics programs such as CorelDraw, Adobe Illustrator, and PowerPoint. This feature is useful if you want to add or change labels on internal and external nodes of your tree (e.g., highlight or label taxonomic groups within the nested hierarchy of the tree), which is especially important in delineating sister taxa and monophyletic groups within a nested hierarchy.

You also may save your tree as a Newick file, which can be viewed in the alternative tree drawing program called TreeView (Page, 2003). This program can be downloaded from the website http://taxonomy.zoology.gla.ac.uk/rod/treeview.html.

TreeView does not create trees, but uses files created by alignment programs to display and print the trees. You also may convert TreeView files to the Postscript format, which can be converted to PDF, JPG, or TIF formats (described above) for importing to graphics programs.

To print your phylogenetic tree on multiple pages directly from TreeView: In TreeView, select Print Preview from the File menu. Once in the Preview window, you should see in the middle of the tool bar a drop-down menu for Pages. The default setting is 1. You may select 2 or 3 pages (maybe more, depending on how many taxa you include in your tree).

To convert TreeView files to a format compatible with other graphics programs: In TreeView, select Print from under the File menu (or within the Print Preview window), and then define your parameters:

Experiment continues

Printer Name > Select Adobe PDF.

Under Properties > Deselect "Do not send fonts to Adobe PDF", then press OK to return to the Print window.

Select Print to File, then press OK.

Output File Name: "FILENAME.ps" to create Postscript file.

6. Print a copy of your tree in PDF format for your laboratory notebook and/or final report.

REFERENCES

Cole, J. R., B. Chai, R. J. Farris, Q. Wang, S. A. Kulam, D. M. McGarrell, G. M. Garrity, and J. M. Tiedje. 2005. The Ribosomal Database Project (RDP-II): sequences and tools for high-throughput rRNA analysis. *Nucleic Acids Res.* **33:**D294–D296 (doi:10.1093/nar/gki038).

Page, R. D. M. 2003. Visualizing phylogenetic trees using TreeView, Unit 6.2. *In Current Protocols in Bioinformatics.* John Wiley & Sons, Inc., New York, NY.

Students will be using Molecular Evolutionary Genetics Analysis software version 4.0 (MEGA4) to create an optimized, multiple-nucleotide sequence alignment required to generate an evolutionary distance-based phylogenetic tree (Kumar et al., 2008; Tamura et al., 2007). **ClustalW** is the primary multiple sequence alignment program in MEGA4, which is available free at the website http://www.megasoftware.net/.

A multitude of alternative packages (>300) and servers (>50) are available for creating multiple sequence alignments and building phylogenetic trees—some are free while others must be purchased. Please consult the website http://evolution.genetics.washington.edu/phylip/software.html for the most comprehensive list currently available.

For Experiment 7.2, students will use MEGA4 to construct an NJ tree with an analysis that includes 2,000 bootstrap replicates (review bootstrapping in Section 7.2). If there are many taxa in the tree, as may be required to strengthen bootstrap support for various branches in the tree, it may be necessary to collapse branches and create subtrees in order to simplify presentation of the tree. If possible, include an outgroup in the rooted tree (refer to Section 7.1 for a review of the nested hierarchy and outgroup concepts). However, it is appropriate to present the results as an unrooted tree. The following procedure details the steps required to build a rooted NJ tree with bootstrap values. More advanced manipulations of the tree require consultation of the literature (Hall, 2008). Ultimately, the goal is to create a tree that will support or refute the hypothesis generated by previous nucleotide BLAST or RDP-II Classifier results.

METHODS

Part I Uploading FASTA files to MEGA

1. Download MEGA4 onto your computer (the website URL is provided above).

2. Retrieve or create a single FASTA file containing all 16S rDNA sequences that will be included in the MEGA4 phylogenetic analysis. Specifically, this file should contain the 16S rRNA gene sequences for isolates or clones, as well as the gene sequences for the best hits from your NCBI-BLAST search (Experiment 6.2), for nearest neighbors obtained by using RDP-II SeqMatch (Experiments 6.3 and 7.1), and for additional sequences (i.e., outgroup, intermediate lineages) acquired using RDP-II Hierarchy Browser (Experiment 7.1). Confirm that the description line for FASTA sequences in this file refers to your isolate or clone sequence by "Sample ID #":

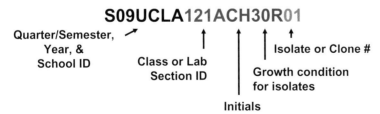

Important: Avoid using the pound sign (#) in the description of your sequences since MEGA uses this symbol as part of the header information. In addition, there should be no spaces in the Sample ID. Replace spaces with the underscore character (_).

3. Before a phylogenetic tree can be generated by MEGA, the FASTA file must be converted to MEGA format (.meg), which is an aligned series of sequences generated by ClustalW. Before the sequences can be aligned by MEGA, the FASTA file extension

Experiment continues

must be changed from .txt or .fa to .fasta, otherwise it will not be recognized by MEGA.

If the file extensions are not visible, open the folder containing the FASTA file. Select Tools from menu bar, and then click Folder Options as shown:

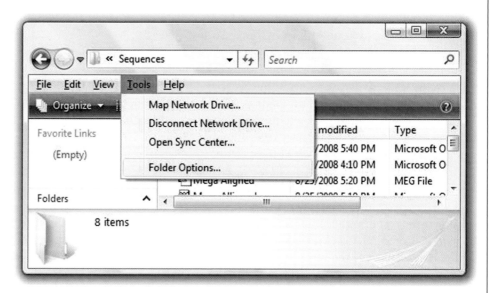

Under the View tab, deselect the box next to Hide Extensions for Known File Types, and then press the Apply button and OK to close the window. You now should be able to view the extension for all files in that folder.

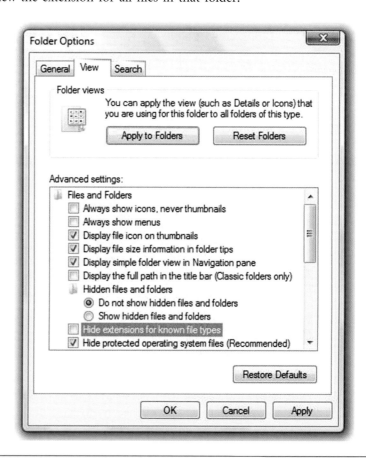

Change the file extension from .txt or .fa to .fasta, noting a difference in the appearance of the icon as follows:

myrdp_downlo
ad_59_seqs.fa

myrdp_downlo
ad_59_seqs.fa
sta

Note: For additional instructions about file interconversions, students are encouraged to consult the literature (i.e., see Appendix I of Hall [2008]).

4. The FASTA file with .fasta extension eventually will be converted to a MEGA file (.meg). This manipulation is done within the MEGA program itself once an optimized multiple sequence alignment has been constructed.

a. Open MEGA4, and select Alignment Explorer/CLUSTAL from the Alignment menu option as shown:

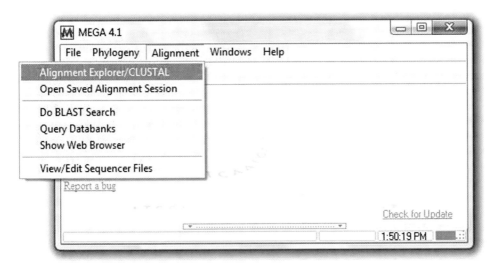

Select Create a New Alignment from the option list, and press OK.

A confirmation window will appear, asking: "Are you building a DNA [Yes] sequence alignment (otherwise choose [No] for Protein)?" Select Yes for DNA sequence alignment.

Experiment continues

b. In the Alignment Explorer window, select Insert Sequence From File under the Edit menu options as shown:

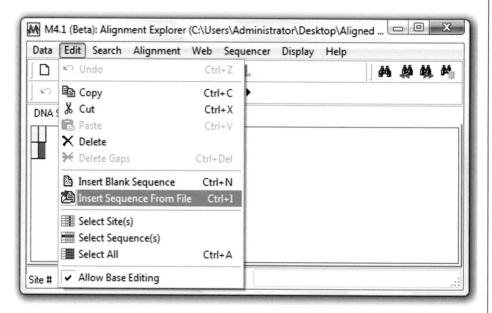

Notice that Allow Base Editing is selected, a feature that permits modification of your aligned sequences in Part II.

Choose FASTA from the drop-down menu under Files of Type. Then select your .fasta file created above in step 3, and press Open.

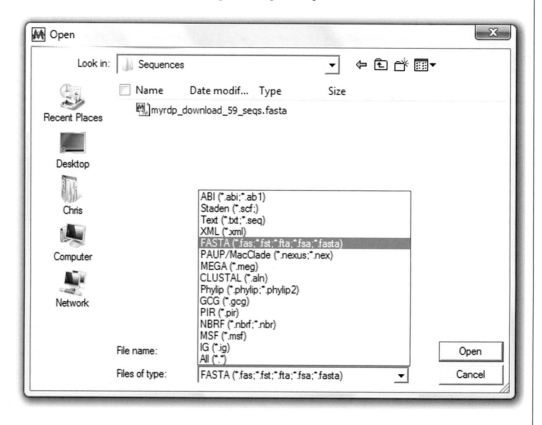

c. Once the FASTA file is uploaded to MEGA, the sequences will be displayed in the Alignment Explorer window. Shorten the identifiers for each sequence, leaving only the Sample ID for isolates and clones, or the genus and species name for type strains (e.g., eliminate GenBank accession numbers or RDP identifiers). This modification will make it easier to recognize the names of organisms in your phylogenetic tree. However, *before deleting this information from the MEGA window,* make sure that the GenBank accession numbers or RDP identifiers have been recorded in a table in your laboratory notebook together with the shortened identifier so that the data can be cross-referenced and the original sequence information can be independently verified.

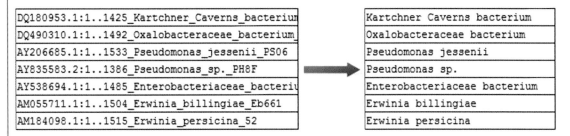

DQ180953.1:1..1425_Kartchner_Caverns_bacterium	Kartchner Caverns bacterium
DQ490310.1:1..1492_Oxalobacteraceae_bacterium	Oxalobacteraceae bacterium
AY206685.1:1..1533_Pseudomonas_jessenii_PS06	Pseudomonas jessenii
AY835583.2:1..1386_Pseudomonas_sp._PH8F	Pseudomonas sp.
AY538694.1:1..1485_Enterobacteriaceae_bacterium	Enterobacteriaceae bacterium
AM055711.1:1..1504_Erwinia_billingiae_Eb661	Erwinia billingiae
AM184098.1:1..1515_Erwinia_persicina_52	Erwinia persicina

d. SAVE the MEGA session, which includes all current work, by selecting Save Session from the Data menu options. Provide a file name (*hint:* it may be helpful to note in the file name that this session produced an unaligned MEGA file). Make sure the file type indicates that the file is a MEGA alignment session (Aln Session), which has a .mas file extension.

Part II Multiple Sequence Alignment in MEGA by ClustalW

1. Open the saved MEGA session from Part I above by selecting Open Saved Alignment Session from the Alignment menu in the main MEGA4 window. Choose the appropriate file name with the .mas extension.

2. Select all the sequences by choosing Select All from Edit menu. Start creating an alignment with selected sequences by choosing Align by ClustalW from the Alignment menu.

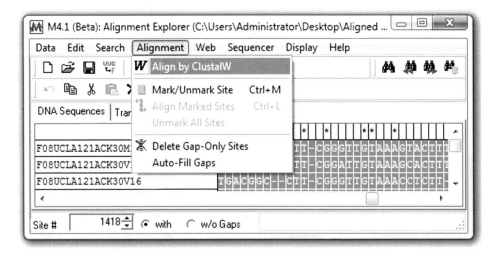

Experiment continues

a. The default gap opening and extension penalties may be changed for the pairwise alignment used to make a guide tree as well as the multiple alignment parameters used to make the final tree. For now, change the Gap Opening Penalty under Multiple Parameters from 15 to 3 and the Gap Extension Penalty from 6.66 to 1.8 (Hall, 2008). Then press OK to continue with the alignment.

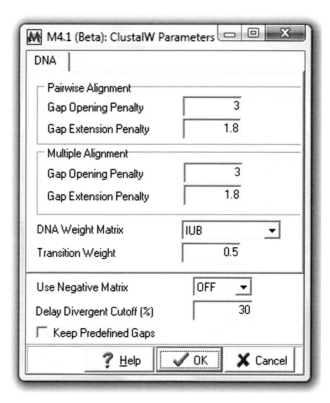

b. A MEGA window showing the progress of the ClustalW pairwise alignment (guide tree construction) and multiple alignment will appear. The time needed to complete this process varies, depending on the number of sequences and settings selected for alignment parameters.

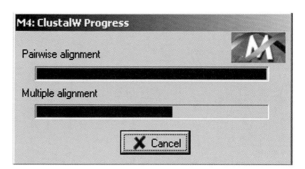

3. The sequences need to be edited *after* the alignment. For the isolate sequence reads generated from the 519R primer, the 16S rRNA sequences are probably much shorter than the 16S rRNA sequences obtained from NCBI-BLAST or RDP-II. The extra nucleotides at the beginning and end of each database sequence have to be deleted. These manipulations can be done within the MEGA Alignment Explorer widow. Simply highlight any nucleotides you wish to remove from the alignment, and press the Delete key.

Note: If you make a mistake, you may cancel your action by selecting Undo from the Edit menu.

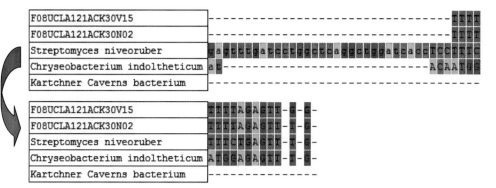

4. Save the current MEGA session by selecting Save Session from the Data menu options. Provide a new file name (*hint:* it may be helpful to note in the file name that this session produced the first aligned MEGA file). Confirm that the file has a .mas file extension.

Part III Acquiring Additional Sequences for the Alignment

During the MEGA alignment sessions, it may be determined that additional nucleotide sequences are needed. For example, a different outgroup sequence or homologous sequences from alternative lineages may be desired, depending on the outcome of the phylogenetic analysis in Part IV. To acquire additional sequences for the alignment, follow the next series of steps.

1. Access the RDP-II home page: http://rdp.cme.msu.edu/.

2. Open the Browser link at the top of the page. Select a sequence from the list of organisms on Hierarchy Browser. Place the cursor over the ID number next to the species name. As shown below, a small window will appear, giving you two options for the DNA sequence format: FASTA or GENBANK. Click on FASTA, and a new tab will open with the sequence displayed in FASTA format. Copy the URL for this page.

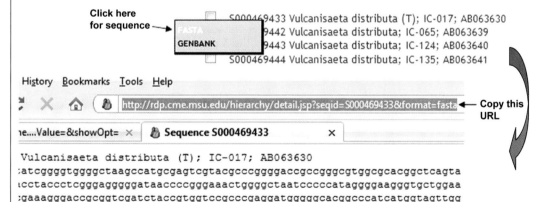

Experiment continues

3. Return to the Alignment Explorer window in MEGA, and select Show Browser from Web menu options.

4. Paste the URL from step 2 in the Address window. Then press the Add to Alignment button.

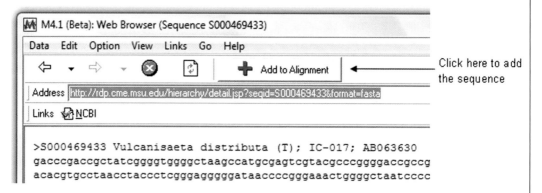

The nucleotide sequence will be added to your alignment (highlighted blue in the example shown below).

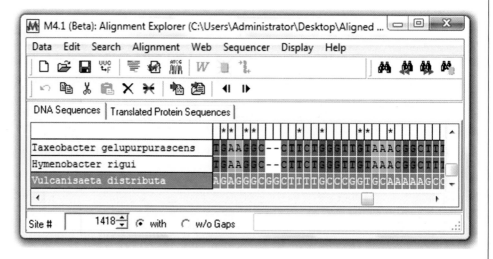

5. Because you are adding unaligned sequences to an aligned series of sequences, you will need to repeat the alignment procedure described in Part II (step 2), except that you should select the box next to Keep Predefined Gaps in the ClustalW Parameters window for assigning gap penalties. By doing so, the addition of new sequences will not drastically change the multiple sequence alignment, which can become problematic when the new sequences are of a different length or have a significantly different sequence composition (as may be the case with a new outgroup, for example).

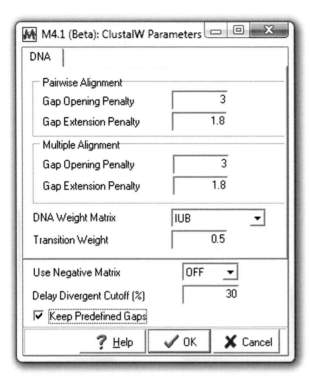

Caution: The decision to add new sequences in a manner that differs from the original sequences could introduce a systematic bias into the phylogenetic analysis (see Section 7.2 for other examples of bias in tree construction). Keeping predefined gaps makes the process of adding new sequences to the alignment faster but does not necessarily make it better. If the addition of a new sequence destroys the original alignment, a different sequence (i.e., one that is more similar to the sequences in the original alignment) should be used instead. This strategy achieves an entirely new alignment that is subject to less bias than one that maintains predefined gaps.

6. Once the alignment has finished, delete the portions of the new sequences that extend beyond the length of the original aligned sequences, similar to what was done in Part II (step 3).

7. Save the current MEGA session by selecting Save Session from the Data menu options. Provide a new file name (*hint:* it may be helpful to note in the file name that this session produced the second or third aligned MEGA file). Confirm the file has a .mas file extension.

Part IV Building a Phylogenetic Tree from a Multiple Sequence Alignment in MEGA

1. Once you are satisfied with the quality of the multiple sequence alignment, you are ready to build a phylogenetic tree. However, the aligned sequences first must be converted to MEGA format (.meg) as follows:

Experiment continues

a. Open the saved MEGA session from Part II or III above by selecting Open Saved Alignment Session from the Alignment menu in the main MEGA4 window. Choose the appropriate file name with the .mas extension.

b. To convert the data set to MEGA format, select Export Alignment from the Data menu options, choosing the MEGA Format. Provide a file name, and confirm that the file has a .meg file extension before pressing Save.

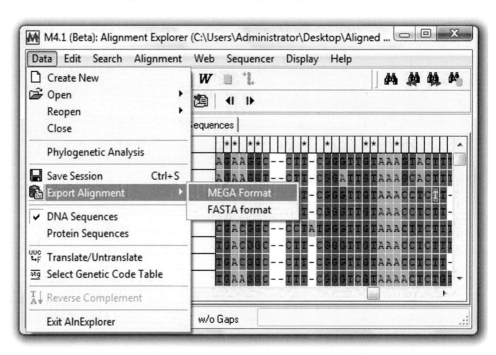

Provide a short title for your aligned data set when prompted by the window requesting Input Title of the Data, then press OK.

Since the 16S rRNA gene does not encode protein data, press No when asked "Protein-coding nucleotide sequence data?" in the next window prompt.

Note: Please be aware that unexpected conversion errors may occur, preventing immediate conversion of FASTA files to the MEGA format. To troubleshoot such problems, students must use the Text Editor, which can be found under the File menu option in the original MEGA4 window. For a detailed description of MEGA file specifications, students are encouraged to read the *MEGA4 Manual,* available in PDF format from the MEGA website (see Section 5, part 3, of the manual).

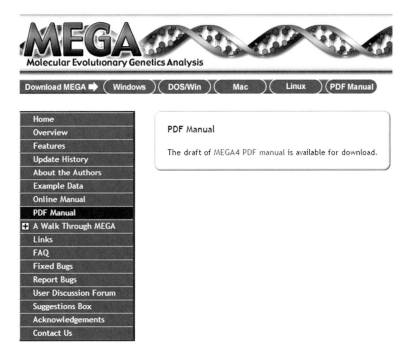

Briefly, the MEGA file converter looks for a line in the FASTA file that begins with a greater-than sign (>) and replaces it with a pound sign (#). It assumes that the first word (i.e., the string of characters comprising the sample ID) following the greater-than sign (>) in the FASTA file is the name of the sequence and deletes the rest of the description line. All information in the description line following the first word is deleted. The following line is taken as the nucleotide sequence data, ending with the next line that began with a greater-than sign (>) in the FASTA file.

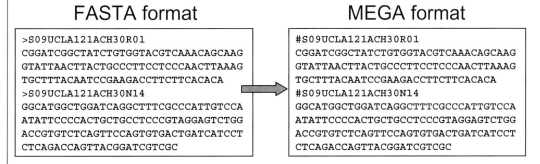

FASTA format

```
>S09UCLA121ACH30R01
CGGATCGGCTATCTGTGGTACGTCAAACAGCAAG
GTATTAACTTACTGCCCTTCCTCCCAACTTAAAG
TGCTTTACAATCCGAAGACCTTCTTCACACA
>S09UCLA121ACH30N14
GGCATGGCTGGATCAGGCTTTCGCCCATTGTCCA
ATATTCCCCACTGCTGCCTCCCGTAGGAGTCTGG
ACCGTGTCTCAGTTCCAGTGTGACTGATCATCCT
CTCAGACCAGTTACGGATCGTCGC
```

MEGA format

```
#S09UCLA121ACH30R01
CGGATCGGCTATCTGTGGTACGTCAAACAGCAAG
GTATTAACTTACTGCCCTTCCTCCCAACTTAAAG
TGCTTTACAATCCGAAGACCTTCTTCACACA
#S09UCLA121ACH30N14
GGCATGGCTGGATCAGGCTTTCGCCCATTGTCCA
ATATTCCCCACTGCTGCCTCCCGTAGGAGTCTGG
ACCGTGTCTCAGTTCCAGTGTGACTGATCATCCT
CTCAGACCAGTTACGGATCGTCGC
```

There are a number of other features common to all files in MEGA format. The first line must contain the keyword #mega to indicate that the data file is in the proper format for the program. On the second line, the data file should include a title, which is a succinct description of the data. On the third line, the data file may include a description statement. The title and description must be written according to a specific set of rules. The title may not occupy more than one line of text. It must begin with !Title and end with a semicolon (;). The description may occupy multiple lines of text. It must begin with !Description and end with a semicolon. Neither the title nor the description may contain semicolons inside the statements.

Experiment continues

```
#mega

!Title  This is an example title;
!Description  This is detailed information the data file;
```

2. Once the sequences are in MEGA format, a phylogenetic tree can be constructed from the alignment. In the main MEGA4 window, click on the "Click me to activate the data file" link. Open the alignment file that you exported in MEGA format in step 1. MEGA automatically opens the Sequence Data Explorer window, which can be minimized or closed.

   After the alignment file has been activated, return to the main MEGA4 window and select Bootstrap Test of Phylogeny and then Neighbor-Joining from the Phylogeny menu options.

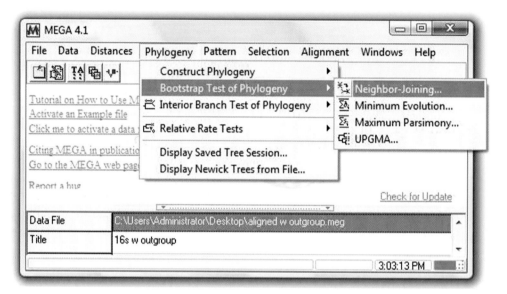

In the Analysis Preferences window, open the Options Summary tab. Choose Pairwise Deletion instead of Complete Deletion for the Gaps/Missing Data option (Hall, 2008).

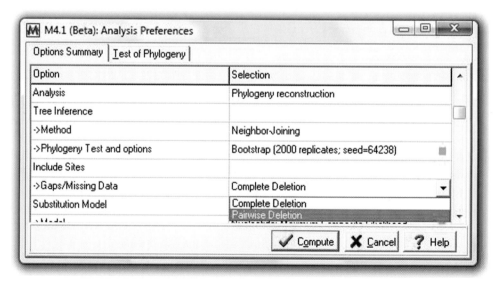

Note: The Pairwise Deletion option is only a good choice if all the sequences in the alignment are nearly the same length and the edges of the alignment have been trimmed back to a well-aligned section of the gene sequence.

Next, open the Test of Phylogeny tab. Make sure Bootstrap is selected under Test of Inferred Phylogeny, and change the number of Replications from 500 to 2,000 (Hall, 2008). Click on the red check mark.

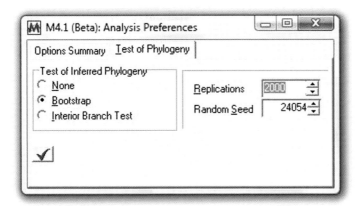

Press Compute at the bottom of the Analysis Preferences window to construct a phylogenetic tree.

3. After MEGA finishes making the tree, it may appear very crowded, making it difficult to actually read the names of the organisms at the branch tips or the bootstrap values at the internal nodes of the tree. To make it easier to inspect the tree details, select Topology Only under the View menu options or click the button as shown below:

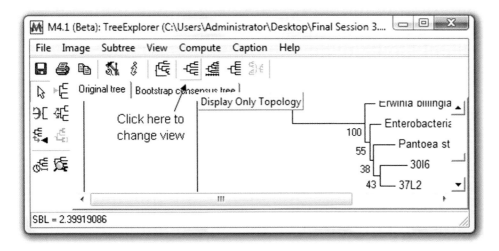

At this point, every tree produced looks different because the sequences for every tree are unique to a project.

Experiment continues

Save your tree by selecting Save Current Session under the File menu options. Provide a file name, and confirm that the file has a Tree Session File extension (.mts) before pressing Save. This option allows you to return to the tree analysis later. If you instead select Export Current Tree (Newick) from the File menu options, your tree may be viewed in the alternative tree-drawing program called TreeView (see Experiment 7.1 for details).

It is advisable that you print the tree as a record for your laboratory notebook by selecting Print from the File menu. However, you also may save the image as a TIFF file (.tif) or copy the image to the clipboard by selecting the appropriate options under the Image menu. These file formats allow you to import the tree images into graphics programs such as PowerPoint, CorelDraw, and Adobe Illustrator as well as text editing programs such as Microsoft Word. You may then print the tree on a single page, since the image size can be reduced to fit the page. In addition, this feature is useful if you want to add or change labels on internal and external nodes of your tree (e.g., highlight or label taxonomic groups within the nested hierarchy of the tree), which is especially important in delineating sister taxa and monophyletic groups within a nested hierarchy.

4. For each tree produced by MEGA, obtain the caption for the figure legend detailing the specifications of your analysis. Click on Caption, and a new window will appear with the text modified in accordance with the analysis performed. Print the caption for your laboratory notebook. In addition, copy the text to the clipboard and paste the caption in a text editing program (e.g., Microsoft Word). Save this information with your tree for subsequent reference or presentation purposes.

5. Make sure sample ID numbers and taxonomic assignments for all isolates and clones have been recorded in the "I, Microbiologist" database (CURL Online Lab Notebook).

Part V Troubleshooting the Analysis

1. Carefully inspect your phylogenetic tree, noting poor bootstrap values at internal nodes, multifurcating branches, any species for which there appears to be redundant representation, or organism sequences (e.g., nearest neighbors) that do not cluster with the isolate or clone sequences as expected. For example, in the tree figure below, several species (red circles) are slated for removal from the tree for one or more of these reasons.

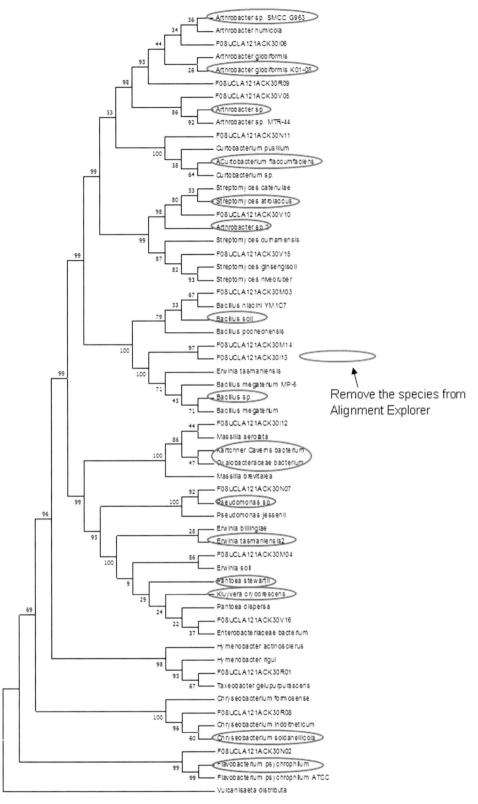

Remove the species from
Alignment Explorer

Experiment continues

2. Remove undesired sequences by opening the relevant alignment session (*hint:* it should have a .mas file extension) in the Alignment Explorer window. Click on the sequence you would like to remove, and simply press the Delete key. After those sequences are eliminated from the alignment, repeat the tree construction process (Part IV). Notice in the example that the tree now is simplified and redundancy is minimized.

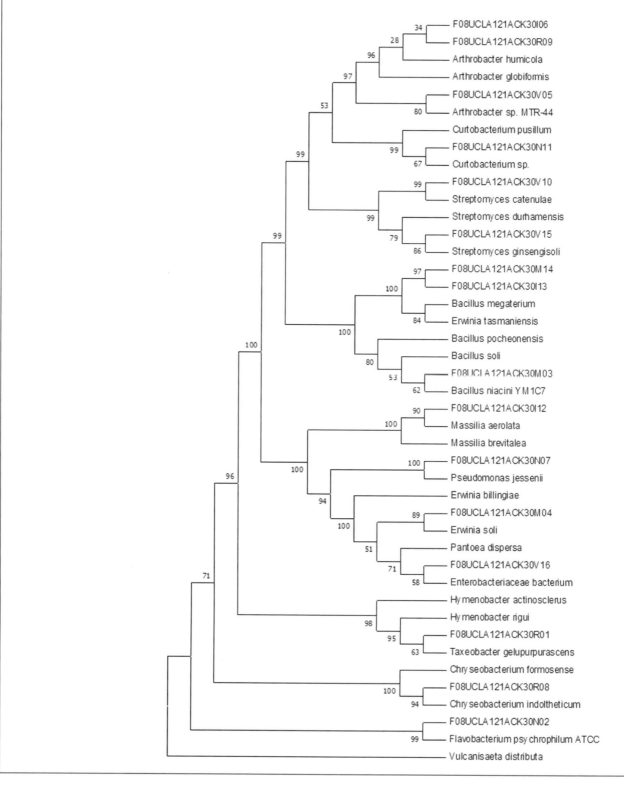

This type of manipulation will have to be done over and over again, ideally until a tree is obtained with few to no multifurcating nodes, few to no redundancies, and well-nested isolate and clone sequences such that a taxonomic assignment can be made to the lineage comprising the monophyletic group encompassing your isolate and clone sequences.

3. To improve bootstrap values for the internal nodes in the phylogenetic tree, it may be necessary to choose additional sequences from the same genus, family, order, class, or phylum as your isolates or clones, depending on the level of resolution desired in making a taxonomic classification. Remember that an internal node defines the lineage for which the taxonomic assignment is being made. Each node can delineate any level within the hierarchy. The goal is for your sequence to be nested within a taxonomically defined clade, which can be constructed by the inclusion of sequences intermediate to the outgroup and nearest neighbors. (*Hint:* Think of an intermediate sequence as a bridge linking two distantly related sequences because it has greater similarity to each of the two sequences than the two sequences have to each other.) It is impractical to expect that all isolate and clone sequences will be classified at the genus or species level, but adding sequences representing transitional positions in the lineage may help define an internal node, allowing you to identify the phylum, class, order, or even family to which a sequence belongs.

   To acquire additional sequences for the alignment and tree, follow the procedure provided in Part III above.

4. For the cultivation-independent part of the "I, Microbiologist" project, building an informative phylogenetic tree can be more challenging. It is difficult to find matches in NCBI-BLAST or RDP-II SeqMatch that are type strains because the sequences are derived from the uncultivated majority (recall the discussion of microbial bottlenecks to microbial ecology in Unit 2). In fact, the nearest neighbors may include only 16S rRNA genes from uncultured or uncharacterized bacterial species. It is less likely that this will be a problem for the cultivation-dependent part of the project. Thus, it is probable that the phylogenetic tree derived from cultivation-dependent sequences (isolates) will reflect taxonomic categories at the genus or species level, whereas the tree resulting from cultivation-independent sequences (clones) will reveal categories at only the higher taxonomic levels in the hierarchy (e.g., phylum, class, order, and maybe family).

5. It will not be a surprise if your phylogenetic tree has more than 50 sequences represented by the end of the analysis. As more and more sequences are added to the tree, branching order and branch length become obscured. One feature in MEGA that allows you to see details within clades composed of a large number of sequences is the set of drawing options under the Subtree menu options on the Tree Explorer page. Using this tool, you can collapse certain clades and then display them as separate tree figures, or subtrees. This way, you can easily visualize the relationships among clades without losing the necessary details.

Experiment continues

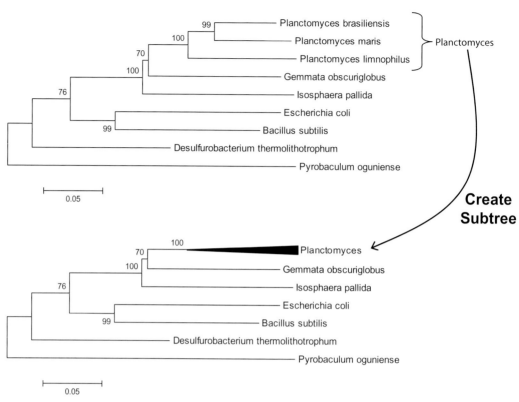

REFERENCES

Hall, B. G. 2008. *Phylogenetic Trees Made Easy: a How-To Manual,* 3rd ed. Sinauer Associates, Inc., Sunderland, MA.

Kumar, S., M. Nei, J. Dudley, and K. Tamura. 2008. MEGA: a biologist-centric software for evolutionary analysis of DNA and protein sequenes. *Briefings Bioinformatics* **9:**299–306.

Tamura, K., J. Dudley, M. Nei, and S. Kumar. 2007. MEGA4: Molecular Evolutionary Genetics Analysis (MEGA) software version 4.0. *Mol. Biol. Evol.* **24:**1596–1599.

APPENDIX
MEDIA, SUPPLIES, and EQUIPMENT

Part A
RECIPES AND PREPARATION NOTES FOR MEDIA

1. LB agar (also available commercially) (Miller, 1972)

   Tryptone .... 10.0 g
   Yeast extract .... 5.0 g
   NaCl .... 10.0 g
   Agar .... 15.0 g
   Distilled water .... up to 1.0 liter

   pH 7.0 at 25°C (may need to add up to 0.5 ml of 5 N NaOH)

   Autoclave at 121°C for 15 minutes to sterilize.

   For LB broth, omit agar.

   For LB + antibiotics, filter sterilize antibiotic solutions, and add to LB medium after autoclaving. The medium should be ca. 55°C. If antibiotics are added when the medium is too hot, they will be deactivated.

   - For LB-Kan plates (TOPO cloning): Add 50 mg of kanamycin per ml to a final concentration of 50 μg/ml.
   - For LB–Kan–X-Gal plates (TOPO cloning): Make a fresh 20-mg/ml solution of X-Gal by dissolving 400 mg of X-Gal in 20 ml of dimethylformamide. Then add 4.0 ml of the 20-mg/ml solution per liter of medium.

2. R2A agar (Reasoner and Geldreich, 1979)

   R2A agar .... 18.2 g
   Distilled water .... up to 1.0 liter

   The agar solution must be heated with frequent agitation before being autoclaved, as follows. Weigh out the appropriate amount of dry R2A agar into a 2-liter beaker. Add 1 liter of distilled water. Using a magnetic stirrer/hotplate, heat the solution while rapidly stirring. Continue heating until the dry component dissolves. Autoclave at 121°C for 20 minutes to sterilize.

   R2A agar (final pH of 7.2 at 25°C) contains (*The Difco Manual*, 11th ed.):

Yeast extract	Sodium pyruvate
Proteose Peptone no. 3	Potassium phosphate, dibasic
Casamino Acids	
Dextrose (glucose)	Magnesium sulfate
Soluble starch	Agar

3. 2× VXylA base (Modified VL55 base) (Davis et al. 2005)

   2-(*N*-Morpholino)ethanesulfonic acid
   (MES) .... 3.9 g
   20 mM MgSO$_4$ · 7H$_2$O .... 20 ml
   30 mM CaCl$_2$ · 2H$_2$O .... 20 ml
   20 mM (NH$_4$)$_2$HPO$_4$ .... 20 ml
   Selenite/tungstate solution .... 2 ml
   Trace-element solution SL-10 .... 2 ml
   Distilled H$_2$O .... 920 ml

   Adjust pH to 5.5 with 200 mM NaOH–100 mM KOH (about 12 ml). Autoclave at 121°C for 20 minutes to sterilize. After cooling to ca. 55°C, add the following:

   10% (wt/vol) xylan from birchwood
   (Fluka) .... 10 ml
   Vitamin solution I[c] .... 2 ml
   Vitamin solution II[D] .... 6 ml

4. 1× VXylA agar medium

   Phytagel .... 16 g
   Distilled H$_2$O .... 1.0 liter

   Autoclave at 121°C for 15 minutes. After cooling to ca. 55°C, add the Phytagel to equal volume (1 liter) of 2× VXylA base at ca. 55°C to make 2 liters of agar media.

   - For streak plating, increase the amount of Phytagel to 32 g/liter.
   - You should add the gel to the medium, not the medium to the agar, to avoid bubble formation.
   - For VXylA[hex] plates, add 1 ml of 20-mg/ml cycloheximide to a final concentration of 20 μg of cycloheximide per mil in a total volume of 1 liter.

5. Vitamin solution I[c]

   4-Aminobenzoate .... 40 mg
   (+)-Biotin .... 10 mg
   Nicotinic acid .... 100 mg
   Hemicalcium D-(+)-pantothenate ..... 50 mg
   Pyridoxine hydrochloride .... 150 mg
   Thiamine hydrochloride .... 100 mg
   Cyanocobalamin .... 50 mg
   Distilled H$_2$O .... 1.0 liter

   Mix well, and filter sterilize through a 0.22-μm-pore-size filter.

6. **Vitamin solution II**[D]

DL-6,8-Thioctic acid	10 mg
Riboflavin	10 mg
Folic acid	4 mg
Distilled H_2O	1.0 liter

Mix well, and filter sterilize through a 0.22-μm-pore-size filter.

7. **Selenite-tungstate solution**

NaOH	0.5 g
$Na_2SeO_3 \cdot 5H_2O$	3 mg
$Na_2WO_4 \cdot 2H_2O$	4 mg
Distilled H_2O	1.0 liter

Autoclave at 121°C for 20 minutes to sterilize.

8. **Trace-element solution SL-10**

25% HCl	10 ml
$FeCl_2 \cdot 4H_2O$	1.5 g
$CoCl_2 \cdot 6H_2O$	190 mg
$MnCl_2 \cdot 4H_2O$	100 mg
$ZnCl_2$	70 mg
H_3BO_3	6 mg
$NaMoO_4 \cdot 2H_2O$	36 mg
$NiCl_2 \cdot 6H_2O$	24 mg
$CuCl_2 \cdot 2H_2O$	2 mg
Distilled H_2O	1.0 liter

Autoclave at 121°C for 20 minutes to sterilize.

9. **200 mM NaOH–100 mM KOH**

NaOH	0.80 g
KOH	0.56 g
Distilled H_2O	100 ml

10. **20 mM $MgSO_4 \cdot 7H_2O$**

$MgSO_4 \cdot 7H_2O$	0.49 g
Distilled H_2O	100 ml

11. **30 mM $CaCl_2 \cdot 2H_2O$**

$CaCl_2 \cdot 2H_2O$	0.44 g
Distilled H_2O	100 ml

12. **20 mM $(NH_4)_2HPO_4$**

$(NH_4)_2HPO_4$	0.26 g
Distilled H_2O	100 ml

13. **Potassium phosphate buffers (1 M Na_2HPO_4 and 1 M NaH_2PO_4)**

K_2HPO_4 (dibasic formula weight [FW], 174.18 g/mol)	17.4 g
Distilled water	Up to 100 ml

Adjust to pH 6.3 with KOH or phosphoric acid.

KH_2PO_4 (monobasic FW 136.09)	13.6 g
Distilled water	up to 100 ml

Adjust to pH 6.3 with KOH or phosphoric acid.

Prepare 1 M phosphate buffer (PB) stock solution (100 ml) as follows. Mix 23.5 ml of 1 M K_2HPO_4 (dibasic, pH 6.3) with 76.5 ml of 1 M KH_2PO_4 (monobasic, pH 6.3). Confirm that the pH is still 6.3. Adjust with KOH or phosphoric acid as needed.

Filter sterilize (do not autoclave phosphate buffers).

14. **Na_2 pyruvate (1 M)**

Sodium pyruvate (FW 110.04)	11 g
Distilled water	up to 100 ml

Filter sterilize (do not autoclave).

15. **FeNaEDTA (20 mM)**

$FeSO_4 \cdot 7H_2O$	5.56 g
Na_2EDTA	7.45 g
Distilled water	up to 1.0 liter

Adjust the pH to 6.3 with NaOH. Dispense 100-ml aliquots into screw-cap bottles. Autoclave at 121°C for 20 minutes to sterilize.

16. **Micronutrients**

H_3BO_3	2.86 g
$MnCl_4 \cdot 4H_2O$	1.81 g
$ZnSO_4 \cdot 7H_2O$	0.22 g
$CuSO_4 \cdot 5H_2O$	0.08 g
$Na_2MoO_4 \cdot 2H_2O$	0.025 g
$CoSO_4 \cdot 7H_2O$	0.001 g
Distilled water	up to 1.0 liter

Filter sterilize, and aseptically dispense 100-ml aliquots into screw-cap bottles.

17. **Vitamins**

Thiamine HCl	0.1 g
Pyridoxine HCl	0.5 g
Nicotinic acid	0.5 g
Folic acid	0.1 g
Biotin	0.225 g
Riboflavin	0.1 g
Calcium pantothenate	0.1 g
Distilled water	up to 1.0 liter

Filter sterilize, and aseptically dispense 100-ml aliquots into screw-cap bottles.

18. **N$_2$-BAP agar (modified BAB medium) (Murry et al., 1984)**
    1 M NH$_4$Cl (FW 53.49) .... 5.0 ml
    80 mM MgSO$_4$ · 7H$_2$O (FW 246.47) ... 2.5 ml
    70 mM CaCl$_2$ · 2H$_2$O (FW 147.02) .... 1.0 ml
    20 mM FeNaEDTA .... 1.0 ml
    Micronutrients .... 1.0 ml
    Vitamins .... 1.0 ml
    Agar .... 15.0 g
    Distilled water .... up to 980 ml

Adjust to pH 6.3 dropwise with dilute KOH or NaOH. Autoclave at 121°C for 20 minutes to sterilize. After cooling to ca. 55°C, aseptically add the following sterile stock solutions:

    1 M PB (pH 6.3) .... 10.0 ml
    1 M Na$_2$ pyruvate .... 10.0 ml

The total volume should now be 1.0 liter.

19. **Cycloheximide (20 mg/ml)**

In a fume hood, weigh out 0.02 g of cycloheximide. Dissolve in 1 ml of 95% ethanol.

20. **Benomyl (50 mg/ml)**

In a fume hood, weigh out 0.05 g of benomyl. Dissolve in 1 ml of chloroform.

21. **ISP medium 4 agar (Shirling and Gottlieb, 1966)**
    ISP medium 4 agar .... 37.0 g
    Distilled water .... up to 1.0 liter

The agar solution must be heated with frequent agitation before being autoclaved, as follows. In a fume hood, weigh out the appropriate amount of ISP medium 4 agar into a 2-liter beaker. Add 1 liter of distilled water. Using a magnetic stirrer and a hot plate, heat the solution while rapidly stirring. Continue heating until the dry component dissolves. Then autoclave at 121°C for 20 minutes to sterilize.

For ISP 4 + antibiotics, filter sterilize the antibiotic solutions and add them to ISP medium 4 after autoclaving; the medium should be ca. 55°C. If antibiotics are added when the medium is too hot, they will be deactivated.

  - For ISP 4[hex, ben] plates: Add 1 ml of 20-mg/ml cycloheximide and 1 ml of 50-mg/ml benomyl to final concentrations of 20 μg of cycloheximide per ml and 50 μg of benomyl per ml in a total volume of 1 liter.

22. **Mueller-Hinton (M-H) agar (available commercially) (Mueller and Hinton, 1941)**

Add 38 g to 1 liter of distilled water. Boil to dissolve completely. Autoclave at 121°C for 15 minutes to sterilize.

M-H agar (final pH of 7.3 at 25°C) contains (*The Difco Manual*, 11th ed.):

Infusion from beef	Starch
Casamino Acids, technical	Agar

23. **Rhizobium defined medium (RDM) (Vincent, 1970)**
    RDM A stock .... 100.0 ml
    RDM B stock .... 100.0 ml
    Biotin stock (0.25 mg/ml) .... 4.0 ml
    Thiamine stock (10 mg/ml) .... 1.0 ml
    Sucrose .... 5.0 g
    Distilled H$_2$O .... up to 1.0 liter

Autoclave at 121°C for 20 minutes to sterilize.

For RDM agar plates, add 15 g of agar per liter of medium.

24. **RDM A stock (10×)**
    CaCl$_2$ · 2H$_2$O .... 1.0 g
    KNO$_3$ .... 6.0 g
    MgSO$_4$ · 7H$_2$O .... 2.5 g
    Distilled H$_2$O .... up to 1.0 liter

Autoclave at 121°C for 20 minutes to sterilize.

25. **RDM B stock (10×)**
    K$_2$HPO$_4$ .... 10.0 g
    KH$_2$PO$_4$ .... 10.0 g
    FeCl$_3$ · 6H$_2$O .... 0.1 g
    Distilled H$_2$O .... up to 1.0 liter

Autoclave at 121°C for 20 minutes to sterilize.

26. **SOC medium**
    Tryptone .... 20.0 g
    Yeast extract .... 5.0 g
    NaCl .... 0.58 g
    KCl .... 0.19 g
    MgCl$_2$ .... 2.03 g
    MgSO$_4$ .... 2.47 g
    Glucose (dextrose) .... 3.6 g
    Distilled water .... up to 1.0 liter

Autoclave at 121°C for 20 minutes to sterilize.

27. **Mineral salts (MS) base**

NH_4Cl .... 1.0 g
$NH_2HPO_4 \cdot 2H_2O$ .... 2.14 g
KH_2PO_4 .... 1.09 g
$MgSO_4 \cdot 7H_2O$ .... 0.2 g
Trace salts solution .... 10.0 ml*
Distilled water .... up to 1.0 liter
pH 7.0

*Trace salts solution is prepared in 0.1 N HCl as follows. It is added to the base before sterilization (autoclave at 121°C for 20 minutes).

$FeSO_4 \cdot 7H_2O$ .... 300.0 mg
$MnCl_2 \cdot 4H_2O$ .... 180.0 mg
$Co(NO_3)_2 \cdot 6H_2O$ .... 130.0 mg
$ZnSO_4 \cdot 7H_2O$ .... 40.0 mg
H_2MoO_4 .... 20.0 mg
$CuSO_4 \cdot 5H_2O$ .... 1.0 mg
$CaCl_2$ .... 1.0 g
HCl (0.1 N) .... up to 1.0 liter

28. **Minimal salts broth (MSB)**

MS base + 0.1% carbon source

29. **Glucose minimal salts broth (Glucose MSB)**

MS base + 0.1% carbon source + 0.1% or 0.2% glucose

30. **1-mg/ml carboxymethyl cellulose (CMC) overlay medium or hard agar (Teather and Wood, 1982)**

Carboxymethyl cellulose
   7H3 SXF .... 1 g
Agar .... 6 g (15 g for hard agar)
MSB .... 1.0 liter

CMC will be at a final concentration of 0.1%. Agar will be at a final concentration of 1.5% for hard agar and 0.6% for overlay medium. Autoclave at 121°C for 20 minutes to sterilize.

31. **Congo Red (1 mg/ml)**

Make solution fresh before use; properly discard as hazardous chemical waste after 3 months.

32. **1 M HCl (preparation depends on concentration and purity of vendor stock)**

Make solution fresh before use; discard after 3 months.

33. **1 M NaCl**

Dissolve 0.5844 g of NaCl in 100 ml of deionized water. Autoclave at 121°C for 20 minutes to sterilize. Make solution fresh before use; discard after 3 months.

34. **5 M guanidinium thiocyanate buffer (Pitcher et al., 1989)**

Guanidine thiocyanate .... 59.08 g
$Na_2EDTA \cdot 2H_2O$ .... 3.72 g
Sarkosyl (N-lauroylsarcosine) .... 0.5 ml
Total volume with deionized H_2O ..... 100 ml

Mix guanidine thiocyanate and EDTA while heating at 65°C until dissolved. After cooling to ca. 55°C, add Sarkosyl. Bring up the total volume with deionized water to 100 ml. Filter sterilize using a 0.45-μm filter. Store at ambient temperature.

35. **7.5 M ammonium acetate**

Ammonium acetate .... 57.81 g
Total volume with deionized H_2O ..... 100 ml

36. **Chloroform–2-pentanol (24:1, vol/vol)**

37. **Glycerol (80%, vol/vol)**

Prepare in a 1-liter graduated cylinder. Pour 100% glycerol into the cylinder to the 80-ml mark, and add 20 ml of distilled water. Cover, seal the top of the cylinder with Parafilm, and invert to mix. Autoclave at 121°C for 20 minutes to sterilize. Store at 4°C.

38. **Ethanol (70% vol/vol)**

Mix 70 ml of absolute ethanol (200 proof) with 30 ml of deionized water. Store at 4°C.

39. **0.5 M KCl**

Dissolve 0.7455 g of KCl in 100 ml of deionized water. Autoclave at 121°C for 20 minutes to sterilize.

40. **Dimethyl sulfoxide (DMSO)**

Sterilize with a nylon filter.

41. **Crystal violet**

Crystal violet .... 10.0 g
Ethanol (95%) .... 100.0 ml
Ammonium oxalate .... 4.0 g
Total volume with distilled H_2O .... 500 ml

42. **Gram's iodine**

I_2 .... 1.0 g
KI .... 2.0 g
Total volume with distilled H_2O .... 500 ml

43. Alcohol

Ethanol (95%)

44. Safranin

Safranin .... 1.0 g
Ethanol (95%) .... 40.0 ml
Distilled H$_2$O .... 400 ml

45. TE buffer (10 mM Tris-HCl, 1 mM EDTA) (pH 8.0)

1 M Tris-HCl (pH 8.0) .... 1.0 ml
0.5 M EDTA (pH 8.0) .... 0.2 ml
Sterile distilled water .... 90.0 ml

Prepare stock solutions of 1 M Tris-HCl and 0.5 M EDTA. Check the pH, and adjust each to pH 8 with NaOH as necessary. Autoclave at 121°C for 20 minutes to sterilize. Mix the indicated volumes of sterile stocks, and add sterile water such that the total volume is 100 ml.

46. 1× TAE buffer

First prepare a 50× TAE buffer stock as follows:
Tris base .... 242.0 g
Glacial acetic acid .... 57.1 ml
0.5 M EDTA (pH 8.0) .... 100.0 ml
Distilled water .... up to 1.0 liter

Autoclave at 121°C for 20 minutes to sterilize. Then dilute the stock solution 1:50 into deionized water for use as 1× working stock solution for agarose gel electrophoresis.

47. 6× TAE load dye

Xylene cyanol (XC) ... 0.25 g (0.25%, wt/vol)
Bromophenol blue
(BϕB) .... 0.25 g (0.25%, wt/vol)
Sterile 80% (vol/vol)
glycerol .... 37.5 ml (30%, vol/vol)
50× TAE buffer ...... 12.0 ml (6×)
Sterile distilled
water ..... up to a final volume of 100 ml

Mix glycerol, buffer, and water first. Add XC and BϕB dyes last. If the pH is correct, the load dye solution should be blue (not blue-green). Store at ambient temperature.

48. 6× TAE load dye with 6% SDS

Xylene cyanol (XC) ... 0.25 g (0.25%, wt/vol)

Bromophenol blue
(BϕB) .... 0.25 g (0.25%, wt/vol)
Sterile 80% (vol/vol)
glycerol .... 37.5 ml (30%, vol/vol)
50× TAE Buffer ...... 12.0 ml (6×)
Sodium dodecyl sulfate
(SDS) .... 6.0 g (wt/vol)
Sterile distilled
water ..... up to a final volume of 100 ml

While wearing a protective mask and gloves and using a fume hood, weigh out SDS first and pour it into a glass beaker that has been precalibrated with a 100-ml mark. Add water, buffer, and glycerol next. Stir on a hotplate. If SDS resists going into solution, heat the solution to dissolve SDS. Add XC and BϕB dyes last. If the pH is correct, the load dye solution should be blue (not blue-green). Let the solution cool to ambient temperature, and pour it into a 100-ml screw-cap bottle. Store it at ambient temperature.

References

Davis, K. E. R., S. J. Joseph, and P. H. Janssen. 2005. Effects of growth medium, inoculum size, and incubation time on culturability and isolation of soil bacteria. *Appl. Environ. Microbiol.* 71:826–834.

Difco Laboratories. 1998. *The Difco Manual,* 11th ed. Difco Laboratories, Sparks, MD.

Miller, J. H. 1972. *Experiments in Molecular Genetics.* Cold Spring Harbor Laboratory, Cold Spring Harbor, NY.

Mueller, J. H., and J. Hinton. 1941. A protein-free medium for primary isolation of gonococcus and meningococcus. *Proc. Soc. Exp. Biol. Med.* 48:330–333.

Murry, M. A., M. S. Fontaine, and J. G. Torrey. 1984. Growth kinetics and nitrogenase reduction in *Frankia* sp. HFP ArI3 grown in batch culture. *Plant Soil* 78:61–78.

Pitcher, D. G., N. A. Saunders, and R. J. Owens. 1989. Rapid extraction of bacterial genomic DNA with guanidium thiocyanate. *Lett. Appl. Microbiol.* 8:151–156.

Reasoner, D. J., and E. E. Geldreich. 1979. A new medium for the enumeration and subculture of bacteria from potable water. *Abstr. 79th Annu. Meet. Am. Soc. Microbiol.,* abstr. N7.

Shirling, E. B., and D. Gottlieb. 1966. Methods for characterization of *Streptomyces* species. *Int. J. Syst. Bacteriol.* 16:313–340.

Teather, R. M., and P. J. Wood. 1982. Use of Congo red-polysaccharide interactions in enumeration and characterization of cellulolytic bacteria from the bovine rumen. *Appl. Environ. Microbiol.* 43:777–780.

Vincent, J. M. 1970. IBP handbook no. 15. *A Manual for the Practical Study of Root-Nodule Bacteria.* Blackwell Scientific Publications, Oxford, United Kingdom.

Part B
LABORATORY SUPPLIES AND EQUIPMENT

LABORATORY STRAINS

Strain designation	Identity	Description	Experiment
UCLA #1010	*Serratia marcescens*	Gram-negative rod	1.2
UCLA #1003	*Escherichia coli* B	Gram-negative rod, Nonmotile strain	4.3
UCLA #1218	*Staphylococcus epidermidis*	Gram-positive cocci clusters	4.3
ATCC 533	*Micrococcus luteus*	Gram-positive indicator	4.4
UCLA #1246	*E. coli* (wild-type)	Gram-negative indicator	4.4
UCLA #1247	*E. coli fis tolC*	Indicator	4.4
UCLA #1248	*E. coli smpA surA*	Indicator	4.4

UCLA laboratory strains are available upon request; however, a material transfer agreement (MTA) may be required. Please contact Erin Sanders (erinsl@microbio.ucla.edu) to request strains. The American Type Culture Collection (ATCC) strain can be purchased from the Bacteriology Collection at the ATCC Global Bioresource Center (http://www.atcc.org/).

STANDARD SUPPLY AND EQUIPMENT LIST

Many of the following materials also are listed with the experiment(s) for which they will be used. Note that there are additional experiment-specific supplies not listed here.

✔ Deli refrigerator or cold room for storage at 4°C

✔ Frost-free freezer for storage at −20°C

✔ Freezer for storage at −80°C

✔ Incubator or warm room set at 30°C

✔ Incubator or warm room set at 37°C

✔ Shaking water bath set at 37°C

✔ Water bath set at 100°C

✔ Water bath set at 55°C

✔ Water bath or heat block set at various temperatures: 42°C, 45 to 48°C, 50°C, and 60°C

✔ Thermal cycler (PCR machine)

✔ Digital timer

✔ Analytical scale

✔ Weighing paper or boats

✔ Vortex mixer (a Vortex Genie II is recommended because it is compatible with MO BIO soil kit adaptors)

✔ Clinical centrifuge (or equivalent for lower-speed spins)

✔ 13- and 18-mm glass culture tubes

✔ 15-ml sterile conical tubes with screw caps

✔ 1.8- and 2.0-ml clear microcentrifuge tubes

✔ 1.8-ml clear, "boil-proof" microcentrifuge tubes

✔ PCR tubes with caps

✔ Tube racks

✔ Sterile plastic pipettes (5, 10, and 25 ml)

✔ Pipette pump or bulb and/or automatic pipette gun

✔ Pipettors (P10, P20, P200, and P1000) with pipette tips

✔ Gloves

✔ Kimwipes

✔ Paper towels

✔ Adsorbent pads

✔ Disinfectant (Amphyl, 10% bleach, or 70% ethanol)

✔ Labeling tape

✔ Ethanol-resistant marker (sold by VWR)

✔ Styrofoam ice buckets

✔ Bunsen burner with tubing and striker

✔ Long, sterile sticks in 18-mm tube covered with metal cap

✔ Wire inoculating loop

- ✔ Sterile flat toothpicks in 50-ml beaker covered with aluminum foil
- ✔ Sterile swabs
- ✔ Cryogenic storage vials with caps
- ✔ Microcentrifuge
- ✔ 125-ml Erlenmeyer flask
- ✔ Glass petri dish (use as lid for flask)
- ✔ Spatula
- ✔ Hot mitts
- ✔ Power supply
- ✔ Mini-gel electrophoresis chamber and lid
- ✔ Mini-gel casting tray and carrier
- ✔ 13-well combs (two) and wide-well comb (one)
- ✔ Small, white metal ruler (with metric markings)
- ✔ Safety glasses
- ✔ Computers with printers and access to internet

SPECIALTY SUPPLIES OR EQUIPMENT

The following materials are necessary for experiments conducted in the "I, Microbiologist" project but may not be considered standard laboratory supplies or equipment.

- ✔ Digital camera
- ✔ Drying oven (105°C)
- ✔ Desiccator
- ✔ Light microscope (with bright-field and phase optics; Olympus CX41 recommended)
- ✔ Digital microscope camera
- ✔ UV/visible spectrophotometer (Thermo Scientific NanoDrop recommended)
- ✔ 96-well UV microplate (new, clean) or 0.1-ml quartz cuvette if NanoDrop is not available
- ✔ DNA sequencing service or instrument
- ✔ Gel documentation system

Index

Page references followed by f denote figures; those followed by t denote tables.

429